Engineering Materials

This series provides topical information on innovative, structural and functional materials and composites with applications in optical, electrical, mechanical, civil, aeronautical, medical, bio- and nano-engineering. The individual volumes are complete, comprehensive monographs covering the structure, properties, manufacturing process and applications of these materials. This multidisciplinary series is devoted to professionals, students and all those interested in the latest developments in the Materials Science field, that look for a carefully selected collection of high quality review articles on their respective field of expertise.

Indexed at Compendex (2021)

More information about this series at https://link.springer.com/bookseries/4288

Arief Suriadi Budiman

Principles of Extreme Mechanics (XM) in Design for Reliability (DfR)

With Special Emphasis on Recent Advances in Materials Characterization and Experimentation Techniques

Springer

Arief Suriadi Budiman
Xtreme Materials Laboratory (XML)
Engineering Product Design (EPD)
Singapore University of Technology and Design
(SUTD), Singapore, Singapore

Department of Manufacturing and Mechanical
Engineering and Technology (MMET)
Oregon Institute of Technology (Oregon Tech),
Klamath Falls, USA

ISSN 1612-1317 ISSN 1868-1212 (electronic)
Engineering Materials
ISBN 978-981-15-6719-3 ISBN 978-981-15-6720-9 (eBook)
https://doi.org/10.1007/978-981-15-6720-9

This Springer imprint is published by the registered company Springer Nature Singapore Pte Ltd.
The registered company address is: 152 Beach Road, #21-01/04 Gateway East, Singapore 189721, Singapore

Preface

This book addresses issues pertinent to mechanics and stress generation, especially in recent advanced cases of technology developments, spanning from micrometer interconnects in solar Photovoltaics (PV), next-gen energy storage devices to multilayers of nanoscale composites enabling novel stretchable/flexible conductor technologies. In these cases, the mechanics of materials have been pushed to the extreme edges of human knowledge to enable cutting-edge, unprecedented functionalities and technological innovations. Synchrotron X-ray diffraction, *in-situ* small-scale mechanical testing combined with physics-based computational modeling/simulation have been widely used approaches to probe these mechanics of the materials at their extreme limits due to their recently discovered distinct advantages. The techniques discussed in this manuscript are highlights specially curated from the broad body of work recently reported in the literature, especially ones that the author had led the pursuits at the frontier himself. Extreme stress generation in these advanced material leads to often new failure modes and hence, reliability of the final product is directly affected. From the recent topics and various advanced case studies covered in this book, the reader gets an updated knowledge of how new mechanics can and have been applied in Design for Reliability (DfR) for some of the latest technological innovations known in our modern world. Further this also helps in building better designs, which may avoid the pitfalls of the current practiced trends.

Singapore, Singapore | Arief Suriadi Budiman

Contents

About the Author

Arief Suriadi Budiman received his B.S. in Mechanical Engineering from Institute of Technology, Bandung (ITB), Indonesia, his M. Eng Sc in Materials Engineering from Monash Univ., Australia and his Ph.D. in Materials Science and Engineering from Stanford University, CA in 2008. During his doctoral candidacy at Stanford's Department of Materials Science & Engineering under the supervision of Professor William D. Nix (MRS Von Hippel Award 2007), Dr. Budiman received several research awards (MRS Graduate Silver Award 2006, MRS Best Paper 2006) and contributed to several high-impact journal publications (Acta Materialia, Applied Physics Letters, Journal of Electronic Materials). Subsequently, Dr. Budiman has been awarded the prestigious Los Alamos National Laboratory (LANL) Director's Research Fellowship in 2009 to conduct top strategic research for the energy and national security missions of the Los Alamos National Laboratory's. Dr. Budiman has a truly international career being academics in Singapore (2013–2019), Indonesia (2020–2021) and the United States (MIT – Massachusetts Institute of Technology in 2014/5, and currently at Oregon Tech – Oregon Institute of Technology/OIT), as well as collaborating with researchers in Europe (ANR), China (Chinese Academy of Sciences) and Japan (Tokyo University). He has published a book "Probing Crystal Plasticity at the Nanoscales – Synchrotron X-ray Microdiffraction" (Springer 2015). He has two U. S. Patents and one pending.

Chapter 1
Introduction

Mechanics is always an important factor in designing next generations of advanced and emerging micro/nanoscale devices and systems. It certainly has been paramount for the success of the advanced semiconductor microelectronics industry and the next generations of 3D Integrated Circuits (IC) design and packaging technologies. Mechanics of materials—in particular, stress, elasticity, plasticity and fracture insights—have not only enabled design of robust and reliable electronics devices, but often also crucially influenced the next revolutions of the technology, in terms of performance and extreme functionality. Today's advanced semiconductor microelectronics are the testament of how the mechanics of materials against electromigration and low-k dielectric breakdown, for instance, are crucial factors for their reliability, while some of the performances (for instance, in terms of computing speed) were due to strain engineering of silicon, and extreme functionality (in advanced memory technology, for example, that has been enabled by 3D integration) has been increasingly adopted only by full control of the 3D state of the stresses in the devices/systems.

Stress continues to play ever-increasingly critical roles in next-generation electronics devices and systems, as novel materials and multi-length scales and dimensions of technology are the only constant in the industry (nanophotonics, nanoplasmonics, MEMS/MicroElectroMechanical Systems, and many others). Yet emerging advanced micro/nanoscale industries—in next-generation energy systems (battery, solar photovoltaics, etc.), novel fabrication and manufacturing technologies (additive manufacturing, artificial intelligence in materials design), and innovations in new, unprecedented functionalities of materials (novel metallic flexible/stretchable conductor for electronic skins, etc.)—are arising where the environments become ever more challenging and extreme in complexities. These new technological needs and trends have been unraveled only in the past decade or so. The case in point here lies, for instance, in the next-generation energy storage materials (Li-ion battery) that are both smart and sustainable. The mechanics aspects in the next-generation Li-ion battery have recently led to the importance of intercoupling between mechanics and electrochemistry in allowing the design of novel electrode materials for future

A. S. Budiman, *Principles of Extreme Mechanics (XM) in Design for Reliability (DfR)*, Engineering Materials,
https://doi.org/10.1007/978-981-15-6720-9_1

Li-ion battery technology. This is in addition to the rising trends in other energy technologies, such as novel photovoltaic cells/systems and next-generation nuclear energy system, not to mention the breakthroughs promised by the vision of emerging flexible electronics and extreme stretchability in wearable devices and technologies. The essence of stress is even in the word "stretchable" in the technology that allows next-generation flexible and wearable materials and electronics.

Mechanics and Design for Reliability (DfR) in engineering devices and systems are always intertwined. Most certainly for mechanical devices, but it is even so for electronics, photonics or, for that matter, biological devices. No matter for which of the applications the devices and systems are, the materials under operational conditions would eventually fail and when they do, failures are almost always controlled by the mechanics of the materials. Moreover, with the recent advances in nanotechnologies and novel fabrication methodologies, the mechanics of new/advanced nanomaterials are no longer just the classical laws of mechanics of materials (i.e. elasticity, plasticity, and fracture mechanics). New, ground-breaking mechanisms of elasticity and plasticity of materials, for instance, are—as we speak—being discovered by nanomechanicians all over the world right now! These new mechanics are what we call here the "eXtreme Mechanics (XM)." These new insights hold ever more importance in the Design for Reliability (DfR) of next-generation micro/nanoscale devices and systems. Device makers have more and more now realized that until they ship robust and reliable devices, they will not yet achieve commercial success. In a design-focused research theme like ours, it should be our unique proposition that we are not just great nanoresearchers and scientists, but that we are primarily great nanodesigners—able to offer complete and elegant engineering solutions from ideas to beautifully working, robust and reliable nanodevices; from invention to the most valuable company on earth (think Apple, Inc. and Tesla, Inc.)!

This book consists of the most recent updates of the mechanics of materials (i.e. the "eXtreme Mechanics") for the cutting-edge design in the advanced micro/nanoscale materials. How they have crucially affected the state of the art of design and technology of today's most important and tomorrow's emerging devices and systems will be highlighted in this book through real-world cases in several major areas of the modern technological world. Three emphasized clusters of industries have been identified as our focus in this book—advanced nano/microelectronics (3D IC, advanced packaging, MEMS, etc.), smart energy technology (battery, photovoltaics, etc.), and emerging flexible electronics (and other extreme technologies).

Overall background as well as the motivation for this book have been described in this chapter. The description of the eXtreme Mechanics (XM) principles as new insights as well as updates of the classical mechanics laws is presented in Chap. 2: **Extreme Mechanics in Advanced, Emerging Materials**. Much of these new insights of XM principles were only recently discovered in the past decade or so, only with the advent of novel, ground-breaking experimentation and materials characterization techniques, such as the synchrotron X-ray nanodiffraction and *in-situ* mechanical testing inside an Scanning Electron Microscropy (SEM). These research methodologies used in probing the eXtreme Mechanics (XM) and enabling the Design for Reliability (DfR) cases/issues in the following chapters of this book are presented in Chap. 3: **Probing Mechanics at the Extremes**.

Chapters 4–6 represent the more practical aspects of the new mechanics (XM) and the extreme probing capability of the techniques to make a difference in the advanced and emerging materials at small scale and in real (industry-relevant), advanced devices and structures. Chap. 4: **Latest Updates in Advanced Micro/Nanoelectronics, Microsystems, and MEMS (Micro-Electro Mechanical Systems) Devices and Technologies** consists of recent cases in which new insights of XM have continued to enable Design for Reliability (DfR) in the Through-Silicon Vias (TSV) technology for 3D Integrated Circuits (IC) design and in the fabrication and plasticity of low-melting metal nanostructures for advanced nanotechnological system design. In addition, unique mechanoelectrical behaviors of Cu/Nb nanolayered structures were revealed with potential applications in Structural Health Monitoring (SHM) and yet another important DfR application in enabling next technology nodes in the semiconductor industry roadmap (ITRS/International Technology Roadmap for Semiconductors).

Latest important trends in the smart and sustainable energy industries are presented in Chap. 5: **Recent Cases in Next-Generation Energy Technologies and Systems**. Recent issues in the novel silicon-based nanostructure anode design for the emerging Lithium ion battery technology as well as how to enable the thin silicon solar cell design for the next-generation Solar PV (photovoltaics) technology will be discussed in this chapter. In addition, yet another very recent trend in the PV materials (especially with the ever-critical encapsulant polymer layers) with important implications in the silicon solar PV industry will be highlighted.

Further, Chap. 6: **Emerging Trends in Additively Manufactured Materials and Novel Flexible/Stretchable Conductor Technologies** explores further the unique, distinctive ways XM principles have allowed additive manufacturing lead to the development of 3D architectured materials with unprecedented fracture and impact properties in one aspect, and in the other, how a clever interfacial design/engineering allowed by a new fabrication path (using severe plastic deformation—a form indeed of extreme mechanics in the metal forming technologies) in the metallic nanolayer Cu/Nb materials may enable novel metallic conductor technologies for extreme stretchability in wearable and biomedical applications.

While Chaps. 4 and 5 provide concrete cases and real-world industry-relevant examples of how XM has extended DfR capabilities and limitations in important high-technology devices/systems/technologies in microelectronics and energy industries, Chap. 6 elucidates further how the powerful interactions between XM and the newest advances in the manufacturing methodologies like AM (Additive Manufacturing) have opened up exciting, new and unprecedented possibilities for Design for Reliability (DfR) in emerging devices and technologies.

These chapters were born out of the research activities and investigations led by the author himself during the last 5–7 years especially with his research group at Singapore University of Technology and Design (SUTD). The cases highlighted here are intended as showcases and real-world industry-relevant examples for how XM principles have indeed extended the DfR capabilities in today's engineering design and will certainly also in tomorrow's world. They have been curated carefully by the author to demonstrate and to show as cases in point of the main thesis of this book. These pieces have individually been published in leading journal publications in each own field as original contributions, but the versions included in this book have incorporated significant updates, edits, and additional recent insights as well as been put in new lights and contexts of XM for DfR in the newest technological world (with Additive Manufacturing, atomic engineering of interfaces, etc.) from the unique perspectives of the author. The former members of the research groups "Xtreme Materials Laboratory (XML)" and "Xtreme Photovoltaics (XPV)" at SUTD, Singapore are fully acknowledged for their contributions through the pieces of highlights in the book.

Finally, conclusions to this body of work are provided in Chap. 7. In this book, we have captured a portion of the broad spectrum of research, representing some of the most important and recent advances in the probing of the principles of XM and in extending the DfR capabilities along with enabling extreme design and innovations. This book provides readers up-to-date information on the impact of these recent advances on XM principles and DfR innovations. We hope the readers will find this text to be a significant collection of important research highlights representing the rapid advancements in these exciting research fields.

Chapter 2
Extreme Mechanics in Advanced, Emerging Materials

Plasticity and fracture of materials at the nanoscales can deviate significantly from the classical laws of the same phenomena in bulk properties. This could have important implications if the materials are to be used effectively in real-world engineering systems. Nanoscale materials and composites have been known to have important effects related to size, but today many other emerging materials—resulting from, or enabled by, novel manufacturing routes (additive manufacturing, or bottom-up approaches such as atomic or layer-by-layer deposition methodologies) combined with 3D multi-hierarchical microarchitecturing—have been reported to approach extreme limits of materials properties.

There are at least three aspects where the new insights of the mechanics of materials have been considered distinctively to be departing the trajectories predicted by the classical laws of elasticity, plasticity, and fracture mechanics—the principles we present here as the "eXtreme Mechanics" or XM. First, "Smaller is Stronger" (which is the subject of Sect. 2.1: **Smaller is Stronger—Extending the Elasticity Limit of Nanoscale Crystalline Materials**) has been considered the most significant and important aspect of XM and has become widely accepted in the scientific communities in the world recently. It is considered as an update of the elastic part of the classical laws of mechanics—the most fundamental perhaps for engineering materials and applications, as it governs not only deformation in failure events, but also the functionality (and thus performance) of many devices/systems (e.g. piezo-devices/sensors, for instance, perform their functions based on elasticity of the materials). However, the new insights here involve not the actual magnitude of Young's modulus itself, but rather on the extension of the elastic regime. The definition of strength of materials is represented by the yield strength parameter. The yield point is the upper limit of the elastic regime—that is, further deformation of the material will carry it to the plastic regime (i.e. yielding). So this is an update of XM in Elasticity of Materials.

The second aspect is indeed how plasticity of the materials at the nanoscales deviates from the classical laws of strain hardening, for instance, and the old dichotomy

A. S. Budiman, *Principles of Extreme Mechanics (XM) in Design for Reliability (DfR)*, Engineering Materials,
https://doi.org/10.1007/978-981-15-6720-9_2

of "strong but brittle" (or "ductile, but not strong") in materials development/design. We call it "old dichotomy" as with the new insights of XM in plasticity here, we can now have strong materials with very high ductility or large amount deformability. By controlling the defect dynamics in the crystalline nanometals, for instance, we could create microstructures that could self-heal and remove the defects from accumulating in the materials (which is the primary reason why strain hardening leads to stronger but brittle metals). This is described in detail in Sect. 2.2: **Smaller is More Deformable—Controlling Defect Dynamics at the Nanoscales**. This is hence an update of XM in Plasticity.

Finally, fracture events have also been found in emerging materials to depart the trajectories as predicted by classical fracture mechanics. Nanoscale materials and composites have been known to have important effects related to size, including fracture events and parameters leading to final end of the service/lifetime of the materials. However, today many other emerging materials—resulting from, or enabled by, novel manufacturing routes (additive manufacturing, or electrospinning methodology, along with atomic engineering of interfaces)—have been reported to approach extreme limits of materials lifetime. By controlling the dynamics and interactions of the crack tip in a 3D architectured polymer composite material (enabled by electrospinning as an additive manufacturing methodology), for instance, we could create 3D nanostructures that could very effectively and efficiently keep blunting the crack tip so that it does not propagate so easily anymore, thus delaying the crack from achieving the critical length in the material significantly. Thus, we could now do away with yet another difficult dilemma in materials design and development—"strong, but not tough" (in terms of fracture events), or "strong but more prone to impact." Sect. 2.3: **Extreme Fracture Limits—Fracture at the Small Scales and Emerging 3D Architectures** will have some details for this new, important update of XM for the design of emerging lightweight, strong but also fracture and impact-resistant materials. Finally, this in an update of XM in Fracture Mechanics.

2.1 Smaller is Stronger—Extending the Elasticity Limit of Nanoscale Crystalline Materials

The mechanical behavior of materials in small dimensions (submicron and nanoscale) often deviates considerably from the behavior of bulk materials. In the macroscale (bulk), the mechanical properties of materials are commonly described by single valued parameters (e.g. yield strength, hardness, etc.), which are largely independent of the size of the specimen. However, as specimens are reduced in size to the scale of the microstructure, their mechanical properties deviate from those of bulk materials. For example, in thin films—where only one dimension, the thickness, reaches the micron-scale and below—the flow stress is found to be higher than its bulk value and becomes even higher as the film gets thinner. These size effects are usually attributed to grain size hardening [1–6] and to the confinement of dislocations within the film by the presence of the substrate and in some cases the passivation [7–13].

Yield point has been considered as the onset of plastic deformation in materials, but in this section, we will discuss delaying or pushing further the yielding onset point as extending the elastic regime of the materials. Or in other words, increasing the resistance of the crystalline materials (which are the focus of the materials system in this book) for plastic flow, by making defect movements harder. This of course has been the known strategy for strengthening crystalline materials (especially metals, such as in precipitation hardening), thus extending the elastic regime. The increase in the yield point here could still be explained by classical law of flow mechanics. But, as nanotechnologies enable materials development with smaller and smaller scales of microstructures or specimen geometries, such as thin films and nanostructures (like nanopillars, nanowires, etc.), they are known to withstand much higher stresses, then the known yield strength for the bulk version of the materials. For instance, stresses in metallic thin films (like Au or Cu) were measured in high hundreds of MPa, which are at least one order of magnitude higher than their bulk yield strength values [6–9]. These high strains were measured using advanced X-ray diffraction methods [1, 7, 8], thus we know the interplanar distances between atomic layers have changed beyond their normally observed elastic regimes.

In nanoindentation, this extension of elastic regime (or yield strength) has been known to depend inversely with indentation depth—the smaller the depth of indentation, the harder/stronger the crystal. This effect has been attributed to strain gradients or geometrically necessary dislocations (GNDs) [14–28]. In other small-scale structures, like submicron single-crystal pillars subjected to uniaxial compression, however, this size effect has been observed even though the deformation imposes no strain gradients [29–31]. These different kinds of size effects observed as the scale of deformation tends toward the submicron and nanoscales, thus suggest that the onset of plastic flow is no longer governed by the classical plasticity model.

In order to provide a useful framework for the work presented in this book, the principles of the Taylor relation between flow stress and dislocation density [29] are first presented; this is then followed by a discussion of the limitation of this classical flow-stress relation. Two accounts of size effects in hardening/strengthening—the Nix and Gao model of strain gradient plasticity [19] and the dislocation-starvation plasticity—are reviewed in this section.

2.1.1 *Classical Flow-Stress Relationship: The Taylor Relation*

The Taylor relation states that the flow stress of a material is proportional to the square root of the density of dislocations. By recognizing that the shear stress resulting from the self-stress field of a dislocation varies as the inverse of the distance from the dislocation and that the average spacing of dislocations is defined by the reciprocal of the square root of the dislocation density, ρ, Taylor argued that the stress required to move a dislocation past another dislocation, τ, is given by

$$\tau = \frac{\mu b}{2\pi}\sqrt{\rho} \tag{2.1}$$

where μ is the shear modulus of the material, b is the Burgers vector.

The more general form of this relation for the case of a dislocation moving in a forest of dislocations is given by

$$\tau = \alpha\mu b\sqrt{\rho} \tag{2.2}$$

where α is the Taylor coefficient, which depends on the character of the dislocation forest [32–36]. The important implication of this relation is that the flow stress is always enhanced by the presence of dislocations.

However, this statement cannot always be true. As the scale of deformation tends toward the scale of microstructural features (the spacing of dislocations, grain sizes) this classical relation may no longer hold. This has been illustrated clearly by Lilleodden using Fig. 2.1 (courtesy of Lilleodden [19]). Figure 2.1a illustrates that when the volume is large relative to the spacing of dislocations, the Taylor relation is expected to fully describe the flow stress. In Fig. 2.1b, the volume is close to the spacing of dislocations, such that the averaging implied by Eq. 2.1 is no longer applicable, and consequently a deviation from the Taylor relation is expected. Finally, in the limit of a deformed volume that is less than the cube of the dislocation spacing, as in Fig. 2.1c, plasticity may be controlled by the creation of dislocations, which requires near-theoretical stresses. Thus, the Taylor relation would obviously not be

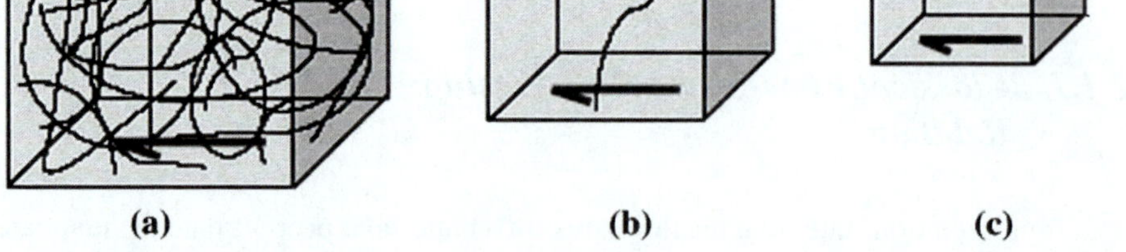

Fig. 2.1 Illustration of the limitation of Taylor relation: **a** deformed volume >>$\rho^{-3/2}$, **b** deformed volume ~$\rho^{-3/2}$, **c** deformed volume <$\rho^{-3/2}$; in each case, the same dislocation density is assumed (courtesy of Lilleodden [19])

applicable in this case. The flow stress in the case described in Fig. 2.1c should be very high. This is one limitation of the classical flow-stress relationship described by Taylor relation. Clearly, the length scale of the deformation needs to be considered when describing the flow stress.

2.1.2 The Nix and Gao Model of Strain Gradient Hardening

Indentation size effects (ISE) have been widely observed. Crystalline materials display a strong size dependence when they are indented over a scale from a fraction of a micron to tens of microns: the smaller the indentation depth, the higher the hardness, such as shown in Fig. 2.2a. Such behavior cannot be explained using classical plasticity as we have discussed in the last sub-section (Taylor relation). Following the work of Stelmashenko et al. [15] and De Guzman et al. [16], Nix and Gao [19] provided a simple explanation for this depth-dependent hardness, in terms of the geometrically necessary dislocation density as a function of indentation depth. Durst and Goeken [27] as well as Feng [55] later modified the model to account primarily for the fact that the plastic zone radius is not equal to the contact radius, as Nix and Gao had assumed. Still the revised model takes the form:

$$\frac{H(h)}{H_0} = \sqrt{1 + \frac{h_0}{h}} \tag{2.3}$$

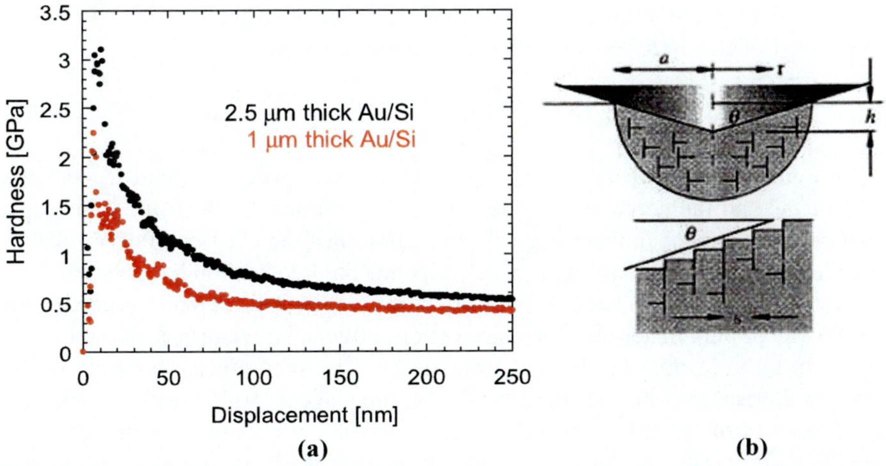

Fig. 2.2 Indentation Size Effects (ISE): **a** Hardness versus displacement data for two thicknesses of Au thin films on Si substrate (courtesy of Lilleodden [19]); **b** geometrically necessary dislocations created by a rigid indentation

which can also be shown equivalently

$$\frac{H(h)}{H_0} = \sqrt{1 + \frac{\rho_G}{\rho_S}} \tag{2.4}$$

where

$$\rho_S = \frac{H_0^2}{3C_H^2 \alpha_t^2 \mu^2 b^2} \tag{2.5}$$

$H(h)$ is the hardness as a function of h, the depth of the indentation, while H_0 is the limit of the hardness when the indentation depth (h) becomes indefinitely large, and h_0 is a material length scale. In Eq. 2.5, C_H is a constant associated with the plastic zone size, α_t is the Taylor constant, μ is shear strength and b is the magnitude of Burgers vector.

Equation 1.5, thus, implies depth-dependent ρ_G (as ρ_S is nominally constant), or in other words depth-dependent strain gradients. The correlation between the hardness numbers and the associated ρ_G had been derived in a complete manner elsewhere [37], and with some rearrangements, the final form such as shown in Eq. 2.6 can be used

$$\rho_G = \frac{H_0^2}{3C_H^2 \alpha_t^2 \mu^2 b^2} \left\{ \left(\frac{H}{H_0} \right)^2 - 1 \right\} \tag{2.6}$$

Mechanism-based strain gradient (MSG) plasticity is a model of hardening in which the Taylor relation is adopted as a founding principle:

$$\tau = \alpha \mu b \sqrt{\rho_{total}(h)} = \alpha \mu b \sqrt{\rho_G(h) + \rho_S} \tag{2.7}$$

The total dislocation density, ρ_{total}, is the sum of the statistically stored dislocations (SSDs), ρ_S, and the geometrically necessary dislocations (GNDs), ρ_G, which is a function of h. This phenomenological model modifies the classical Taylor relation by adding the strain gradient term which depends on deformation length scales.

Plastic deformation in small volumes requires higher stresses than are needed for plastic flow of bulk materials. There are various effects, both extrinsic and intrinsic, that seem to be responsible for this observation. The size effects observed in thin metal films arise from the constraints of surrounding layer or from the microstructural characteristics of the thin films. These kinds of size effects have been called the extrinsic effects [14]. The intrinsic effects, in contrast, are those size effects that arise in small, unconstrained single crystals under deformation. There are two possible sources of size effects that have been identified: strain gradients and dislocation starvation.

In nanoindentation experiments, where the length scale of the deformation reaches the microstructural length scale of the material, the governing relations between stress and strain deviate from the classical laws that apply to bulk materials. At small depths of indentation the hardness of crystalline materials is usually higher than that of large indentations, as illustrated in Fig. 2.2a. This indentation size effect is an intrinsic one and has been explained using the concept of geometrically necessary dislocations (GNDs) and strain gradients [15–28].

According to this picture, the hardness increases with decreasing depth of indentation because the total length of geometrically necessary dislocations forced into the solid by the self-similar indenter scales with the square of the indentation depth, while the volume in which these dislocations are found scales with the cube of the indentation depth (Fig. 2.2b). This leads to a geometrically necessary dislocation density that depends inversely on the depth of indentation. The higher dislocation densities expected at smaller depths lead naturally to higher strengths through the Taylor relation [32], and this leads to the indentation size effects (ISE).

A different kind of intrinsic size effect is observed when single crystalline materials are deformed homogenously, without strain gradients. Recently Uchic et al. [29] and others [30, 31] have shown that micropillars of various metals with diameters in the micron range, subjected to uniaxial compression, are much stronger than bulk materials. For example, micropillars of gold ranging in diameter between 200 nm and several microns have been found to have compressive flow strengths as high as 800 MPa, a value ~50 times higher than the strength of bulk gold [30, 31]. The accounts of strain gradient plasticity, as discussed above, appear to break down for the case of micropillar compression because the deformation is essentially uniform.

Greer and Nix [31] suggested that the high strengths of submicron pillars of gold might be controlled by a dislocation-starvation hardening process. In this mechanism and for small enough pillars, the mobile dislocations have a higher probability of annihilating at a nearby free surface than of multiplying and being pinned by other dislocations. When the starvation conditions are met, plasticity is accommodated by the nucleation and motion of new dislocations rather than by motion and interactions of existing dislocations, as in the case of bulk crystals.

With the increasing need from industry to develop materials of high mechanical performance at small length scales, a good understanding of such material response under deformation has become important since many of these responses, as we have seen, are dependant both extrinsically as well as intrinsically on the behavior of structural entities at this scale (grain boundaries, inclusions, intrinsic intra- and inter-granular stress distribution). There is a limited amount of experimental data at these scales due to the lack of suitable techniques. This has long prevented modeling material behavior at these length scales, which in turn has prevented progress in developing a systematic link of materials properties from the macroscopic to the microscopic. Understanding and controlling plasticity and the mechanical properties of materials on this scale could thus lead to new and more robust nanomechanical structures and devices.

2.1.3 Dislocation-Starvation and Dislocation Nucleation-Controlled Hardening

Unlike the size effect that we observed and studied in the earlier section (strain gradient hardening), a different kind of intrinsic size effect appears to have also been observed [38–40] when single crystalline materials in the form of micropillars are deformed homogenously, without strain gradients. Recently Uchic et al. [38] and others [39, 40] have shown that micropillars of various metals with diameters in the micron range, subjected to uniaxial compression, are much stronger than bulk materials. For example, micropillars of gold ranging in diameter between 200 nm and several microns have been found to have compressive flow strengths as high as 800 MPa, a value ~50 times higher than the strength of bulk gold [39, 40]. This suggests that in spite of much progress on size effects on strength there is still no unified theory for plastic deformation at the submicron scale. The accounts of strain gradient plasticity, as illustrated in the previous section appear to break down for the case of micropillar compression because the geometry of the micropillar compression is not expected to include externally imposed plastic strain gradients that might lead to extra hardening for small samples.

While significant macroscopic strain gradients are not expected to develop during the uniform compression of micron-sized pillars, we cannot preclude microscopic strain gradients from forming. Nevertheless, even the presence or the absence of macroscopic strain gradients has not been directly and experimentally observed, especially in the case of metallic pillars. Since strain gradients and GNDs are directly related to the local lattice curvature, the technique of Scanning X-ray Microdiffraction (μSXRD) using a focused polychromatic/white synchrotron X-ray beam can be used to determine the density of GNDs. This has proven to be useful in the study of the early stages of electromigration failure in interconnect lines, wherein lattice bending and GNDs are created by electromigration processes [41, 42]. This capability is related to the continuous range of wavelengths in a white X-ray beam, allowing Bragg's Law to be satisfied even when the lattice is locally rotated or bent, resulting in the observation of streaked Laue spots.

Using this approach, we can monitor the change in the Laue diffraction peaks before and after uniaxial compression of a submicron single-crystal Au pillar. A quantitative analysis of the Laue peak widths then allows us to estimate the density of GNDs in the pillar. The absolute number of geometrically necessary dislocations in the crystal can then be determined using the dimensions of the pillar. A comparison of the numbers of geometrically necessary dislocations before and after the uniaxial compression provides information about the change in microstructure associated with plastic deformation. The synchrotron technique of scanning white-beam X-ray microdiffraction has been described thoroughly elsewhere [43, 44]. The power of this technique to study local plasticity and mechanical behavior of materials at small scales stems from the high brilliance of the synchrotron source, as well as the recent advances in X-ray focusing optics (allowing sample mapping at the submicron level).

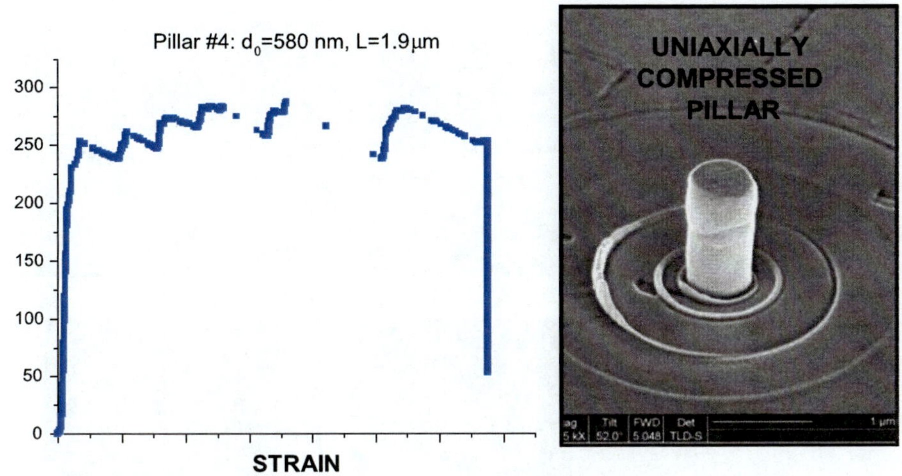

(a) (b)

Fig. 2.3 Stress–strain behavior of <111>-oriented Au single-crystal submicron pillar: **a** flow stress increases significantly beyond its typical bulk values; **b** SEM image of a uniaxially compressed pillar after deformation. Reproduced with permission from Elsevier

Figure 2.3a shows the stress–strain curves of the 0.58-μm Au pillar obtained during the compression testing. Uniaxial loading in the <111> direction of the Au crystal/pillar, corresponding to a high-symmetry orientation, would result in the activation of multiple slip systems, with the pillar deforming uniformly around its diameter as it is compressed. The flow stress reaches value as high as 280 MPa. This is close to 10 times the yield stress of gold in bulk and falls consistently in the flow stress versus pillar diameter chart described by Greer and Nix [40]. In this <111> loading orientation, and despite the presence of the end constraints, the pillar remains centrally loaded and preserves its cylindrical shape throughout the deformation process as shown in Fig. 2.3b. While upon further inspection we could not observe slip steps on the surface of the deformed pillar, we do however observe a visible slip marking (not shown very clearly in this particular SEMimage in Fig. 2.3b) that appears to be consistent with the trace of a {111} plane of the pillar crystal. The angle made by the plane causing the trace was within a few degrees of the expected trace of a {111} plane, after correcting for subsequent plastic deformation of the pillar. The final diameter of the pillar after the uniaxial compression is 0.67 μm, which represents a total strain of close to 35%.

A diffraction scan was again taken covering the deformed pillar. We subsequently select a particular Laue diffraction spot (in the case shown here, $(\bar{3}11)$ diffraction spot) associated with the location of the deformed pillar for further quantitative analysis and comparison. We subtracted the background intensity and checked that there is no Au crystal rotation involved (which would have manifested in the shift of the position of the Au Laue peak with respect to the Laue diffraction pattern of the chromium substrate reference), in order to be able to directly compare the

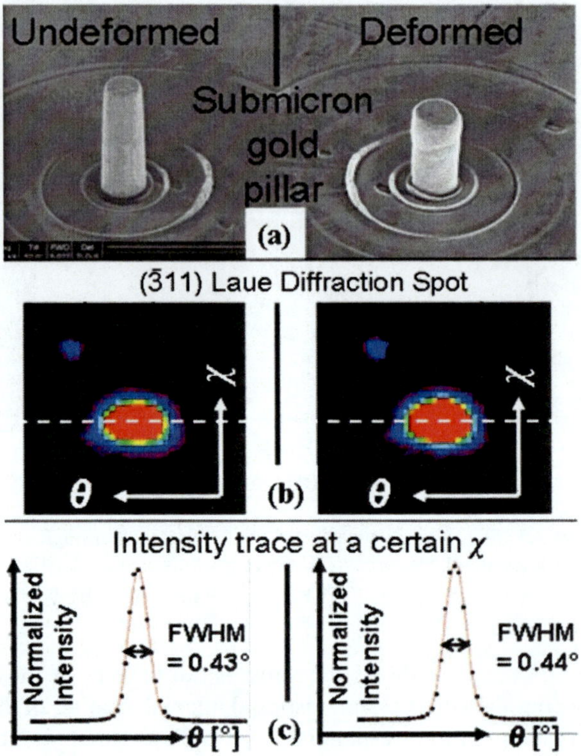

Fig. 2.4 Side-by-side comparison between undeformed and deformed states of the pillar crystal; **a** SEM image of the pillar, **b** Laue diffraction spot ($\bar{3}$ 11), and **c** quantitative analysis. Reproduced with permission from Elsevier

($\bar{3}$ 11) Laue spots before and after deformation and infer what happened to the pillar crystal during the deformation process. Figure 2.4 shows the data of the pillar crystal (SEM images, Laue diffraction spots, and Intensity profiles) for both undeformed and deformed states side by side.

Figure 2.4b shows that both ($\bar{3}$ 11) Laue spots have the same shape and that they are both rounded—not broadened toward a certain direction (streaked). This rounded shape is typical of an undeformed crystal, whereas broadening of a Laue diffraction spot in a certain direction (streaking) would have been associated with the presence of strain gradients in the deformation volume. A more detailed treatment of the streak length and its correlation with the curved, or in general, plastically deformed crystal has been described elsewhere [41, 42]. In addition, streaked Laue diffraction spots have also been observed recently in as-fabricated silicon micropillars [46].

In Fig. 2.4c we take the intensity traces along a particular χ to study the Laue diffraction peak profile more quantitatively (here the χ angle is simply the angle orthogonal to the 2θ angle). The profiles were fitted with Lorentzian curves. The measured FWHMs (full width half maximum) of both profiles show that there is an increase of 0.01° in the angular width. However, this difference is still within the experimental error bar of the instrumentation [7, 10] rendering the two measurements statistically identical. The angular resolution of this technique was calculated using a few assumptions on the experimental sample setup with respect to the CCD camera and on the capability of the indexing code [47], which are applicable to our micropillar compression experiments. The technique is sensitive to local lattice rotation, and thus this angular detection limit is applicable to geometrically necessary dislocations (GND).

In terms of the capability of this technique to detect GNDs, it is also limited by the instrumental broadening inherent in the observed Laue diffractions spots. One practical way to estimate the extent of instrumental broadening in our experiments is to take FWHM measurements of Laue diffraction spots coming from a silicon substrate/wafer, as such single crystals are very close to being defect-free and 100% pure. Based on measurements on such silicon wafer substrates, for similar experimental settings, the instrumental broadening contribution to the observed FWHM of Laue diffraction spots is 0.06°. This places a limit on our technique corresponding to the number of GNDs associated with a peak broadening of 0.06°.

This limitation indicates that relative lattice rotations smaller than $\Delta\theta = 0.06°$, or $\Delta\theta = 10^{-3}$ radians, which might be produced by compressive deformation could not be detected. Converting that measurement to the possible number of dislocations that could be left in the crystal after deformation depends on how the dislocations are distributed. To make these calculations we consider the model shown in Fig. 2.5.

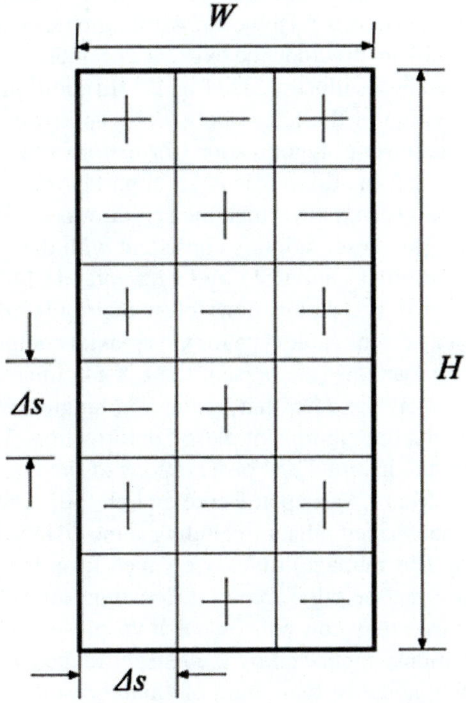

Fig. 2.5 A model of the pillar single crystal consisting of different domains of size Δs, each assumed to be occupied by like-signed edge dislocations, leading to a local lattice curvature of magnitude $|\kappa|$. Reproduced with permission from Elsevier

Figure 2.5 shows a model where different domains of size Δs are each assumed to be occupied by like-signed edge dislocations, leading to a local lattice curvature of magnitude $|\kappa|$. Thus within each domain there are only geometrically necessary dislocations. Taken over the whole crystal the dislocations can be regarded as statistically stored dislocations. We know from the Cahn-Nye relationship [11, 12] that the local geometrically necessary dislocation density is $\rho = |\kappa|/b$ where b is the magnitude of the Burgers vector. But the local curvature is $|\kappa| = d\theta/ds \approx \Delta\theta/\Delta s$, so that $\rho = \Delta\theta/b\Delta s$. With this model the total number of dislocations in the entire crystal would be $n = \rho WH = WH\Delta\theta/b\Delta s$. Taking the dimensions of the upper crystal to be $W = 600\,\text{nm}$ and $H = 1100\,\text{nm}$, and the Burgers vector to be $b = 0.3\,\text{nm}$ and using the X-ray broadening resolution of $\Delta\theta = 10^{-3}$ we find $n = 2.2 \times 10^3/\Delta s$. When the domain size is the same as the width of the crystal, $\Delta s = 600\,\text{nm}$, this leads to 3–4 dislocations left in the crystal after deformation. The expected number of dislocations that could be left in the crystal after deformation naturally increases with decreasing domain size. In the limit, a very high density of statistically stored dislocations cannot be ruled out by these experiments, though escape of these dislocations from the nearby free surfaces makes this unlikely.

Thus, the fact that the Laue diffraction spot in the deformed pillar is not streaked suggests that there might not be significant macroscopic strain gradients created during the uniaxial compression of the pillar (only 3–4 GNDs left in the crystal after deformation when the domain size is taken as the same as the width of the crystal, $\Delta s = 600\,\text{nm}$). However, within the resolution and limitation of our technique, we could not preclude the existence of microscopic strain gradients within the inspected volume, as illustrated in Fig. 2.5. In addition, as we find neither a shift in the absolute position of the Laue spots, nor a split, we can further infer that there is no crystal rotation or polygonization (formation of subgrain structures) in the pillar crystal upon the deformation. This is an important observation considering the huge amount of strain (~35%) to which the crystal was subjected.

This observation is consistent with the earlier TEM observations on a deformed gold pillar conducted by Greer and Nix [40]. Their TEM results showed that there were only 2 dislocations left in their pillar after deformation and that they were both of a non-movable type for the uniaxial compressive loading of their experiment. Both observations (the present Laue X-ray microdiffraction and the earlier TEM results) support the idea that submicron single-crystal gold pillars are nearly defect-free even after significant plastic deformation. This view was recently further supported by the in situ TEM observations of the compression of single-crystal nickel pillars by Minor's group at Berkeley Lab [50]. They found that the dislocations present in undeformed pillars (including some dislocation loops near the pillar surfaces created by FIB damage) quickly escaped from the pillar during compressive deformation, leaving the pillar free of dislocations after compression.

We may now conclude that the present white-beam X-ray microdiffraction observations, supported by the closely related TEM results, [40, 50] are consistent with the model of hardening of small crystals by dislocation starvation and dislocation nucleation or source-controlled plasticity, as suggested by Greer and Nix [40]. In ordinary plasticity (i.e. in typical, bulk samples), dislocation motion leads to dislocation multiplication by various cross-slip processes, invariably leading to softening before strain hardening occurs through elastic interaction of dislocations. However in small samples, such as the submicron Au single-crystal pillar under study here, dislocations can travel only very small distances before annihilating at free surfaces, thereby reducing the overall dislocation multiplication rate. The central idea is that, as dislocations leave the crystal more frequently than they multiply, the crystal can quickly reach a dislocation-starved state. When such a state is reached, continued loading would force other, harder sources of dislocations to be activated in the crystal, leading to the abrupt rise in the measured flow stress (i.e. hardening).

2.2 Smaller is More Deformable—Controlling Defect Dynamics at the Nanoscales

Plasticity of materials following the classical laws was often marked by the "strong but brittle" (or "ductile, but not strong") dilemma in materials development/design. Nanoplasticity or plasticity of materials at the nanoscales has recently been pushed much further to allow advanced crystalline materials to withstand much larger deformability by controlling the defect dynamics in the nanoscale, crystalline multilayered metals (nanolayers), for instance. Atomic design of interfaces or interface engineering has enabled nanolayers to have microstructures that could self-heal and remove the defects from accumulating in the materials (which is the primary reason why strain hardening in bulk crystalline metals leads to stronger but brittle metals). Hence, this section discusses new plasticity mechanisms involving interactions with interfaces in nanolayered structures which have pushed the materials deformability to extreme limits in Cu/Nb nanolayers [51–56].

In order to provide a background for the work presented in this section, the overall review of nanoscale metallic nanolayers, especially the Cu/Nb nanolaminates [52] is first presented; this is then followed by a discussion of the unique roles of interfaces in the case of Cu/Nb nanolayers in enabling the extreme deformability. One account of how the unique interface-based plasticity models could be used in more general nanostructures, beyond the nanolaminates, is also included in the discussion of this section.

2.2.1 Nanocale Metallic Multilayered Materials

Nanoscale metallic multilayer composites have recently attracted significant interests in the research community due to their radically extended performance limits at mechanical, thermal, chemical and irradiation extremes. The Cu/Nb nanoscale multilayer systems, for example, have been shown to demonstrate ultra-high strength and ultra-high radiation damage-tolerance [51–54]. Studying the deformation mechanisms in the nanoscale metallic multilayered composite materials has increasingly become both an interesting topic scientifically as well as an important subject technologically during these recent years especially with the oncoming development of the new generations of nuclear energy systems as well as aerospace and space technologies. These unique mechanical and radiation damage properties have been proposed to result from the interface-dominated plasticity mechanisms in nanoscale composite materials [55, 56]. The Cu/Nb nanoscale metallic multilayer composites, for instance, have demonstrated that interfaces serve to block dislocation movements as well as play roles as sinks that attract, absorb and annihilate point and line defects including radiation-induced bubbles [51, 55]. As a result, the nanolayered materials can exhibit over an order of magnitude improvement in strength and resistance to damage from particle irradiation. Studying how the dislocation configurations and densities evolve due to deformation will be crucial in highlighting the yield, work hardening and recovery mechanisms in the nanolayered materials.

The majority of these studies on deformation of metallic nanolayered materials however have been largely done on the Cu/Nb systems [51–56] and a few studies were done on Al/Nb systems [57]. Both are basically FCC/BCC systems but even then significant difference was already observed [56, 57] representing a wide range of design space just within the different materials combination following the FCC/BCC coordination. There are certainly research opportunities and scope for new nanolayered materials involving novel interfaces formed of additional crystalline materials (such as HCP) as well as amorphous materials and graphenes that are within the current capability of the experimental research group. This promises new knowledge and potential discoveries, as well as represents possibilities for further opening up of design spaces for nanolayered materials design and integration into real devices/systems.

These materials are thus well suited for fundamental studies of the effects of interfaces at the nanoscales on the strengthening mechanisms in layered metallic composites [58–64]. The mechanisms in which the materials become stronger have been shown to invariably involve interfaces and their interactions with the dislocation processes in most cases [51, 54, 55]. These materials and deformation experiments are thus ideal candidates for experimental studies using synchrotron white-beam X-ray nanodiffraction as a nanoprobe of plasticity. The fundamental understanding of the origins of these nanomechanical behaviors of metallic multilayered materials and the fundamental understanding of the roles of interfaces would eventually permit us to build and control useful engineering nanodevices and future integrated nanotechnologies.

Nanolayered materials have thus shown promise as candidates for next-generation structural materials because of their concurrent high strengths and, as plasticity is recovered through activities in the interfaces, high elastic limits. However, the limited amount of post-elastic deformability, which can lead to sudden and catastrophic material failure, especially in the tensile mode of deformation as illustrated by Aydiner et al. [60, 61] earlier, may have prevented their widespread use in integrated commercial applications. This represents a significant research gap in the attempts to design and integrate new nanomaterials into real-world applications. An appropriate fracture mechanics characterization technique for nanolayered materials and for nanostructured films/materials in general is needed to provide fundamental insights into how interactions between interface and other microstructural features lead to macroscale fracture and fatigue events in nanomaterials enabling their integration for potential widespread use in commercial applications.

2.2.2 *Extreme Deformability as Revealed by Successive Micropillar Compression Experiments*

Among these, nanoscale multilayer composites of Cu/Nb are being keenly explored due to their ultra-high strength (~2.5 GPa) without compromising its ductility (~30%) [62–65]. This idea of using nanolayers where Frank-Read source does not operate in a single layer, to build a strong solid was proposed by Koehler [66]. The ability to reach near-theoretical strength and large plastic deformation has made these materials a serious contender for applications involving high radiation damage, temperature and mechanical loading. The strengths of these nanoscale materials are strongly derived from the interface structure compared to its bulk counterpart [67, 68]. Hirth [69] also reported the role of interfaces and its effects on mechanical properties of metals. The underlying strengthening mechanisms at this incoherent (large lattice parameter mismatch) interface of the face-centered cubic (FCC) Cu, and body-centered cubic (BCC) Nb structure has been investigated using experimental techniques (ex/in situ) [65, 70–72] and simulations (continuum/atomistic) [73–84]. Due to the recent technological advances in fabricating nanoscale, multi-material films of (few to tens) nanometers thick are possible, [85] which are useful to interpret the interface-dominated plasticity phenomenon of these materials. The large strain deformation behavior of the rolled nanoscale Cu/Nb multilayer (each layer of 75 nm thickness) foils was reported by Misra et al. [86]. The transmission electron microscopy (TEM) observations of the samples revealed no dislocation cell structure formation and insignificant rotations of the interface normal as compared to the meso-scale (each layer of 4 μm thickness) multilayers, which showed both these dislocation mechanisms [67]. They concluded that large plastic deformation of the nanoscale materials results in symmetric slip. This mechanism was further investigated by measuring the flow strength including length scale dependency of the Cu/Nb bilayer with thickness each ranging from micrometer to less than a nanometer [87]. It was observed that layer thickness h greater than 100 nm were able to accommodate dislocation pile-ups and hence, yield strength σ_Y with respect to layer thickness could relate to Hall–Petch equation: $\sigma_Y = 1/\sqrt{h}$. As the thickness reduces to few (5) to tens ($\leq$50) of nanometers, there is no dislocation loop formation and pile-ups and hence, the Hall–Petch equation for flow strength breaks down. The threading dislocations, i.e. surface defects formed during the layer growth was predicted to be the plasticity mechanism at this nanoscale. When the deformation reaches a threshold (onset of plastic region), these dislocations accumulate at the interfaces leading to an Orowan bowing. The movement of these dislocations within a nanolayer is termed as confined layer slip or confined layer plasticity (CLS/CLP). Further, when the thickness of the nanolayers is in order of one to few (≤ 5) nanometers, the strength saturates and even drops. The interface resistance to dislocation crossing across nanolayers was believed to be the phenomena for this behavior.

To elucidate quantitatively the interfacial strength of the Cu/Nb nanolayers, ex-situ pillar compression and in situ indentation tests were performed on these samples by Li et al. [65]. They reported the interface strength to be in the range of 0.3–0.55 GPa. In situ nanoindentation of these samples was also performed by Li et al. [65] to observe the confined layer slip phenomenon. Their conclusions on these Cu/Nb nanolayers agree well with the insights of atomistic simulations which shows that, due to incoherent interface being weak in shear, they act as potent dislocation traps, limiting dislocation motion and leading to their unusually high strength [63, 88, 89].

In this section, we highlight three pillar samples (each deformed with a different strain schedule up to 35%) of similar dimensions which were compressed using a flat punch. The measured load–displacement results of the samples show variation within the nominal experimental scatter. Based on this affirmation, throughout this section, the results of only one pillar are discussed. The mechanical response of the pillar samples deformed to varying compressive strain is presented and its relation to confined layer slip is proposed.

The true stress–strain mechanical response of the pillar sample from the compression tests is shown in Fig. 2.6. The substrate effect is neglected due to no significant change in the results. The constant volume and homogeneous deformation model are used to interpret the load–displacement data. In this case, the volume measured from initial cross-sectional area A_o (based on d_o) and length L_o of the undeformed pillar is equivalent to the final volume of the deformed sample, therefore, $A_o{:}L_o = A_P{:}L_P$, where A_P and L_P denotes final pillar area and length. To aid the understanding, at each strain ramp, the micrograph of the pillar with its topography taken after each

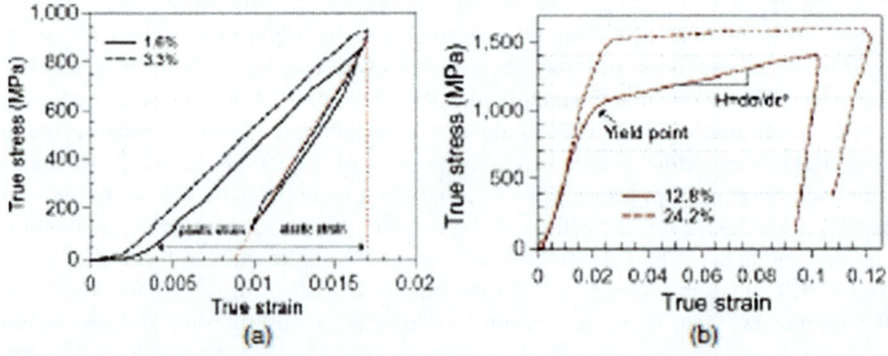

Fig. 2.6 True Stress versus Strain of the pillar obtained from compression tests: **a** cumulative "of 1.6 and 3.3%" and **b** cumulative "of 12.8 and 24.2%". Reproduced with permission from Elsevier

cycle is shown in Table 1. It is seen from the stress–strain response (Fig. 2.6a) that, pillar subjected to total true strain ε^T ($\varepsilon^{elastic} + \varepsilon^{plastic}$) of 1.6 and 3.3%, the single-crystal Cu/Nb nanolayers deforms mostly linearly throughout the loading regime. The unloading of the pillar from a maximum strain of 1.6% results in a true plastic strain ($\varepsilon^{plastic}$) of ~0.9%. Similarly, analyzing the unloading response of the Cu/Nb nanolayer subjected to a maximum strain of 3.3%, it can be clearly observed that, approximately 44% elastic strains are still present. From the unloading curves, the composite elastic modulus obtained for ε^T of 1:6% and 3:3% is ~103.5 and ~105.8 GPa respectively. Young's modulus obtained from the experimental data reasonably agrees to the calculated value of ~116:5GPa based on rule of mixtures. The true stress as seen from Fig. 2.6a shows a peak value of ~933 MPa for 3:3% strained pillar as compared to sample with 1.6% which has a value of ~845 MPa. This slight increase of maximum true stress at higher cumulative strain is obviously expected in mostly elastic regime. When the cumulative compressive strain increases to 12.8%, the Cu/Nb nanolayers yield, i.e. onset of plastic deformation at a stress of ~1016 MPa and thereafter shows a linear hardening behavior in the plastic zone until unloading with a peak stress of ~1404 MPa. Tensile experimental response of a similar Cu/Nb free-standing nanoscale multilayers with a [111] Cu [110] Nb texture monitored through a synchrotron X-ray source confirms this clear elastic–plastic transition behavior [90–92]. A plausible explanation for this kind of mechanical response from this in situ study is Cu nanolayer alone deforms plastically whereas Nb behaves predominantly elastic. The dislocation arrays (both parallel and oblique) formed due to the interaction of dislocations during the glide and some deposited at the interface, as the nanolayer has been subjected to plastic deformation, can cause significant work hardening. There will also be a decrease in dislocation array distance λ as the plastic strains increase, which is related to the in-plane Burgers vector b through this relation: $\lambda = b/\varepsilon$. Subsequently, the stress field of the array which produces a force to repel a dislocation glide will eventually increase which will lead to higher flow stress as reported here. From the unloading regime, the elastic modulus is seen to be ~115.5 GPa, which agrees with the previous observed values.

From the plastic flow curve, the modulus $H = d\sigma^P / d\varepsilon^P$, an equivalent measure of strain hardening rate, is found out to be ~8.4 GPa. Since the dislocations lie within the interface between the two individual nanolayer, during deformation they react with the other similarly deposited dislocations from the other nanolayer leading to crystal recovery/annihilations, which may be the cause for this behavior. Further straining the pillar to 24.2% shows a much higher yield value of ~1480 MPa and an elastic modulus of ~98 GPa noted from the unloading curve is similar to the previously observed values. Moreover the nature of the plastic flow curve shows a considerable drop in strain hardening rate with a H value of ~0.9 GPa with a maximum strength of ~1590 MPa at unloading. The local stresses at the interfaces may have build to a level that dislocations can transmit across the interfaces to yield the Nb layer, which may be the reason for this lower work hardening rate. These results of decrease in plastic strain hardening rate with respect to increase in compressive strain are consistent with experimental observations of Al/TiN multilayer composite [93]. Another interesting observation from this stress–strain curve is elastic strain still

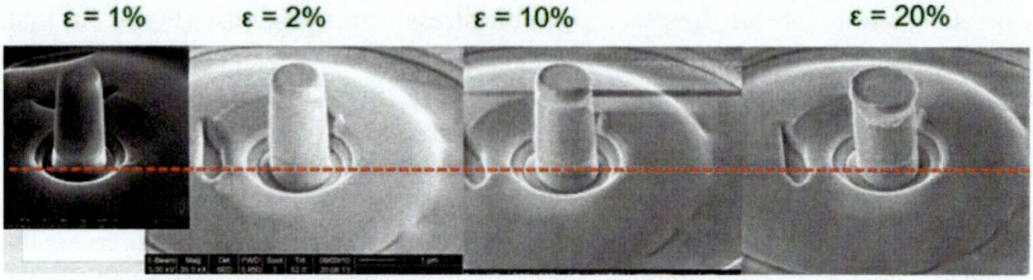

Fig. 2.7 Topography of the pillar after each strain ramp. Reproduced with permission from Elsevier

being slightly present even during this large plastic deformation. A possible recovery of dislocation content at the interfaces could have slightly increased the dislocation array distance λ, resulting in this residual elastic strain. To intuitively understand this limit, one of the pillars was over-strained until 35% of plastic deformation, which showed near zero elastic strain in the nanoscale composite. At this plastic strain, the dislocation emitted from the interface would have completely penetrated into the adjacent layer, leading to slip transmission which has been observed from molecular dynamics simulation [75, 76]. The micrographs of the deformed pillar after each unloading as seen from Fig. 2.7 does not show any shear bands even at this large strains, while a bulge is noticed after a compressive strain of 12.8%. The elastic and plastic deformation behavior of the pillar obtained from these successive loading pattern confirms that indeed the Cu/Nb nanolayer has a high flow strength before yield and is stable until very large plastic strain of 35%.

Synchrotron X-ray microdiffraction Laue studies [94] have confirmed further an increase of statistically stored dislocations in Cu nanolayer, while Nb showed a negligible increase at a plastic strain of 14%. A possible crystal recovery of the Nb nanolayer was proposed for this zero net plastic deformation. Based on these studies, which showed possible crystal annihilation mechanisms involving interface interactions in the nanolayered structure, we describe the unique roles of interfaces in enabling extreme plasticity in the following section.

2.2.3 Roles of Interfaces in Extreme Plasticity

Interfaces play an important role and define their properties of the materials even though they constitute only a small volume fraction of the material. Interfaces play multiple roles in determining mechanical properties in crystalline materials: sources of defects (e.g. dislocations, phase transformation bands, twins and shear bands), sinks for defects via absorption and annihilation, barriers for impeding the propagation of these carriers and preferred sites for storage, reassembly and reaction of interface defects [95, 96]. These defects such as lattice vacancies, solute atoms, dislocations, stacking faults and grain boundaries create the foundations in the plastic deformation of crystalline materials.

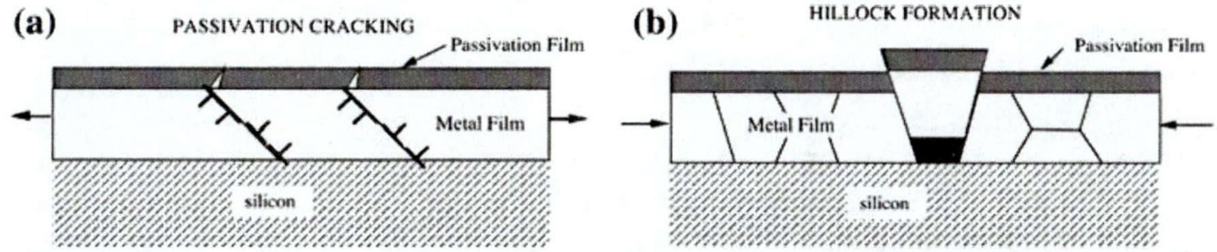

Fig. 2.8 Two mechanisms of inhomogeneous deformation in thin films leading to cracking of the associated passivation film. **a** Dislocation pile-ups can cause passivation cracking, and **b** grain-boundary sliding can cause hillock formation (Reproduced from Demkowiecz et al. [99] with permission from Metallurgical Transactions A)

An example of the influence of interface in the mechanical properties is the growth of thin films on a substrate for commercial applications. The thin films are constrained in one dimension, and the film or layer thickness will usually control the mechanical properties [97–99]. During the growth of the crystalline epitaxial film on a substrate, the film has to be elastically strained to match the lattice spacing of the substrate. When the film becomes sufficiently thick, it becomes energetically favorable to form a periodic array of misfit dislocations. The density of these dislocations is far below that needed to accommodate significant plasticity. Dislocations need to be created or multiplied to support plastic deformation. At low temperatures, the plastic deformation of thin films occurs by dislocation motion with the bowing of pinned dislocations and dislocation pile-up leading to enhanced strength. At higher temperatures, grain-boundary diffusion and grain-boundary sliding become important with the grain morphology playing a significant role in the mechanical properties of the thin films. Sometimes, these same dislocations may cause fracture or delamination. For example, the plastic deformation of passivated oxide on top of a metallic film-substrate caused the fracture to occur either by dislocation slip or by grain-boundary sliding as shown in Fig. 2.8.

Koehler [100] proposed a concept to design a strong solid using the knowledge of dislocations and metallic films at that time. A laminate structure with multiple thin layers of two metals, A and B where one metal has a high dislocation line energy and the other metal has a low dislocation line energy should exhibit resistance to plastic deformation and brittle fracture well in excess to homogeneous bulk metals due to dislocation interaction across the interfaces. This concept was experimentally proven by Matthews and Blakeslee [101, 102] and Lehoczhy [103, 104] where the deformation in the laminate structures is distinct owing to the presence of interfaces.

Metallic multilayered nanocomposites contain a large number of interfaces that give rise to a large number of unique properties. They can be classified into three groups as shown in Fig. 1.6 [105]. Group 1 multilayer nanocomposites are crystalline metal–metal multilayers in combinations of either or both FCC, BCC and HCP crystallographic orientations with similar critical resolved shear stress (CRSS) for slip in both phases. Examples of these multilayer nanocomposites are Cu/Ni, Al/Nb and Ag/W metal–metal multilayers. Group 2 nanocomposites are also comprised of both phases being crystalline and plastically deform via nucleation and propagation

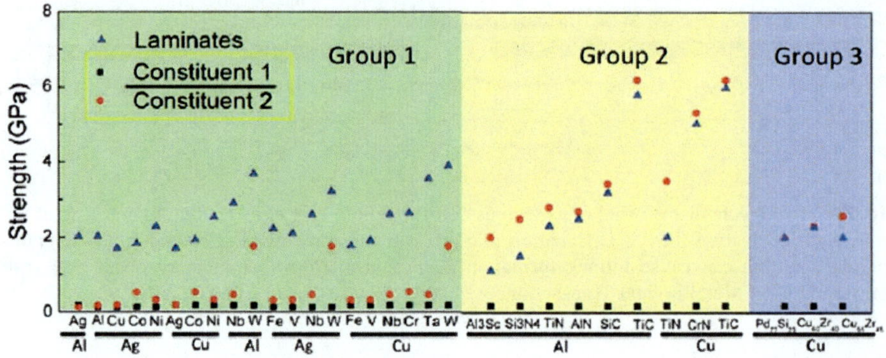

Fig. 2.9 The maximum strength of nanocomposites in accordance to the combination of constituent phases: **a** metal–metal crystalline nanocomposites **b** metal-crystalline hard phase nanocomposites and **c** metal-metallic glass nanocomposites. Black squares indicate the strength of constituent 1 under the black line and red circles represent the strength of constituent 2 above the black line in the horizontal axis. Blue triangles represent the strength of nanocomposites (Reproduced from Beyerlein et al. [105] with permission from Taylor & Francis)

of dislocations, but the CRSS in one constituent phase being much higher than the other. This higher CRSS phase is usually crystalline ceramic or intermetallic together with the metallic nanolayer such as Al/SiC and Cu/graphene multilayered nanocomposites. The last group is comprised of one crystalline phase that plasticity deforms via dislocations and the other amorphous phase where shear transformation zones (STZs) or shear bands accommodate the plastic deformation. The Group 1 multilayered nanocomposites, in particular, Cu/Nb and Al/Nb metallic multilayered nanocomposites, will be the focus of this dissertation (Fig. 2.9).

Interfaces can trap, transmit and annihilate crystalline defects

Multilayered nanocomposites, in particular, metal–metal nanocomposites are known for its ultra-high strength and deformability. Metallic multilayers of soft metals such as copper, nickel and niobium with the strength of a few tens of MPa in bulk can achieve the strength of more than 2–3 GPa [106, 107]. The evolution of film hardness as a function of the layer thickness has been studied through nanoindentation experiments of various types of multilayers where hardness scales inversely with the square root of the individual layer thickness at larger layer thicknesses but tapers off at lower thicknesses as shown in Fig. 2.10 [108, 109]. From these experiments, Misra et al. [106, 110–112] developed dislocation models on metallic multilayers to describe the relationship between multilayer flow strength and individual layer thickness, h for sputter deposited Cu/Nb multilayers. They are the dislocation pile-up model, the confined layer slip model (CLS) and the interface crossing model depending on the range of individual layer thickness one is looking at. At submicron to micron-scale (h > ~50 nm), the Hall–Petch effect occurs where dislocations pile-up at the interfaces til the applied stress plus the stress concentration due to the pile-up exceed the barrier

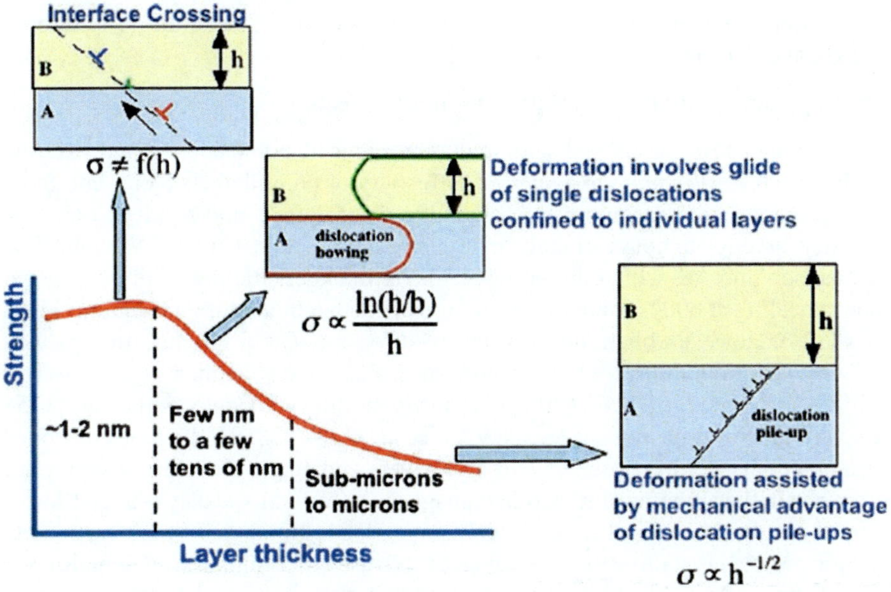

Fig. 2.10 Schematic illustration of the dislocation mechanisms of multilayer strength operative at different length scale (Reproduced from Matthews et al. [110] with permission from Elsevier)

strength and transmit slip across the boundary [114, 115]. Thus, the hardness scales linearly with $1/\sqrt{h}$..

As the layer thickness decreases below 50 nm, dislocation pile-ups may not form, and the deformation behavior is controlled by the nucleation and motion of single dislocations—the confined layer slip (CLS) model and interface crossing. In the confined layer slip (CLS) model, there is a propagation of single dislocation loops confined in the layers parallel to the multilayer interfaces via an Orowan bowing type mechanism [116–121]. The confinement of glide loops to individual layers is because the interface barrier to slip transmission is higher than the confined layer slip stress. Both metallic layers will deform by CLS of Orowan loops to have a slip to propagate through the multilayer nanocomposite.

At very small individual layer thicknesses of $h < 2$ nm, interface crossing by a single dislocation occurs. The critical event that determines the yield strength is the stress needed to push a glide dislocation from one side of the interface to the other [122, 123]. Quantitatively, it is equivalent to the barrier strength of the boundary (in this case, the interface) for slip transmission. Several factors contribute to the interface resistance to single dislocation transmission—(1) Koehler stress [100], (2) coherency stress [122] and (3) other factors such as crystallographic mismatch in the slip systems and stacking fault energy [123]. Hence, if the interfacial structure does not change with layer thickness, the barrier stress also remains independent of the layer thickness. With these dislocation plasticity mechanisms playing a critical role in the mechanical properties of metallic multilayered nanocomposites, interfaces do play an important role in trapping and gliding dislocations resulting in ultra-high strength and deformability.

Interfaces act as sinks for radiation-induced defects

Interfaces can also act as sinks for radiation-induced point defects and external species (such as He), and the enhanced diffusivity of point defects along interfaces could promote rapid recombination of unlike point defects and result in radiation tolerance superior to conventional single-phase bulk metals. Cu/Nb PVD multilayered nanocomposites with 4 nm individual layer thickness did not exhibit blistering upon annealing at 600 °C under the exposure of room temperature irradiation conditions [124]. However, blistering was observed for only Cu or Nb thin films as well as Cu/Nb 40 nm multilayered nanocomposites. Radiation tolerance superiority was similarly observed for Al/Nb, Cu/V and Cu/Mo multilayered nanocomposites [125–127]. Atomistic modeling by Demkowicz et al. [124, 128, 129] using embedded atom potentials (EAM) of the Cu/Nb multilayers and single elements show a clear difference in the amounts of radiation damage in the form of vacancy—interstitial or Frenkel pairs as shown in Fig. 2.11. Collision cascade simulations were carried out on 4 nm perfect crystalline FCC Cu and BCC Nb as well as model Cu/Nb multilayer in Kurdjumov–Sachs (K–S) configuration. Radiation damage in the form of Frenkel pairs was observed for all cases. However, the number of point defects remaining around the Cu/Nb multilayered interface after collisions was significantly lower than in the Cu or Nb perfect crystals. The example in Fig. 1.8a showed no point defects remain in the copper or niobium layers near the Cu/Nb interface as all the vacancies and interstitials were recombined and absorbed into the interface. At these interfaces,

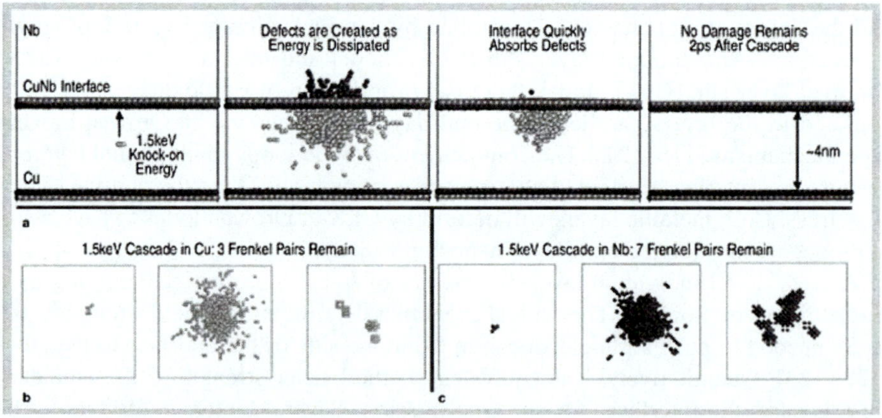

Fig. 2.11 Simulations of 1.5 keV collision cascades performed by Demkowicz et al. in (**a**) Cu/Nb multilayer nanocomposites, (**b**) perfect crystalline fcc copper (light color) and (**c**) perfect crystalline bcc niobium (dark color). Only the defected atomic environments are presented here. The Cu/Nb interface quickly absorbs all of the point defects that would have been created by collision cascades (Reproduced from Mara et al. [124] with permission from JOM)

the vacancy formation energy is much lower, attracting the vacancies and interstitials to get trapped there and undergo accelerated recombination. Thus, interfaces in multilayered nanocomposites exhibit high resistance to ion-irradiation induced damage.

2.3 Extreme Fracture Limits—Fracture at the Small Scales and Emerging 3D Architectures

We have seen in the last section how plasticity at the nanoscales can deviate significantly from the same phenomena in bulk properties. In this section, we will continue with how fracture limits in nanoscale and emerging 3D architectured materials may deviate markedly as compared to the bulk counterparts. Nanoscale metal–metal multilayered materials, where interfaces can be created and modulated in a highly controlled manner and with relative ease, provide a model material system platform to understand fracture deformation based on their interactions with other microstructural features at the nanoscales. Recently, there is a growing trend to understand fracture of multilayered materials to see the interactions between the crack and multilayered interface through novel experimentation techniques. In this section, we will introduce the rationale behind this trend, the current microfracture testing methods to test and analyze multilayer fracture behavior and the challenges faced doing so. Three examples of fracture at the extremes are highlighted in this work—the in situ observation upon tensile testing of film on a substrate, in situ delamination along the layers and the in situ microfracture clamped beam bending technique across the multilayers. Finally, emerging trends in 3D architectures in materials are reviewed especially in terms of the extreme fracture mechanisms.

2.3.1 Fracture at the Nanoscales

Nanoscale metal–metal multilayered materials can be considered a model system to design new interfaces as a possible structural metal. The enabling factor in these multilayered nanolaminates is the interface interaction between the two metals that enables these novel properties of ultra-high strength, sinks for radiation-induced defects, obstacles to deflect and hinder crack growth, and thermal stability up to 800 °C. An example can be seen from the use of the Cu/Nb multilayered nanocomposite, a type of nanoscale metal–metal multilayered composite, as a potential radiation damage tolerant material [130–132]. While multilayered nanolaminates have been studied a lot, the interactions at the nanolayer interface are only understood in detail recently for their mechanical behavior in the elastic and plastic regime.

Fracture and plasticity in nanoscale metal–metal multilayered materials has been observed experimentally. Metallic multilayered nanolaminates of soft metals such as copper, nickel, and niobium with the strength of a few tens of MPa in bulk can achieve the flow strength of more than 2–3 GPa experimentally through nanoindentation experiments [133–135]. The hardness of these nanolaminates scales inversely with the square root of the individual layer thickness, *h*, at larger layer thicknesses but tapers off at lower thicknesses [136, 137]. The flow strength is estimated as the nanoindentation measured hardness divided by a factor of 2.7 [138]. Micropillar compression experiments on nanoscale metal–metal multilayered materials confirmed the ultra-high flow strength values with plasticity mechanisms of the onset of yield, work hardening and shear banding which can be attributed to their interfaces [139–141]. Further, in situ transmission electron microscope (TEM) nanoindentation experiments showed clear evidence of dislocation plasticity behavior along the interfaces of multilayered nanolaminates [142, 143].

Misra et al. [134, 144–146] developed dislocation models on nanoscale metal–metal multilayered materials to describe the relationship between multilayer flow strength and individual layer thickness. They are the dislocation pile-up model, the confined layer slip model (CLS) and the interface crossing model depending on the range of individual layer thickness. Later, these concepts were confirmed by atomistic modeling of the multilayered interfaces to explain the expected plasticity behavior of its constituents [147–151]. Experimentally, Budiman et al. grew nanoscale single-crystal metal–metal multilayered materials [152] and performed synchrotron X-ray microdiffraction (μSXRD) Laue and monochromatic scans on successive ex situ compressions (before and after) at 5 different strain points up to accumulative strain of 35% [153, 154]. The μSXRD technique [155, 156] can characterize the microstructure, plasticity and stress and its evolution in a particular micron/nanoscale location in many crystalline applications in microelectronics, [156–159] solar cells, [156] and nanocrystalline materials [160–162]. For nanoscale metal–metal multilayered materials, the μSXRD technique quantitatively examines the evolution of dislocation densities during the deformation in nanoscale metal–metal multilayered materials through the change in the Laue diffraction peaks before and after each uniaxial compression as geometrically necessary dislocations (GNDs) are directly related to the local lattice curvature. By comparing the X-ray ring width and how it broadens as a function of strains, a relationship between dislocation density and strain in both the Cu and Nb nanolayers was established where a linear relationship was observed up to 3.5% strain followed by the saturation of the ring width broadening at higher strains. This represents an extended regime of linear response in terms of defect density with respect to deformation that was confirmed with resistivity measurements of Cu/Nb micropillar compressions. A linear trend of resistivity with respect to true strain for strains in excess of 3.5% was observed and suggests a significant regime for use for strain sensor/detection/monitoring capability.

So now, in a more systematic way, we want to extend the current knowledge of interfaces in nanoscale metal–metal multilayered materials on their roles in fracture and fatigue. Interfaces in metal–metal multilayered materials restrict the expansion of the plastic zone, resulting in the increase of tensile stresses at the crack tip. Its interaction with this and many other interfaces may result in an enhanced fracture toughness or "cracking resistance" of multilayer by deflection of the crack in the multilayers or even blunting of the crack at one of the interfaces [163–165].

Early work on fracture in nanoscale metal–metal multilayered materials focused on cracks along the interface to understand the damage resistance or delamination between the layers during the fabrication of nanoscale multilayers. Lashmore and Thomson [166] did early experimental work on the fracture of Cu/Ni multilayers of layer thickness 10 nm with a notch along the interface and observed misfit dislocation interactions between the pre-defined notch and multilayer interfaces around it as shown in Fig. 2.12a. The generated misfit stresses amount to a giant Peierls friction stress with the deformation zone highly localized at the crack tip. The dislocations' proximity around the crack tip suppressed continuous emissions, and this scenario favored a transition to a Mode II brittle, cleavage behavior through elastic theory and non-linear atomistic modeling observations by Anderson et al. [168]. Matthew and Blakeslee [58] observed cracks across the interface at certain regions of their fabricated GaAs / $GaAs_{0.5}P_{0.5}$ 25 nm multilayered nanocomposite samples in Fig. 2.12b. One of the cracks stopped at the interfaces by an array of misfit dislocations. During in situ transmission electron microscope (TEM) tensile testing of Cu/Nb PVD 40 nm free-standing multilayered films along the layers, the material exhibits flow stress of 1.55 GPa and elongation to failure of 3.4%. Its fracture surface reveals evidence of ductile fracture with little necking occurring before failure with cracks along the grain boundaries within the layers [169]. Hattar et al. [170] observed the fracture behavior from an initial fabrication crack during the in situ TEM tensile straining experiment of Cu/Nb 60 nm/150 nm multilayers and this fracture was attributed to be due to confined layer slip within the layers containing and preceding the crack tip. They observed four crack hindrance mechanisms—microvoid formation, crack

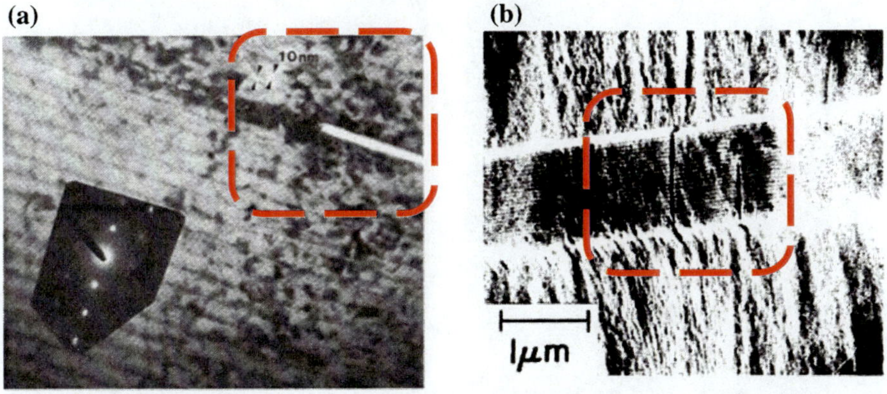

Fig. 2.12 Examples of multilayered nanolaminates with cracks/notches (**a**) along the interface of Cu/Ni multilayered nanolaminates (Reproduced with permission from Cambridge University Press [166]) and (**b**) across the interfaces of GaAs/GaAs0.5P0.5 25 nm multilayered nanocomposite (Reproduced with permission from North-Holland Publishing Co. [167]). The area of interest is highlighted inside the red dotted box

deviation, layer necking and crack blunting as shown in Fig. 2.13. The presence of the Cu/Nb interface influenced the location of the microvoids, redirection of the crack, the extent of necking along the crack path and location of the crack blunting to take place mainly in the Cu layer close to the interface like obstacles.

While the interface effects were known to affect the strength of nanoscale metal–metal multilayered materials via well-known strengthening mechanisms of Hall–Petch effect, confined layer slip and dislocation transmission across the interface [171], their effects in delaying fracture events (crack initiation, propagation) have not been widely reported experimentally. Interfaces have indeed been known to

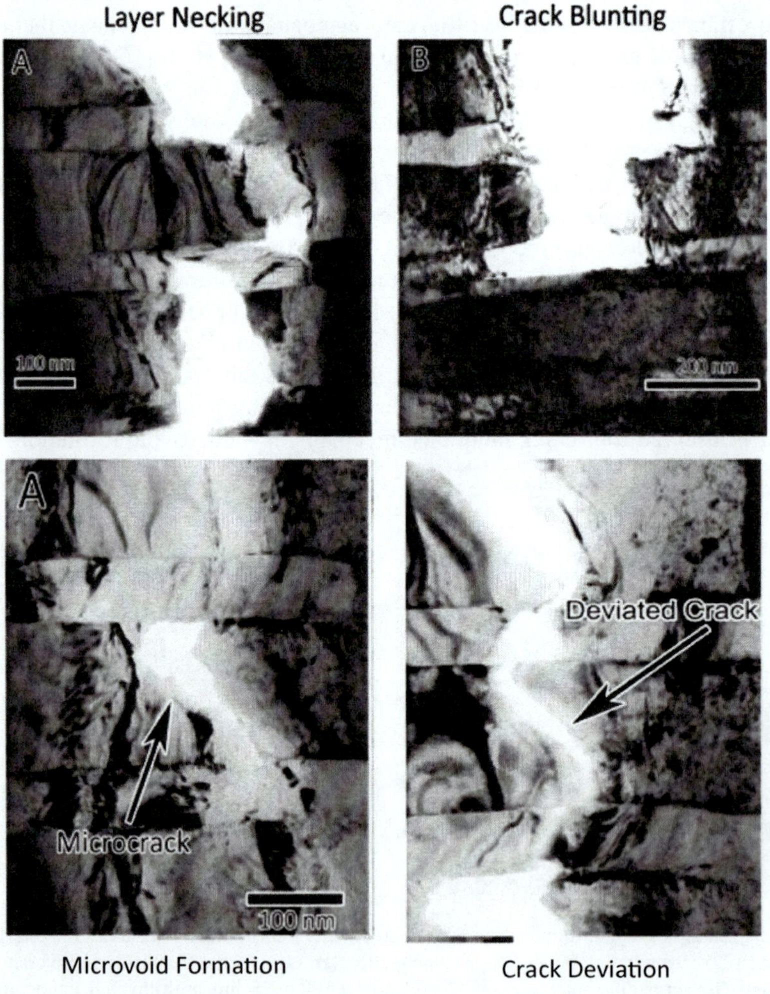

Fig. 2.13 Four crack hindrance mechanisms observed by Hattar et al. in Cu/Nb multilayered nanolaminates (Reproduced with permission from ASME [170])

blunt crack tip in larger layered composite materials, but at the nanoscales, the exact mechanisms through which interfaces interact with the crack tips, and especially in how the interplay of other microstructural features (such as grain boundaries) affects the events leading to the final catastrophic failure has not been much studied. It is only in the last few years that researchers have started looking in detail into the mechanisms behind the fracture along and across the multilayered interfaces. It would be important and crucial to study interface interactions with a propagating crack in real-time during fracture through a small-scale microfracture testing technique. The effects of the constituent layer interfaces and grain boundaries on plasticity, fracture and fatigue of the multilayers can be examined experimentally and will be further evaluated theoretically or through simulations. Current real-time observations of a crack or notch propagating along or across the multilayered nanolaminates in a controlled setting are limited to mainly metal/ceramic multilayered systems [172–174].

So now, in a more systematic way, we want to extend the current knowledge of interfaces in nanoscale metal–metal multilayered materials on their roles in fracture and fatigue. That is because we have known plasticity of multilayered nanolaminates through existing literature of tensile and compression experiments. Now we can embark to learn more about fracture in multilayered nanolaminates through the evaluation of how the multilayered interface interacts in the presence of a crack. Further, fatigue is a combination of both fracture and plasticity. Thus, through the knowledge of what we know about plasticity and will soon know about the fracture, this also means we can have the understanding of fatigue damage mechanisms and especially how they can interact with the interface interactions.

Some Experimental Considerations

The small-scale in situ microfracture technique allows the simultaneous and synchronous recording of the load–displacement information, in addition to crack initiation and propagation events. The fracture load can be precisely noted, and the crack opening displacement (COD) can be determined via direct observation or digital image correlation (DIC) techniques [175]. The development of instruments [176] to both apply and measure loads and displacements accurately and observe the changes and record in situ [69] has enabled fracture testing of materials at micrometer and nanometer sizes to become a reality. Further, the availability of focused ion beam (FIB) machining, lithography, deep reactive ion etching (DRIE) and laser micromachining to prepare samples at small dimensions have encouraged the growth in this area.

Uniaxial loading in both tensile and compression modes have been done in situ for nanoscale metal–metal multilayered materials inside a scanning electron microscope (SEM) as well as transmission electron microscope (TEM) [170, 174, 178–180]. Results from these experiments have provided many insights into the plasticity mechanisms of multilayers at the nanoscale due to interface interactions. However, these methods are not suitable for fracture studies. In the tensile mode, the maximum tensile stress is experienced throughout the entire volume of the testing sample and fracture may not happen at the weakest direction for the crack driving force. If there are sources of weak grain boundaries or voids, lower values of fracture strength may be attained for the same material through tensile mode than through bending. In compression, the crack is aligned parallel to the compression direction. When the crack propagation is the controlling mechanism of failure, a larger amount of stress is required to fail a material in compression compared to tension. However, the maximum tensile stress is conversely concentrated in a small region on the top surface above the neural axis in bending mode, making it apt for fracture. Further, ASTM standard tests to measure crack propagation and fracture toughness in bulk materials are performed in bending only [181, 182].

Consequently, the conventional methods for evaluating fracture in materials in the bulk form cannot be applied such that new experimentation techniques are required in the micron and nanoscale. Researchers have proposed non-conventional solutions of single cantilever bending [183–185], double cantilever beam (DCB) [186] pillar splitting [188, 189] and clamped beam bending [189, 190] to observe fracture in situ and allow for fracture toughness measurements for various kinds of materials. To identify key interface parameters enabling enhanced fracture and fatigue properties in nanoscale metal–metal multilayered materials, we would want to choose a method that is suitable for interfaces, allows us to view interface interactions with the notch and/or crack in situ in the line of sight, scalable at any length scale and a stable crack growth.

In the past, researchers tend to peel away the film from the substrate after deposition to obtain just the multilayered film [169, 191]. This method was prone to create film samples of random sizes which were hard to control and keep for multiple uses. In recent years, the FIB has been used as a machining tool to modify materials at the micro- and nanoscale [192–194]. The focused ion beam, also known as FIB, involves the use of a finely focused beam of ions inside an SEM to mill or sputter off materials or deposit materials in a particular location for specific applications. This technique saved fabrication time and provided more control over the beam and notch dimensions. A liquid metal ion source, which can be gallium, helium or xenon is placed in contact with a tungsten needle, and the heated source wets the tungsten and flows to the tip of the needle. A Taylor cone is formed with the tip radius of ~2 nm. The electric field around this small tip causes ionization and field emission of the metal ions. Based on the beam current setting, the FIB beam can be used for imaging and site-specific sputtering or milling.

FIB is inherently a destructive technique to the specimen. When the high-energy ions bombard the sample, they will sputter atoms of the sample from the surface. The bombarding atoms (e.g. gallium) will also be implanted into a few nanometers from the surface, making the surface amorphous. Care needs to be taken to minimize the exposure of the milling as much as possible with a staggered milling process to make the structures and a final low voltage and current slow milling (or "polishing") in the end. However, just like in any destructive process, the remnants from the milling process have to be deposited elsewhere. The milling steps such that the areas of interest, (where we want to understand the effect of the interface on their mechanical properties) need to be planned such that the key areas do not get deposited with the remnants. Examples of critical areas include the front and back sides of the beam where we view the multilayered interfaces and crack propagation from the notch across the layers. Any presence of amorphous surface due to the deposited milled material on these critical areas affects the results and analysis.

The accuracy of the FIB milling depends on the beam spot size at the surface being milled. An example is how the pre-notch is created by a single cut across the beam of interest. However, this step results in the back notch being smaller in width and height than the front notch. If there is significant beam drift during the FIB milling, an uneven, wide notch formation is formed instead of the expected sharp, narrow notch for that particular beam.

There has been some literature on the effect of gallium (Ga^{+}) induced damage on the FIB milled microstructures and their plastic deformation mechanisms. The relative depth of ion penetration and damage in materials has been observed through experimentation and simulation [94, 95]. For example, the penetration depth of 30 kV Ga^{+} ions in Cu is calculated to be ~24 nm [195]. Xiao et al. [197] recently investigated the effect of gallium as well as xenon ions on single crystalline and ultrafine-grained aluminum cantilevers and pillars. The minimal effect on the yield strength was observed for single crystalline aluminum samples, but gallium deposition along the grain boundaries of ultrafine-grained aluminum was observed and had a significant impact on the mechanical properties of the samples.

However, it is difficult to conclude that the effects observed in the in situ microfracture experiments can be influenced by the existence of the surface damage layer. Burek and Greer [198] did a comparison study of gold and copper nanopillars made by FIB as well as electron beam lithography followed by electroplating and observed similar mechanical behavior from both techniques. Other issues of the FIB milling technique include the rounding at the edges, different milling rates for heterogeneous materials, a non-uniform foot geometry complicating the control and assessment of the final dimensions [199]. Despite these concerns, FIB is being used extensively as the choice method in the fabrication of micron test beams for in situ small-scale microfracture tests [183, 200].

Once a crack starts to propagate in the beams upon loading, fracture mechanic concepts of linear elastic fracture mechanics (LEFM) for brittle materials or elastic–plastic fracture mechanics (EPFM) can be applied for nanoscale bulk materials depending on the load, crack length and geometry of the tested structure [201]. However, for multilayers, this scenario is not as straightforward, taking into account the material inhomogeneity effect [202]. The interface of multilayers can be a potential crack initiation site or a crack arrestor since the crack path depends on the orientation of the original crack with respect to the interface plane. The architecture of the multilayer structure at the nanoscale can be tweaked to enhance the crack resistance by metal combination, texture, grain size, coherency and growth strains and defect distribution.

Take a scenario of a multilayer of two different metals, M1 and M2 where $E_{M1} > E_{M2}$ and $\sigma_{y\_M1} > \sigma_{y\_M2}$. The alternating stacking of M1 and M2 metal layers will bring out sharp gradients in the residual stress along with the elasto-plastic property mismatch along the coating cross section [102–104]. As the crack tip travels across an interface from M1 to M2, the crack is able to propagate across due to the anti-shielding effect. However, if the crack travels to a subsequent layer from M2 to

M1, crack propagation is hindered and the crack tip experiences a shielding effect [202–204].

As the test structures shrink in size especially in the width and notch height, the necessary stress to generate a dislocation at a crack tip becomes very high, significantly larger than the typical yield stress of a coarse-grained metal or alloy. The far-field stress will decrease to a lesser extent compared to the elastic stress field. As a result, dislocation sources become fewer and those dislocations being generated may escape from the sample too.

The way the test structures are loaded and tested will affect the fracture behavior, especially in EPFM. When tested in bending, besides the multilayered structure arrangement, the dislocation movements will be controlled by the dislocation pile-up at the neutral axis of the bending [205]. The extent of dislocation pile-up will not only be dependent on the sample size but also on the orientation of the dislocation glide plane(s) within the sample [206, 207]. The dislocations in front of the crack will be densely packed. However, if tested in tensile, there will be the influence of the multilayers but little on the influence on the crack tip due to the loosely packed arrangement.

ASTM standard fracture mechanics tests have been established for 10 mm and above structures [181, 182]. However, they are not suitable at the micro- and nanoscale. The size of the plastic zone around the crack tip influences the fracture behavior [201]. If the plastic deformation and opening (CTOD) occur within the K-dominated zone, small-scale yielding with the use of LEFM is performed to calculate the fracture toughness of materials. However, if the size of the plastic zone exceeds the K-dominated zone, EPFM is used to describe the stress and strain behavior in front of a crack. One can refer to Dehm et al. [177] for the specific LEFM and EFEM equations that can be used. This technique of the J-integral is established for homogeneous materials [181], but not so much for heterogeneous materials for multilayered systems [163].

Progress is also being made in modeling such as interfaces and studying their fracture behavior via simulations. To quantify the dislocation activities around the crack tip, crystal plasticity or molecular dynamics models are required to analyze the stress the deformation behavior in front of the crack where the relevant fracture processes take place. The dislocations generated usually reduce the elastic stress field at the crack tip—dislocation shielding. Dislocations can also increase the local elastic field depending on their Burgers vector and position—anti-shielding and may induce decohesion at the crack tip. Hence, the crack tip shielding in micro-samples can be different from that of observed in macro-samples. This should result in a sample size dependent fracture resistance controlled by the interplay between the slip distance, specimen dimensions and the extent of the K-dominated zone. The arrangement of the dislocations and the extent of the dislocation-free zone can be affected by the sample size and loading type.

Fractures Upon Tensile Testing of Film on Substrate

With the increasing trend in flexible electronics and paper-like displays, there was a need to study the plastic deformation and cracking behavior of the multilayered film that could be adhered to a polymer substrate. It has been demonstrated that such a film/polyimide composite could be used to investigate plastic deformation and cracking behaviors by applying a certain strain to the composite [208–209]. The group by G.P Zhang investigated on how the plastic deformation is localized, and cracks initiate from the localized damage regions in nanoscale Cu with FCC/BCC/HCP metal (called X) multilayers bonded to flexible polyimide substrates [210–220]. Upon tensile testing of the dogbone sample, the sample is examined in situ to understand the microfracture and mechanisms behind the fracture of the multilayered metallic films.

Equal layers of Cu and X are deposited by radio frequency (RF) magnetron sputtering system under high vacuum base pressure of 1×10^{-7} Torr onto a 125 μm thick polyimide substrate. The total thickness of the deposited multilayer is about 1 μm. They investigated a wide range of individual layer thicknesses from 1 to 250 nm. The composite is then reshaped to a tensile dog bone with a gauge section of 15 mm in length and 5 mm in width as shown in Fig. 2.14b. The tensile test is performed using a microforce testing machine at a constant strain rate of 1.6×10^{-4} s^{-1} up to 5% strain, and then the samples were unloaded. The amount of strain was accurately measured by the extensometer or electrical resistance method.

During tensile testing, the strain and stress were automatically recorded by machine and then Young's modulus of the film/substrate system where

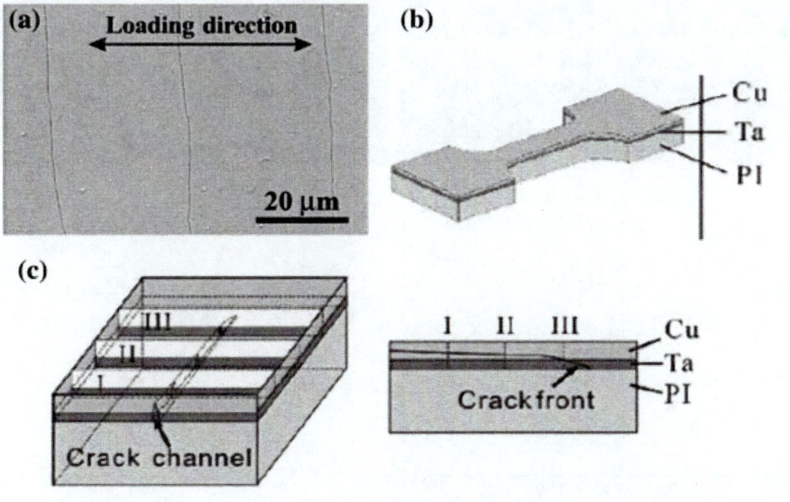

Fig. 2.14 **a** SEM image of the surface of the Cu/Ni multilayered film after the tensile test experiment (Reproduced with permission of Taylor & Francis [215]). **b** Design of the dog-bone-shaped tensile sample of the multilayered film on a polyimide substrate. **c** Schematic illustration of the damage zone ahead of a channel crack in the out-of-plane direction which is cross sectioned at positions I, II and III. To understand the multilayered film fracture, in particular, a position I is chosen ((**b**) and (**c**) Reproduced with permission of Cambridge Core [221])

$$E_f = \left(t_{sb}/t_f\right)(E_{total} - E_s) \tag{2.8}$$

$$\sigma_f = \left(\sigma_{sb}/\sigma_f\right)(E_{total} - E_s) \tag{2.9}$$

A macroscopic strain (ε_c) obtained from the R-ε curves is used to characterize the ductility.

After the tensile test, some cracks formed in the multilayer and grew along the direction almost perpendicular to the loading axis as shown in Fig. 2.14a. To understand the localized deformation across the multilayers, FIB cross sectioning was done at specific locations of interest as shown in Fig. 2.14c. They observed the resultant fracture angle between the macro-cracking direction and the horizontal direction as well as the extent of the crack across the layers. Respective SEM images are often shown thereafter to characterize the fracture behavior as shown in Fig. 2.15a, b.

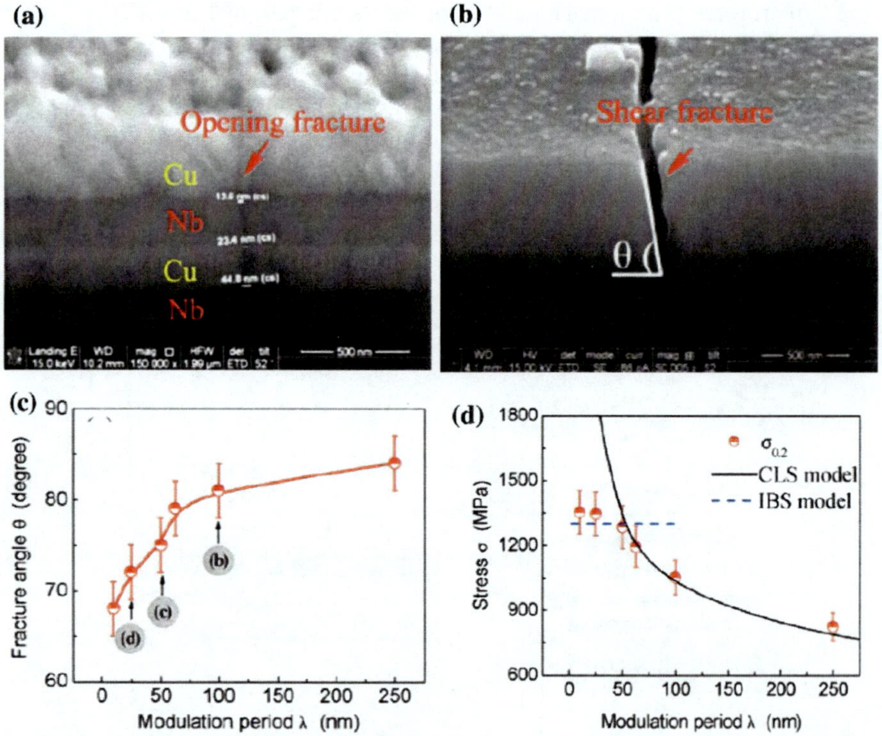

Fig. 2.15 Different fracture profiles as the individual nanolayer decreases for Cu/Nb multilayered nanolaminates from (**a**) an opening fracture at $h = 125$ nm to (**b**) a shear fracture at $h = 5$ nm ((**a**) and (**b**) Reproduced with permission from Elsevier [123]). (**c**) Dependence of theta on λ for the Cu/Cr multilayered films with $\eta = 1$. (**d**) Dependence of stress $\sigma_{0.2}$ on λ for the Cu/Cr multilayers with $\eta = 1$ where the solid curve is calculated from the CLS model and the dash curve is from the IBS model ((**c**) and (**d**) Reproduced with permission from Elsevier [223])

From these studies, they developed a rationale between the trend of fracture at different individual layer thicknesses and theorize them with the known dislocation plasticity mechanisms of Hall–Petch relation, confined layer slip and dislocation crossing across the layers. A large resultant fracture angle with a brittle fractography represented thicker (>50 nm) individual layer thickness whereas a smaller fracture angle represents a ductile shear fracture as shown in Fig. 2.15c, d. They could also quantify the K_{IC} at the notch tip to determine the stresses and trend of fracture. The cracking of brittle films is described analytically with 2D steady-state models, in which the steady-state energy release rate can be described as

$$\xi = \frac{\pi\sigma^2 t}{2E}(1-\nu^2)g(\alpha_1, \alpha_2) \tag{2.10}$$

The fracture toughness (K_{IC}) is expressed as

$$K_{IC} = \sqrt{\frac{E\xi}{1-\nu^2}} \tag{2.11}$$

If strained further until 20%, compressive transverse stresses arise in the film due to a Poisson's ratio mismatch between the substrate and film. The compressive stresses will cause buckling and delamination of the film. The interfacial adhesion energy can be measured from the buckled geometry and film thickness as shown in Fig. 2.16a–d [224].

For Cu/Ta and Cu/Ni on polyimide substrates, fatigue tests were also performed under load control between a maximum tensile load and a constant minimum tensile load of 0.2 N at a frequency of 20 Hz with the same setup. After 1×10^6 cycles were completed at a certain load range, the samples were checked by an SEM. If there was no fatigue damage observed at the specimen surface, the Pm was raised to a higher level until fatigue damage could be observed on the specimen surface by the SEM within 1×10^6 cycles. The trend observed is similar to that of the yield strength of multilayers. The higher the nominal fatigue strength, the lower the individual spacing.

In one of their papers, they have investigated on the use of three-point bending tests on the multilayered film/polyimide substrate system for Cu/Au and Cu/Cr multilayers as shown in Fig. 6e. To ensure substrate cracking prior to multilayer fracture, a brittle 525 μm thick Si (1 0 0) single-crystal was selected as the substrate, and the bending span was set as 6 mm. The width of all specimens was 1.2 mm. Static bending tests were conducted. Manufacturing a pre-crack, loading to induce crack propagation, measuring the critical stress needed to inflict fracture and observing the fracture morphology is followed by a simple procedure to conduct.

During the bending process,

$$\xi_c = -\frac{dU_e}{dA_c} = -\frac{\frac{1}{2}\Delta F^2.dC}{2h.dr} \tag{2.12}$$

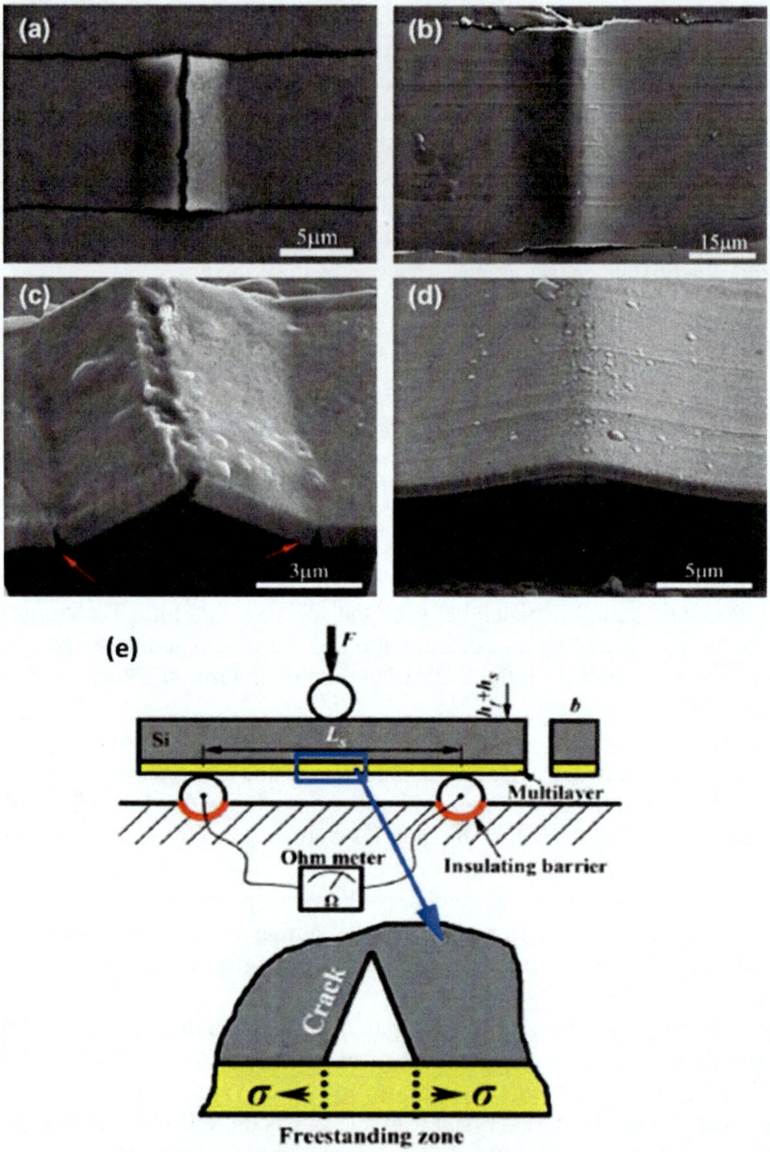

Fig. 2.16 Respective SEM images showing (**a**) cracked buckle in the Cu/Nb multilayers and (**b**) a faultless buckle in the Cu/Zr multilayers, (**c**) The buckles in Cu/Nb multilayers crack not only at the top but also at the base (marked by arrows). (**d**) The buckles in Cu/Zr multilayers are symmetrical and have a smooth arch shape (Reproduced with permission from Elsevier [224]). (e) Schematic illustration of three-point bending apparatus used for localized fracture experiment

$$K_c = \sqrt{E\xi_c} \tag{2.13}$$

The multilayer must be thin enough relative to the substrate in order to eliminate the effects of stress non-homogeneity at the crack tip. The effects of the constituent layer scales, layer interfaces and grain boundaries on plasticity and fracture of the multilayers were examined experimentally and evaluated theoretically.

As expected, a crack was preferentially initiated in the middle of the Si substrate, close to the film/substrate interface due to the maximum tensile stress, and then a local free-standing multilayer was produced ahead of the opening crack. Thus, plastic deformation of the free-standing multilayer subject to tensile stress continued with substrate crack opening until final rupture of the multilayer/substrate system. This method was a slight improvement over the tensile method.

Fractures Upon Clamped Beam Bending of Cu/Nb Nanolayers

As seen in the previous section, performing mechanical experiments in tensile mode until fracture is the easiest way to understand fracture mechanisms in multilayers. However, the formation of cracks in the sample becomes random such that it becomes challenging to analyze the fracture quantitatively. An alternate fracture test geometry needed to be explored to analyze fracture across the multilayers. Clamped beam bending was first demonstrated as a fracture test geometry at the nanoscale on PtNiAl bond coats [172]. It is based on a modification over the three-point bend test by clamping the ends instead of letting them free.

Within elastic limits, simply supported beams deform four times more than rigidly clamped ones [228]. A clamped beam offers a more stable setup with crack arrest after a certain amount of propagation from the notch tip. The position of the maximum principal tensile stress changes from the notch tip to the fixed ends once the *h*/*t* ratio of the crack exceeds ~0.5 [172, 190]. This transition makes the clamped beam bend configuration transform from an unstable to a stable system for crack propagation.

The in situ mechanical experiments are carried out inside the JEOL FESEM (JSM-7600F) using a PI-85 SEM picoindenter (Bruker Inc., USA) with a 20 mN maximum load transducer. Two beams were tested for each material system. The beams were loaded using a conical flat tip of 5 μm diameter and carried out in displacement control mode. A constant loading displacement rate of 5 nm/sec with a Q-control of 0.2, an integral gain of 0.2, a derivative gain of 1 and a specific preload was maintained for all experiments. The load–displacement data and the real-time video of the experiments were synchronized and captured during the experiment using a frame grabber with the TriboScan software (Bruker Inc., USA). Care was taken to perform drift correction before the experiment started and corrected to the final load–displacement measurements.

This geometry offers pure mode I loading as well as controlled mixed mode loading by placing the indenter at fixed offsets from the notch position. The crack can be seen to grow and arrest in stages before complete failure ensues. Such controlled crack growth is important in the measurements of R-curve or fatigue behavior of materials [190, 194] as well as in following crack trajectory and study of toughening mechanisms after the experiment. Jaya et al. [172] did FEM models of the clamped beam geometry for PtNiAl bond coats as shown in Fig. 2.17. The unnotched beam showed maximum tensile stress occurring at both fixed ends and bottom edge center. If the tip was aligned with the notch correctly, the crack was seen to propagate in mode I in the direction of maximum tangential stress toward the loading line. If the tip was offset away from the notch, the crack tended to follow the loading line but did not significantly change the fracture mode based on their XFEM results [172]. Cracks were only observed at the edges when the beams are overloaded.

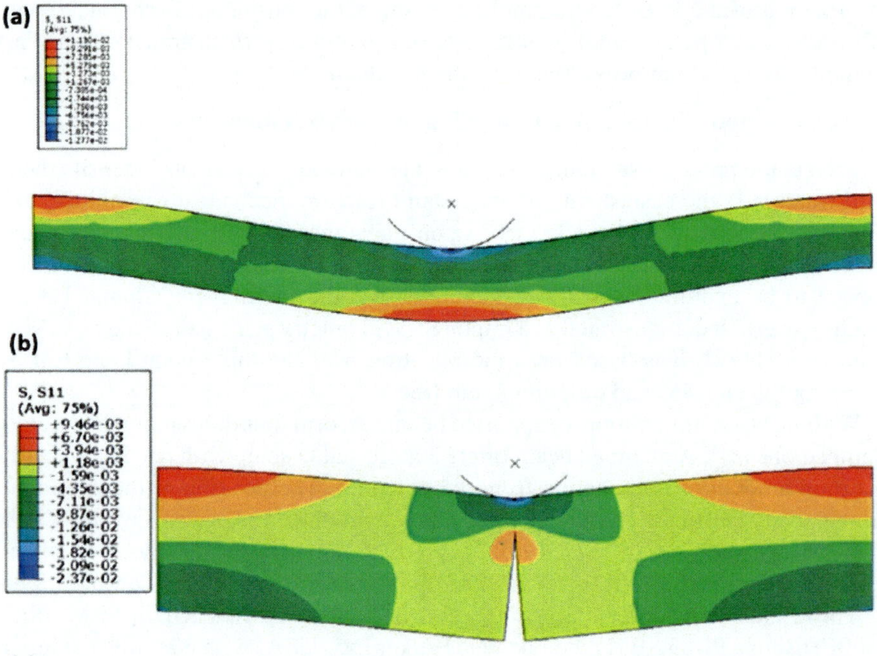

Fig. 2.17 Stress contour plots at a specific load of $P = 0.75$ N using beam specification of PtNiAl bond coats, showing positions of maximum tensile stress S11 and crack trajectories for mode I and mixed mode cases—**a** unnotched beam in bending, **b** 0 μm load offset and **c** 5 μm load offset (Reproduced with permission from Taylor and Francis [172])

The maximum principal stress that opens up the crack acts along the line connecting the notch tip and the indenter. Therefore, the indenter should ideally be positioned precisely across the notch for mode I loading, but misalignments up to 5% of beam length are tolerable. The indenter tip should be sufficiently blunt not to cause indentation impressions or cracking. The test is no more valid if and when edge cracking initiates at the clamped ends. Maintaining larger *l/t* ratios prevents edge cracking. K_I is obtained via an extended finite element method (XFEM) simulations and has a sinusoidal dependence on *h*/*t*. Though the general form of this dependence remains constant for this geometry, each case differing in sample dimensions has to be separately simulated for quantitative determination of K_I in the absence of readily available analytical solutions [189, 190]. If there are cracks greater than the notch height present at one or both edges, the clamped beam might fail prematurely [190].

One might question why not use of single cantilever beams instead of clamped beam geometry to measure the stresses and quantify the resulting K_I. Single cantilever beams are inherently unstable loading systems subjected to maximum shear during the initial loading and the extent of shear to tensile mode drops off at later time steps. Catastrophic failure of the beams occurs, preventing the extraction of information post-mortem. The point of contact does not remain constant since the position is displaced due to the large displacement of the beam in the vertical direction. Hence, the span length becomes variable. Load offsets cannot be quantified easily with single cantilever beams too. However, there have been attempts to use notched cantilever beams to study fracture in multilayered nanocomposite beams [229, 230].

There are challenges in performing fracture toughness values to determine the critical stress intensity factor, K_{IC} or *J*-integral. This is especially so for clamped beam bending geometry where there is lack of analytical formulation for K_{I}. Since K_{I} does not follow a monotonically linear relationship with increasing *h*/*t* ratio, FE modeling is performed to calculate K_{I} [172, 189, 190]. Standard geometries used to determine fracture toughness such as single edge notch bend (SENB) [181] cannot be scaled down easily due to limitations in specimen size and instrumentation. It is complicated to obtain a free-standing thin film specimen mounted on roller supports to replicate SENB test. That is why non-standard geometries such as single cantilever and clamped beam geometries are used.

2.3.2 Fracture Interaction with the Interfaces in Nanolayers—Interfacial Sliding

Another interesting part of Cu/Nb accumulative roll-bonded (or ARB) multilayered nanolaminates is that they have a large anisotropy of the interface shear strength between the rolling (RD) and transverse (TD) directions. To study this effect, Radchenko et al. [225] studied the effect of interface shear strength on the failure in Cu/Nb 63 nm ARB multilayered nanolaminates using the in situ clamped beam bending microfracture technique experimentally based on molecular dynamic results by Demkowicz and Thilly [231]. The results showed a substantial difference between the fracture behaviors along the RD and TD owing to differences in the interface shear strength and grain size. The maximum relative notch width increased before crack initiation was between 50 and 100% for the RD beams and exceeded 400% for the TD beams depended on the material of layer at the notch tip. Through confirmation of the stresses with FEM, [225] they identified that the notch widening is due to plasticity for the RD beams whereas the influence of interface shear is prevalent in TD beams.

The fracture at the nanoscales provided insights into identifying the specific material (Cu or Nb) influencing the notch widening and eventual fracture across the beam as shown in Fig. 2.18. The material was identified from the low-angle backscattering emission (LABE) SEM images. Shear instabilities were formed when the Nb layer was at the notch tip followed by crack initiation. If the Cu layer was at the notch tip, necking and crack happened at the Cu layer followed by slip bands and crack initiation at the subsequent Nb layer. The intricate details of the fracture were captured all

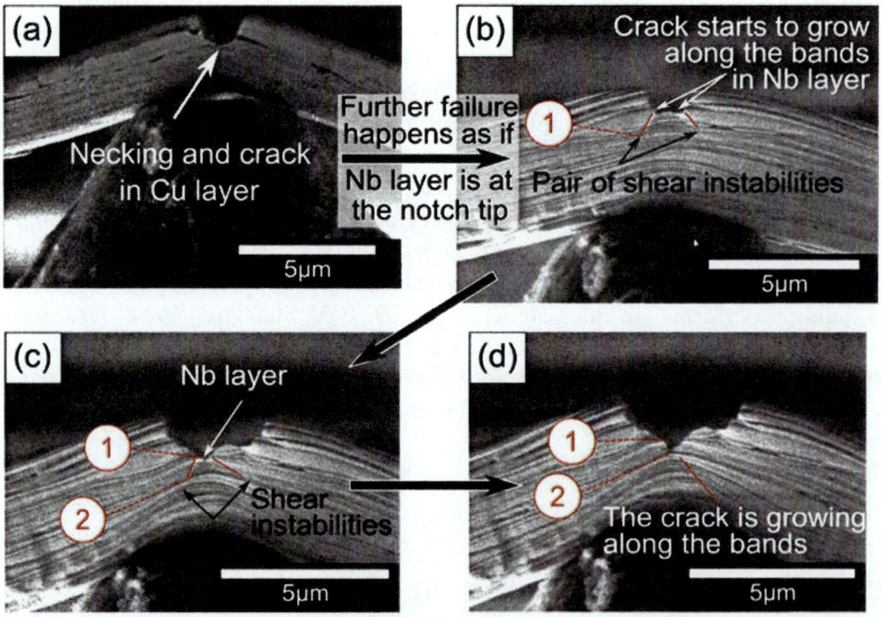

Fig. 2.18 Crack initiation and propagation in the RD beam that varies depending on the layer material (Cu or Nb) on the notch tip. The example shows if Cu was at the notch tip. **a** After some notch widening, necking occurred with a crack initiation. **b** As the crack propagates, a shear instability is formed in the subsequent Nb layer, and the crack starts to grow along the bands in the Nb layer. **c** The crack propagates along one of the shear instabilities until the end. **d** The crack continues to propagate along one of the shear instabilities toward the loading line until it becomes close to the beam surface (Reproduced with permission from Elsevier [225])

due to the in situ technique. Thus, the slip bands originated from the Nb layers at the notch/crack tip in the RD beams followed by crack propagation along these bands.

For the TD beams, the crack propagation is inhibited by interface shear. We suggest that shear bands form subsequently through the beam and lead to the final beam failure. Though the FE simulations for the TD beams show the similar result through complication in the beam stress evolution [225], it could not capture one of the actual experiment beam cases of crack initiation as shown in Fig. 2.19. The influence of the grain boundary at the notch tip resulted in a sharp microcrack formation with limited notch widening to failure. The interface shear in the TD beams could not be inhibited due to the presence of the grain boundaries near the stress concentration zone. The crack growth can be attributed to the formation of microcracks at the grain boundaries besides the main crack. Further, the differences in shear instabilities between RD and TD beams were only understood through the videos of the in situ testing. Post fracture beams might not have captured all the in-between events and result in just speculation.

Nanoscale metal–metal multilayered materials are known to be fabricated by physical vapor deposition (PVD) or sputtering. However, they can also be made by the top-down process of accumulative roll bonding (ARD). Not much work has been done to look into the differences in the interface interactions of crack propagation in real-time in nanoscale metal–metal multilayered materials made by these two fabrication processes. Cu/Nb nanolayers of the similar individual layer thickness of 20 nm are chosen based on the vast literature of its elastic and plastic properties. Fracture of the

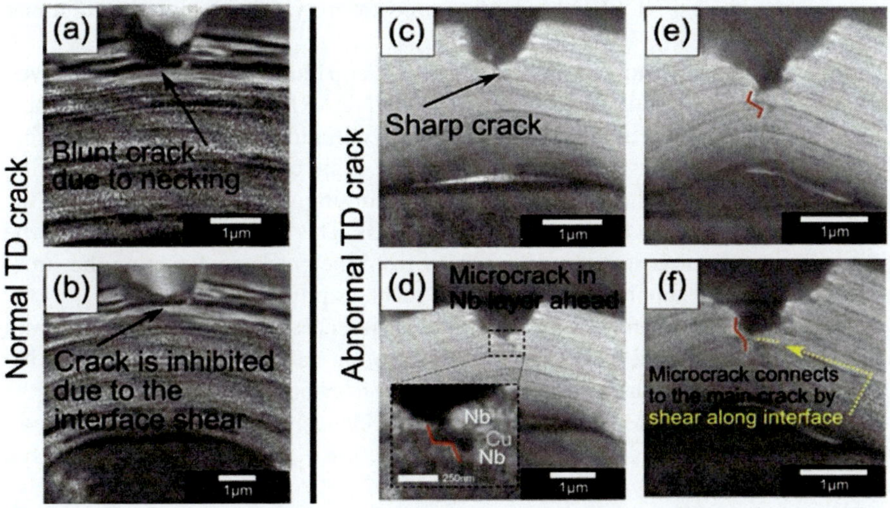

Fig. 2.19 Crack initiation and propagation in the TD beams. **a** The crack initiates in the middle of the notch for most of the tested beams, **b** followed by interface shear at the subsequent layer. However for one of the beams, **c** sharp crack initiation in the layer at the notch tip with no significant interface shear. **d** This is followed by the microcrack formation in the Nb layer besides the main crack [225]

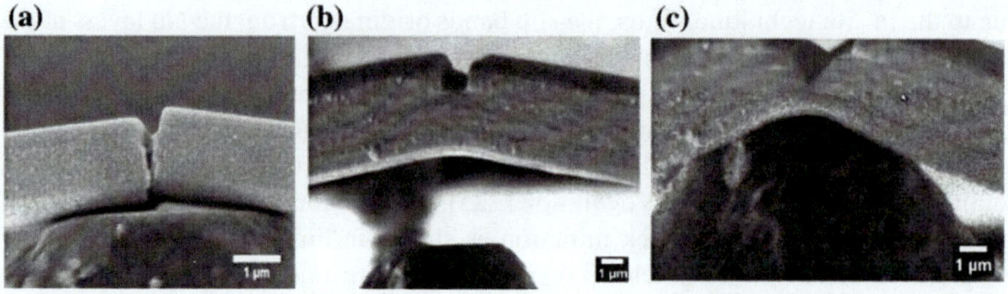

Fig. 2.20 **a** Cu/Nb PVD beam's catastrophic fracture is observed at the end of the test with the crack propagating nearly parallel to the direction of the applied load across the full beams. **b** Notch widening is observed for the Cu/Nb ARB beam. Initial crack from the back of beam while the notch is widening. **d** The crack starts to propagate and tear through the beam with a shear instability is formed at the end of the crack tip, resulting in a ductile fracture (Reproduced with permission from Elsevier [226])

beams across the multilayers can be observed instantaneously. It was unexpected to observe a sudden catastrophic brittle fracture for the Cu/Nb PVD beams as shown in Fig. 2.20a. A linear elastic load–displacement curve was observed up to the max load before a sudden breakage across the layers. Further, from their in situ video, no sign of slip band or shear instability being formed before the fracture is observed unlike what was observed in the Cu/Nb micropillar compression (successive *ex-situ* micropillar compression) of the similar layer thickness [153, 154]. The presence of the multilayers did not attempt to stop the fracture. The fracture only lasted 1–2 frames. Thus, the Cu/Nb PVD beams fractured before yielding along the columnar grain boundaries.

More interesting fracture events leading to crack tip blunting effects were observed for the Cu/Nb ARB multilayered beams. The Cu/Nb ARB beams experience significant fracture events such as notch widening, crack initiation and propagation and final fracture along a shear instability as shown in Fig. 2.2b, c. The large notch widening in the beams is due to the plastic dislocations emitting within the layers below the notch tip, attempting to blunt the crack tip [232]. The dislocations are able to be emitted in the absence of confinement away from the crack tip, lowering the stress at the crack tip and suppressing cleavage as much as possible.

These phenomena can be attributed to the differences in the interfaces and grain sizes and shapes created by the different fabrication paths of the same Cu and Nb multilayers. However, further detailed examinations of both fracture behaviors of the materials at the nanoscales showed competing mechanisms between individual grain structures and interface properties during the fracture of the nanoscale multilayered materials.

The Cu/Nb ARB beams experience significant fracture events such as notch widening, crack initiation and propagation, and final fracture along a shear instability. The large notch widening in the beams is due to the plastic dislocations emitting within the layers below the notch tip, attempting to blunt the crack tip as shown

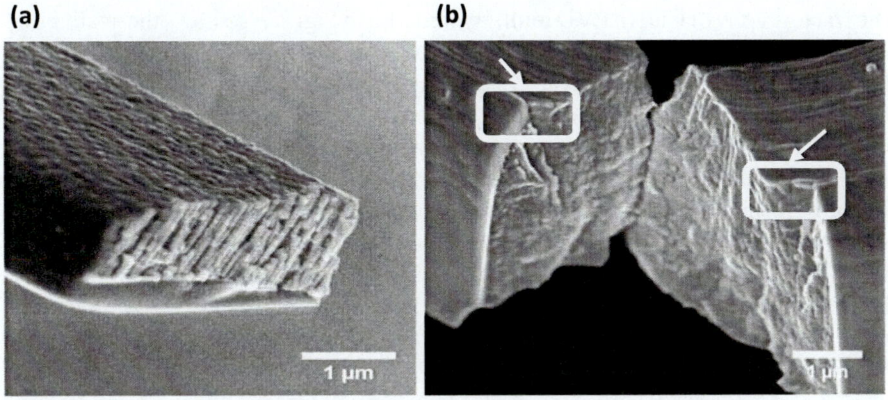

Fig. 2.21 **a** The fractured Cu/Nb PVD beam cross section shows the crack propagating through the columnar grains. **b** Fracture profile of the broken Cu/Nb ARB 4 μm thick beam from the inside. The notch widening is observed at the edges highlighted by the white boxes with arrows (Reproduced with permission from Elsevier [226])

in Fig. 2.21b [232]. The dislocations are able to be emitted in the absence of confinement away from the crack tip, lowering the stress at the crack tip and suppressing cleavage as much as possible. In a layered material, the dislocations are confined by the adjacent, less ductile layers. These confined dislocations send back stress to the crack tip which impedes further dislocation emission [232]. The applied stress intensity is increased after some extent of blunting till it reaches the cohesive strength of the layers in the multilayered beam. The ARB multilayered beam starts to fail by necking with a crack formation and cleavage fracture. Thus, the crack initiated from the backpropagated through the beam. Once the plastic zone approached the bottom of the beam, the tip of the notch started to interact with it. This interaction can be associated with the formation of a shear instability as shown in Fig. 2.20c. The ARB interface, which is known to have a high interface shear strength compared to the PVD interface, permits slip transmission and thus promotes nearly crystallographic slip bands [233]. The crack continued to grow along the shear instability. This crack extended all the way through the remaining beam thickness to its final failure. A similar shear band failure was also observed during the microcompression of Cu/Nb ARB pillars [234]. As a result, a ductile fracture was observed for the Cu/Nb ARB beams due to the dislocation plasticity behavior at the Cu/Nb ARB interface confirmed from the fractography in Fig. 2.21b.

In contrast with the Cu/Nb ARB beams as described above, the Cu/Nb PVD beams experienced a sudden catastrophic fracture event, typical of brittle fracture in the present experiment. In earlier nanoindentation [141] and micropillar compressions [139, 140, 153, 154, 235] experiments using Cu/Nb multilayered nanolaminates, significant plasticity in the Cu/Nb nanoscale multilayered materials was observed due to the influence of the multilayered interface. Thus, at 20 nm layer thickness, the Cu/Nb PVD beams fractured before yielding along the columnar grain boundaries.

The fractography of Cu/Nb PVD multilayered film in Fig. 2.21a shows the availability of grain boundaries through the columnar grain layers for the crack to extend along the grain boundaries. Both layers had a small grain size which made it difficult to deform by dislocation sliding. The columnar grains in Cu and Nb layers were aligned, forming an almost perpendicular fracture path to the tip loading condition [169]. Due to the stress concentration and loss of the interface barrier at the crack tip, the deformation was confined to a local area which severely limited the plasticity of both Cu and Nb layers. Similar observations were made for Cu/Cr multilayered films [213]. The sites of crack initiation became random due to the inhomogeneous distribution of the cohesive strength of the columnar grain boundaries, which lead to the formation of the grain sized steps or cavities at the fracture surface of Cu/Nb PVD beams.

Delamination along the Multilayered Interfaces

The simplest approach to understand the delamination of a particular interface is to perform nanoindentation tests to calculate the interfacial fracture energy by determining the buckle geometry around the indents. However, not all the interfaces will buckle upon indentation. Other explored methods include notched micro-cantilevers [236] and double cantilever beams [186] to understand delamination or adhesion between two layers of the nanoscale metal–metal multilayers. However, all these techniques will test the weakest of all the interfaces in the multilayer stack with little control over the interface of interest. Kiener et al. found a more systematic way to understand the propagation of a crack along the layers using notched cantilever beams [229, 230, 237].

Before the actual in situ fracture experiment, residual stresses and stress gradients were measured through an improved ion beam layer removal (ILR) method [237, 238]. Cantilevers were prepared from single, bi- and tri-layer systems with an individual layer thickness of 500 nm using focused ion beam machining. Subsequently, residual stress profiles were determined with a depth resolution of 50 nm, employing the ion beam layer removal method.

To determine the stress profile, the cantilever is cut free on one side and the thickness of the film in the rear part of the sample, called ILR-area with a length of approximately 15 μm, is reduced gradually using the FIB. Thus, the curvature changes in the ILR-area. The rest of the cantilever stays unaltered and acts as an amplifier for this curvature. The cantilever deflection is measured at the tip. Due to stress redistribution, this deflection changes gradually with each step of thickness reduction [238, 239]. The evaluated film stresses in individual layer depend on the layer order and can be tensile or compressive. An example of a residual stress profile is W-Cu-W done by Treml et al. [237] as shown in Fig. 2.22. As single layers deposited on a silicon substrate at room temperature, both W and Cu exhibit compressive behavior on silicon. However, in the combination of W-Cu-W, the residual stress profile changed from compressive to tensile when transiting from W to Cu with pronounced stress gradients. This technique highlighted the need to consider residual stresses in cantilever beams together with the substrate.

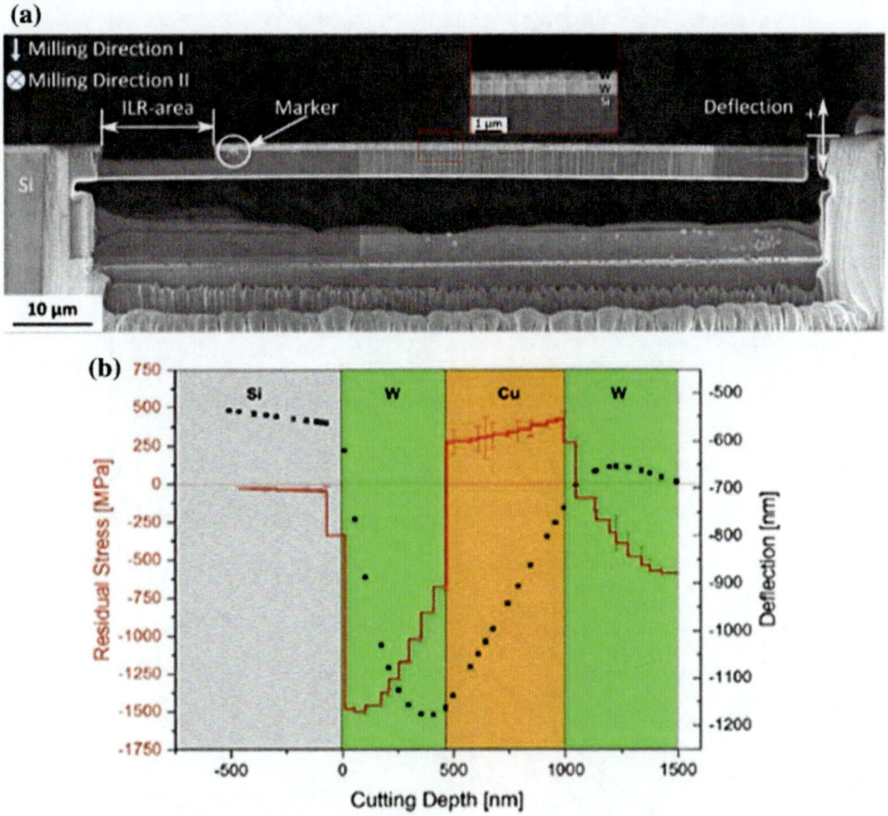

Fig. 2.22 **a** The cantilever beam with the multilayered (W–W) film on top with the indication of the ion beam layer removal (ILR) area close to the support. **b** Deflection and residual stress profile of a W-Cu-W multilayer on a Si substrate (Reproduced with permission from Elsevier [237])

Kozic et al. [229] and Treml et al. [230] used the FIB notched single cantilever bending combined with non-linear FEM analysis to calculate the J-integral at each stage of crack growth as the crack propagated along the W/Cu/W layers under displacement control of 1 µm/min. They observed that a crack tip transition from a softer/more compliant layer into a harder/stiffer layer enhanced the crack tip shielding effect and retarded the crack driving force. The intrinsic fracture toughness of single material layers in complex, multilayered films in terms of J-integral utilizing the concept of configurational forces together with the residual forces.

Another way to initiate stable interfacial delamination within a multilayer system is by the four-point bending of microbridges by Mead et al. [240] Microscale pre-notched bridges with clamped ends were machined into an Al/SiN/GaAs multilayer using FIB milling as shown in Fig. 2.23. A square flat-end indenter was used to induce bending of the bridge by two contact locations. Bridge failure occurred via substrate fracture at the pre-notch, followed by crack deflection, and stable interfacial

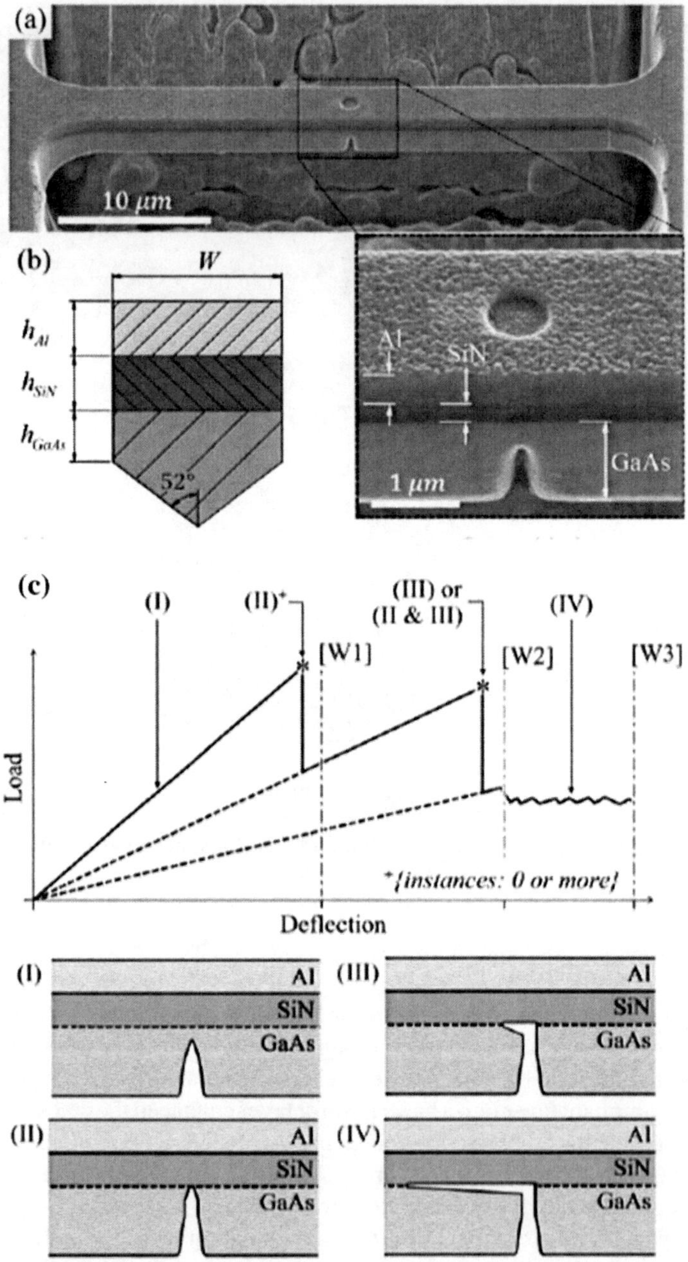

Fig. 2.23 **a** SEM of the clamped beam viewed at 45° tilt. Al, SiN, and GaAs components are labeled. **b** Indicative beam cross section. **c** Indicative load–deflection curve with position labels (I–IV) to explain the changes to the beam during the loading (Reproduced with permission from Elsevier [240])

delamination of the SiN/GaAs interface. Substrate fracture and delamination were identified within the obtained load–displacement curves as a pop-in and region of linear load reduction, respectively.

The in situ delamination along the layers technique highlighted the need to measure the residual stresses of the individual layers before performing the fracture across the layers by the ion layer removal method. The directed notch cantilever beam method of the film on the silicon substrate allows for the crack to propagate from the notch to the interface of interest along the layers with ease of calculation using the J-integral method.

2.3.3 Novel Fracture Mechanisms with Nature-Inspired 3D Architectures

Many natural structural composites are made of relatively weak constituents. However, their hierarchical arrangement results in a great combination of mechanical properties (e.g. toughness and low density), difficult to achieve in man-made materials [242–249]. This has raised the interest of researchers to investigate the structure–property relationships of natural materials. Wood, bone and nacre are interesting natural materials that have high load bearing capacity. In particular, human bone is demonstrated to possess optimal combination of mechanical properties (i.e. stiffness, strength and toughness) [249, 250]. Moreover, they have an ability to deflect crack and absorb high energy before they undergo catastrophic failure [251, 252]. Indeed, bone has been the object of research for many years and currently constitutes a source of inspiration for many nature-inspired materials [249–253].

Mimicking the structure and design mechanisms of the natural composites that are found in connective tissues, bones, exoskeletons [255, 256] can offer fascinating prospects for designing strong and novel synthetic composites. The natural composites are obtained by the self-assembly process and are found to be made of macromolecular building blocks. At the structural levels, the molecules are essentially designed, synthesized and are self-assembled to obtain a higher order micro or macrostructure [255, 256]. This self-assembly process and hierarchical structural design gives superior properties to the material. The proposed study aims to mimic the helicoidal fibrous structures found in the exoskeletons. The remarkable properties of this exoskeleton are attributed to the hierarchical design of the basic building blocks.

A particular study investigated the microstructure of mantis shrimp's dactyl club as shown in Fig. 2.24a to explain the exceptional impact resistance and mechanical properties of these materials [256]. These arms can strike prey with the speed of a 5.56 mm rifle bullet. Impact generated by the force is around 50 kg, which is much more than the weight of the mantis shrimp [257]. Investigations by Weaver et al. [241] have also revealed the complex interplay between microstructure, mechanical properties and impact mechanics of the dactyl club from Odontodactylusscyllarus.

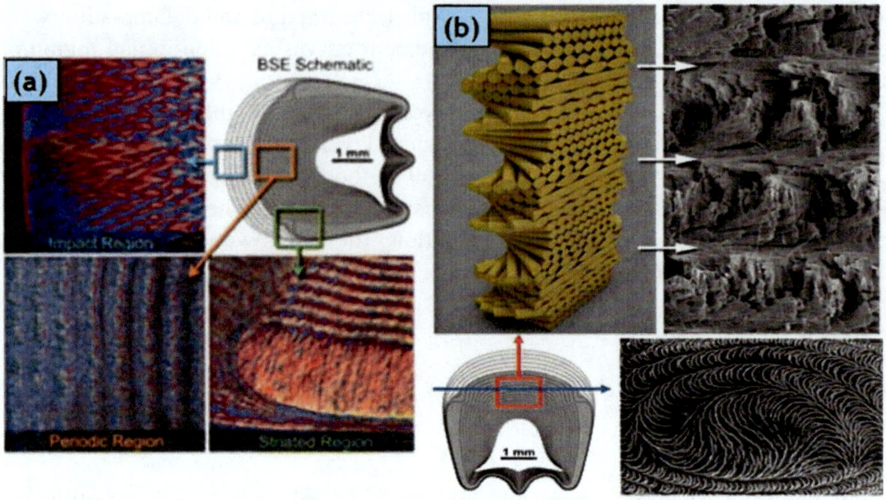

Fig. 2.24 Mantis shrimp a. Club-like arm structure and b. 3D helicoidal architecture of the periodic region which is believed to be the source of the fracture toughening and impact resistance mechanism [241]. Reproduced with permission from Elsevier

Their findings have showed that the club is divided into three distinct regions: (i) the impact surface made of highly crystalline apatite that is near or in direct contact with the external environment, (ii) the periodic region made of a rotated plywood of chitin/amorphous calcium carbonate (ACC) whose main role is to protect the tool against internal damage and (iii) the striated region made of mineralized chitin fibrils running in a nested fashion around the club, which presumably prevents expansion during impact. The periodic mineralized chitin fibril region examined in this study shows a helicoidal architecture (Fig. 2.24b). The advantage of this architecture is, it can dissipate the energy from the impact region through quasi-plastic compressive responses, thereby providing a fracture toughening barrier that hinders catastrophic propagation of micro cracks when subjected to repeated impacts [1, 2]. This damage resistant property explored could lead to potential applications of reinforced ceramics in medical implants and also in defense as soldier body armor materials.

Additive manufacturing (AM) has of late been gaining a considerable amount of attention due to its many possible applications in a wide range of industries; indeed exploring the areas of applicability of the technique is a very active area of research [258]. With the latest 3D printers capable of printing materials with widely contrasting mechanical behaviors simultaneously in complex geometries at micrometer resolutions, the potential of this technology is growing. Using a 3D printer capable of two synthetic polymers in an optimized geometric pattern, Dimas et al. have produced samples of synthetic materials [259] that have fracture behavior similar to bone. More recently, Obaldia et al. [260] has also used 3D printing method to mimic rod-like structures in the teeth of Chitons. These studies demonstrate 3D printing as a means to create fracture-resistant composites inspired by nature.

Learning from nature is an important approach and a prominent scientific endeavor, and especially enabling the nature-inspired design by novel, innovative additive manufacturing (AM) method is an increasingly crucial area for research and technology development for a manufacturing-intensive economes. This research program to investigate detailed mechanisms leading to the enhanced and tunable fracture toughness and fracture resistance of the composite materials will also be significant both scientifically as novel materials science endeavor as well as technologically for the development of novel, unique advanced materials that can be useful for modern societies.

The development of the advanced lightweight, high-performance materials will be highly useful and potentially very productive for human-related or wearable protective technologies such as in soldier body armors and sports/athletic gears, especially since this project could lead to lower cost than existing products in the market (due to the additive manufacturing approach that is highly scalable and economically viable). Since tough and strong composites are used in a wide range of industries such as (a) ***Aerospace***: propellers, aircraft panels, (b) ***Security***: ballistic protection, breathable protective filter media, (c) ***Biomedical***: tissue scaffolds, wound dressing and (d) ***Sports goods***: helmets and protective gears; this technology will build and transform infrastructures for the society.

Electrospinning to Enable 3D Architectures Learned from Nature

Electrospinning is a novel fabrication technique to precisely deposit electrospun fibers in a controlled fashion to obtain 3D hierarchical structures. In this study, layers of controlled geometry and structure were obtained by precise, repetitive deposition of electrospun fibers similar to a 3D printing scheme. This has also demonstrated highly aligned fiber-structure samples having enhanced tensile testing properties [261, 262]. This section combines the aspects of electrospinning and additive manufacturing to obtain 3D hierarchical structures. This approach to obtain 3D structures can bring out similar advantages as 3D printing which is capable of fabricating desired objects irrespective of their complex shapes. Furthermore, this approach addresses some of the limitations that are associated with 3D printing techniques that are capable of constructing complex structures using layer-by-layer approach but lack the smaller filament resolution, which is crucial for mimicking the natural structures composed of micro- and nanoscale fibers. Thus, the integrated, one-step approach of electrospinning in combination with another deposition method offers the most versatile technique by far that enables the fabrication of the hierarchical nanostructured fibers that closely mimic the structures of natural materials. Electrospinning also offers a useful tool to control the fiber size within the 3D hierarchical structure and therefore evaluate the effect of constituent fiber size on the overall mechanical behavior. It is known that the essence of electrospinning is its ability to work at molecular level, atom by atom to create fibers in the sub-micro/nanometer length scale with fundamentally new molecular organizations compared to their bulk counterparts. Therefore, the strength and stiffness of the fibers can be easily controlled by small modifications of their building blocks or in this case the polymer chain conformation and chain orientation within the fibrous geometry.

By using the combined NFES and FDEM, we have comparative advantages in a few ways. Firstly, the techniques are complementary in terms of the length scales of materials that they produce. But also, secondly, one technique could be used in combination with the other to achieve higher level of the mimicry, and thus increasing the possibility to produce a synthetic analogue. Thirdly, the helicoidal structures here, which are thought to be the key to the unique strength and fracture toughness of the material [263] due to its ability to adapt to tensile stress by allowing the plane of fibers to rotate toward or away from the applied tensile load, can be produced rather naturally in an electrospinning process, such as described earlier and shown in Fig. 2.25. The integrated approach is novel and innovative, and thus could lead to allowing comparative advantages for additive manufacturing sectors.

Through the use of novel and innovative, integrated, one-step additive manufacturing approaches (combined NFES and FDM), this research project will provide the common, useful platform to develop technological capabilities that support the growth and competitiveness of advanced manufacturing and engineering sectors, especially in additive manufacturing and advanced materials industries.

The proposed study [264, 265] aims to mimic the helicoidal fibrous structures found in the exoskeletons. This will be enabled using the integrated near-field electrospinning (NFES) and fused deposition modeling (FDM) technique. The NFES

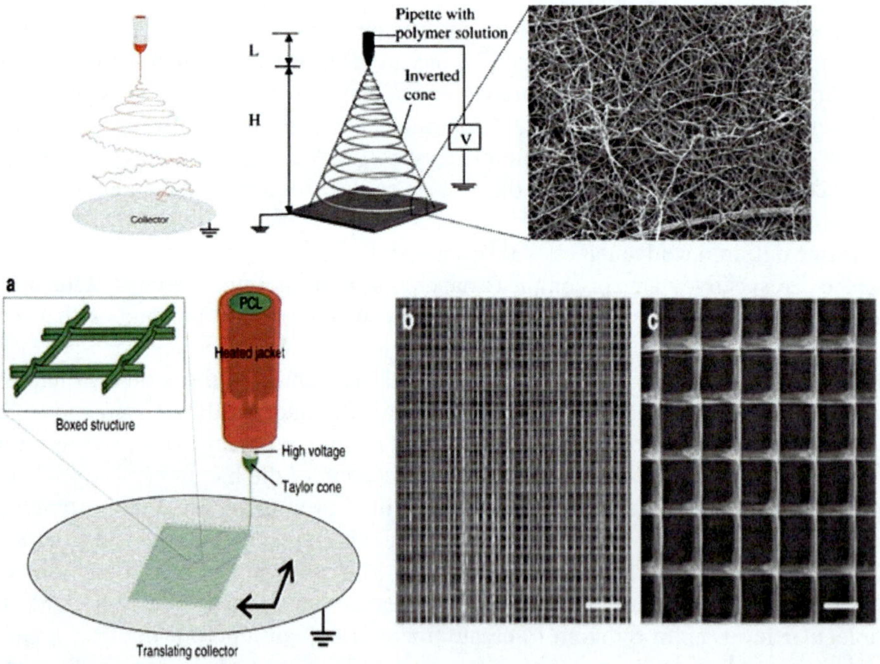

Fig. 2.25 3D structures obtained using melt-electrospinning in the direct writing mode. The stacking of the fibers is evident in the micrographs [260–262] (Reproduced with permission from Hindawi)

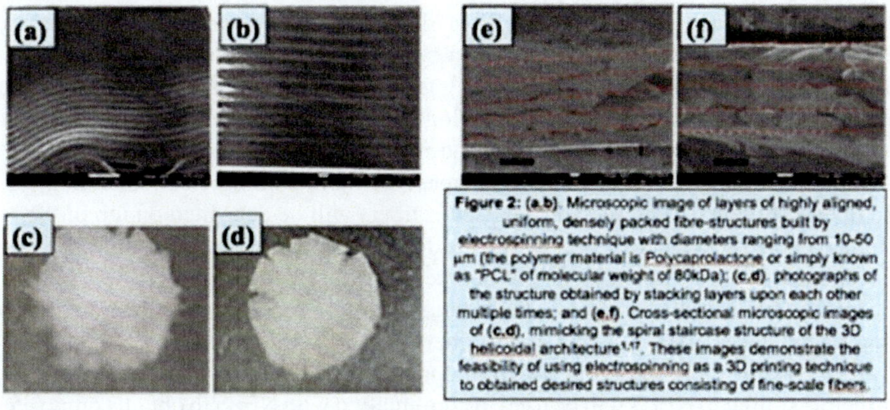

Fig. 2.26 (**a**, **b**). Microscopic image of layers [264, 265] of highly aligned, uniform, densely packed fiber-structures built by electrospinning technique with diameters ranging from 10 to 50 mm (the polymer material is Polycaprolactone or simply known as "PCL" of molecular weight of 80 kDa); (**c**, **d**). Photographs of the structure obtained by stacking layers upon each other multiple times; and (**e**, **f**). Cross-sectional microscopic images of (**c**, **d**), mimicking the spiral staircase structure of the 3D helicoidal architecture [241, 257]. These images demonstrate the feasibility of using electrospinning as a 3D printing technique to obtained desired structures consisting of fine-scale fibers (Reproduced with permission from Hindawi)

technique will be used to precisely deposit and orient the fibers within the composites as shown in Fig. 2.26. Thus, NFES technique will enable us to mimic the helicoidal architecture of the natural materials. FDM technique which has the working principle similar to NFES will be used in combination with NFES to obtain composite structures. In case of FDM, the heated thermoplastic is extruded out through a nozzle. This jet will then be deposited on the computer-controlled collector. The size of the jet extruded out is much larger compared to the fibers obtained using NFES. Thus, it presents an excellent opportunity to obtain multi-scale composite structures.

The conventional setup of electrospinning process relies on the chaotic whipping of liquid jets to generate fibers (see Fig. 2.26), which results in collection of randomly oriented fibers. Recent emergence of near-field electrospinning technique (NFES) demonstrates that fiber deposition and arrangement can be precisely controlled. This technique utilizes the stable liquid jet region by keeping the distance between the electrospinning needle and the collector in the range of 500 μm–5 mm. The voltage used is reduced to 1–3 kV due to the short distance between the electrospinning needle and collector compared to conventional electrospinning which operates in the range of 10–20 kV voltage. NFES technique can be used to deposit the fibers on a motorized collector that is capable of fast motion to allow direct and precise deposition of nearly straight nanofibers. Thus, this tool can be used for direct patterning of fibers and to manufacture novel orderly microstructures as shown in Fig. 2.26.

In this study, the helicoidal architecture found in natural materials will be mimicked using synthetic electrospun fibers [264, 265]. NFES technique will be used to obtain fibers from biodegradable polymers such as polylactic acid (PLA).

The scanning electron microscope (SEM) image that is shown in Fig. 2.26 is an example of the fibers that are obtained in our lab using electrospinning. The SEM image shows that the fibers are highly aligned similar to the fibers that are found in an individual horizontal plane of the biological structures. The length of the fibers will be controlled using the motion of the motorized substrate. Here, the speed of the substrate movement will match the speed of the incoming electrospinning jet. To obtain helicoid structures, the next set of fibers will be deposited on top of these aligned fibers such that the fibers within the domains are misaligned. For this purpose, the substrate will be rotated by few degrees and then will be allowed to oscillate linearly to collect aligned fibers. This process will be repeated until one complete rotation is made. Thus, we shall obtain fibers similar to the natural structures.

Fused Deposition Modeling (FDM) technique [264, 265] which has the working principle similar to NFES will be used in combination with NFES to obtain composite structures. In case of FDM, the heated thermoplastic, polycaprolactone (PCL) of low-melting point will be extruded out through a nozzle. This jet will then be deposited on the computer-controlled collector. The size of the jet extruded out will be much larger compared to the fibers obtained using NFES. Thus, it presents an excellent opportunity to obtain multi-scale composite structures. Figure 2.27 shows the schematic of the integrated, one-step combined NFES and FDM technique. Briefly, a twin spinneret will be constructed, one for NFES and the other for FDM. In the first step, a voltage will be applied to the NFES spinneret to lay down electrospun PLA fibers in a controllable manner on top of the collector. Following this step, the spinneret for FDM will be used to lay the extruded PCL material of low-melting point on top of the deposited fibers. By repeating these steps, we will stack the 2D patterns to obtain 3D structures. However, the computer-controlled stage will be rotated slightly such that the fiber axis of these layer PLA fibers is rotated by few degrees compared

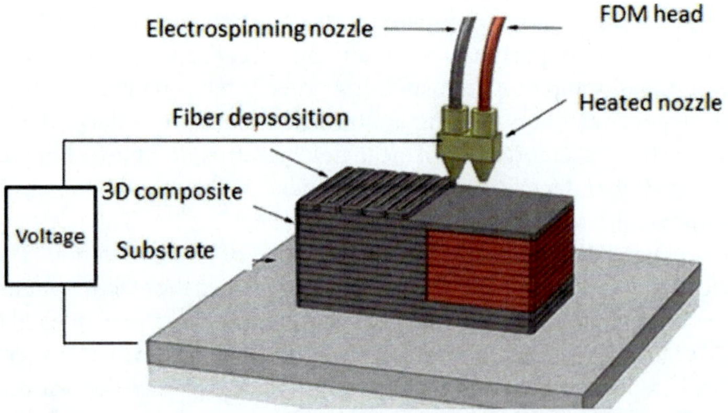

Fig. 2.27 Schematics of integrated, one-step, combined NFES and FDM approach. The two processes will be integrated to first lay down a 2D pattern of electrospun fibers on top a collector. Following this, extruded material from FDM will be deposited directly on top of the electrospun fibers [264, 265] (Reproduced with permission from Hindawi)

to the axis of the previously deposited PLA fibers. This will help us to obtain helicoidal fibrous structure within the composite. Lastly, the obtained sample will be heat pressed above the melting temperature of PCL to obtain a composite.

This research, for the first time, explores integrated NFES and FDM technique for the design and development of high-performance novel composites for advanced structural applications. Several innovations arising from the project can be identified. First, the processing technique using the integrated approach will produce a new class of smart composites. The significance of electrospinning technique to develop high-performance composites is remarkable. Fabrication of fibrous structures using electrospinning is a straightforward and cost-effective process.

Second, fabrication of helicoidal architecture using electrospun fibers is original and their toughness mechanisms have yet to be unveiled. By carefully investigating the effect of fiber diameter and level of hierarchy on their composite's mechanical integrity, the associated strengthening and toughening mechanisms can be elucidated. These results will provide a knowledge base for the design and development of high-performance composites.

Although it is now known that high toughness and strength in natural material can be derived from its unique 3D structures, not many fabrication techniques can be readily used to mimic this architecture. Recently, Grunenfelder et al. [266] mimicked this architecture using a very simple fabrication technique relying on carbon fiber prepregs in epoxy matrix. Yaraghi et al. [267] used 3D printing to produce similar 3D helicoidal architecture but with larger fibers (in millimeters). They demonstrated that the composite is capable of resisting damage from impact. These results are encouraging and motivating for us to go one step further in closely mimicking the architecture and mechanics of natural materials. Our approach is indeed novel as it will enable us to achieve the size of the fibers, the orientation of the fibers and the thickness of the composite as found in the natural materials. Thus, we believe, this study for the first time will provide an in-depth understanding of the toughening mechanisms and thereby lead to the development of high toughness, fracture-resistant composite materials.

Novel design and preparation methods will enable us to systematically investigate and uncover the microstructure and mechanical property relationships of these unique bio-inspired composites. This will facilitate a new design strategy to develop new generation of composite materials. Therefore, new knowledge gained from this research is a major original contribution to the science and technology of nanocomposites and bio-inspired materials research. It will impact this particular field beyond the immediate scope of current project.

Enhanced Fracture Properties with 3D Microarchitectures

As we have described earlier, there have been many researchers working on trying to understand as well as mimicking the molecular structures of natural/biological composites, such as wood, bone and nacre [243, 255, 258–260]. However, for the 3D helicoidal architecture such as found in shells of mantis shrimp dactyl club, a few studies on basic characterization of the natural materials [241, 242, 256, 257, 267] were reported, but mostly how to mimic the unique 3D architecture especially in terms of how it could be enabled by additive manufacturing has not been much studied. Yaraghi et al. [267] reported use of 3D printing to mimic similar 3D helicoidal architecture successfully but the size of fibers is rather large (~millimeters) due to limitation of the 3D printing technique. The following paragraphs are our very recent findings on our investigation. We showed the basic feasibility study of mimicking the 3D helicoidal architecture with finer-sized fibers (in the tens of microns) earlier in this section (Fig. 2.26) using electrospinning (NFES) technique. Furthermore, we have also conducted mechanical testing and shown that our sample with 3D helicoidal architecture indeed exhibited much enhanced toughness as compared to the samples that represent the fracture property of the bulk materials as well as the fiber-structured (all fibers are aligned in the same direction) materials.

The mechanical testing was conducted with Zwick Roell machine as shown in Fig. 2.28a–d on three kinds of samples: (1) bulk PCL, (2) same-orientation PCL fiber sample, (3) helicoidal PCL fiber samples. All samples were made with melt-electrospinning method and have same dimensions. Figure 2.28e showed engineering stress versus engineering strain curves of the tested samples. It clearly shows all the helicoidal samples exhibit qualitatively distinct macroscale crack-delaying behavior compared to the bulk and same fiber orientation samples which exhibited mostly brittle fracture phenomenon. From the areas under the curves, we calculated that the samples with 3D helicoidal architecture have much enhanced toughness compared to the bulk sample, as much as 20 times (for "*Heli t2*" sample).

Furthermore, Fig. 2.29 shows various representative stages of tensile test in the helicoidal PCL fiber samples. While bulk and same-orientation fiber samples exhibited brittle fracture failure mode with crack immediately propagates along the perpendicular direction of loading axis resulting in very little ductility, the helicoidal samples exhibit, as shown in Fig. 2.7, exhibited effective crack-delaying mechanisms. The crack was observed to initiate at the edge near the bottom of the 90° layer, which have no contribution for the total strength (i.e. the weakest part of the sample), then another broken part appeared in the next weakest part of the sample. With further straining ongoing, each layer was observed having the tendency of twisting then dramatically elongating until the last layer of the samples. This represents effective damage delaying mechanisms, which could lead to useful applications in fracture-resistant materials.

Having layers of multi-scale fibers with rotated alignments in each layer, the 3D helicoidal architecture could potentially lead to very efficient (high rate) as well as highly tunable damage dissipation mechanism. We will explore the effect of 3D hierarchy micro-architectured structures and the possible size effects of the substructures on the fracture toughness and fracture resistance properties of the resulting composite materials as well as the associated damage mechanisms. We will vary fiber morphologies (size, surface roughness) as well as inter-fiber and interlayer characteristics (such as adhesion strength). Our findings so far have indeed suggested that we could thus obtain control of deformation mechanisms leading to tunable fracture resistance. Furthermore, the fine size of the fibers could lead to strengthening effects, which will play an important role in the structural performance of the materials, while the high surface area to volume ratio afforded by the nano-sized fiber will be useful

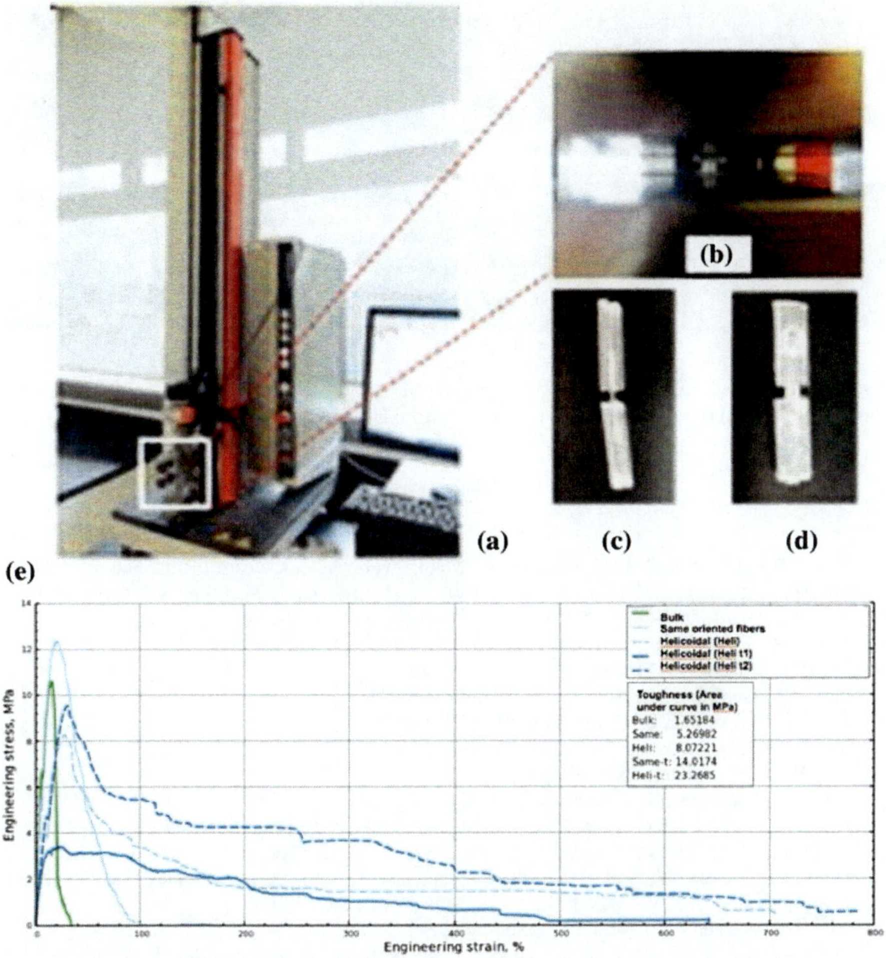

Fig. 2.28 Tensile test of various PCL samples: (**a**) Zwick/roell machine for tensile test (The tensile test was conducted with displacement control of $dL/dt = 100\mu$ m/min and preload of 0.1 N). (**b**) Partial magnification of sample placing; (**c**, **d**) Side and top views of sample setup; (**e**) Engineering stress–strain curves ("*Heli,*" "*Heli t1*" and "*Heli t2*" are all helicoidal samples, but with slightly differing surface treatments) [264, 265] (Reproduced with permission from Hindawi)

to release strain and thus delay fracture events. These deformation mechanisms are likely to be novel and thus could lead to potentially new, breakthrough composite materials technology—strong, yet tough, extremely fracture resistant.

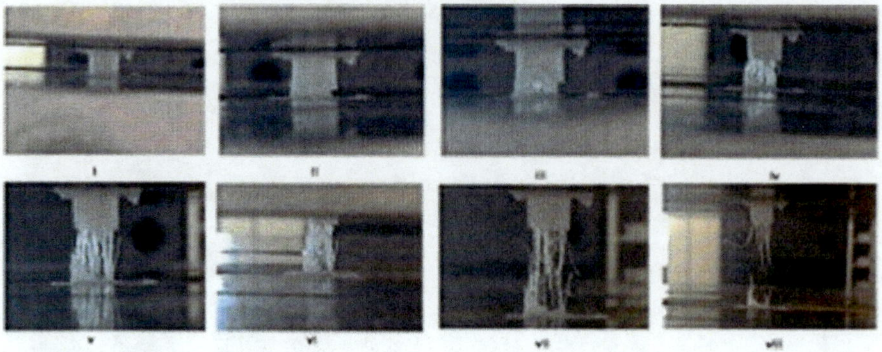

Fig. 2.29 Eight selected representative stages (i–viii) during testing of the helicoidal PCL fiber samples [264, 265] (Reproduced with permission from Hindawi)

References

1. Valek, B.C., Bravman, J.C., Tamura, N., MacDowell, A.A., Celestre, R.S., Padmore, H.A., Spolenak, R., Brown, W.L., Batterman, B.W., Patel, J.R.: Appl. Phys. Lett. **81**, 4168 (2002)
2. Arzt, E.: Acta Mater. **46**, 5611 (1998)
3. Keller, R.M., Baker, S.P., Arzt, E.: J. Mater. Res. **13**, 1307 (1998)
4. Yu, D.Y., Spaepen, F.: J. Appl. Phys. **95**, 2991 (2003)
5. Kraft, O., Hommel, M.: Acta Mater. **49**, 3935 (2001)
6. Huang, H., Spaepen, F.: Acta Mater. **48**, 3261 (2000)
7. Nix, W.D.: Metall. Trans. **20A**, 2217 (1989)
8. Nix, W.D.: Scr. Mater. **39**, 545 (1998)
9. von Blanckenhagen, B., Gumbsch, P., Arzt, E.: Phil. Mag. Lett. **83**, 1 (2003)
10. Arzt, E., Dehm, G., Gumbsch, P., Kraft, O., Weiss, D.: Prog. Mater. Sci. **46**, 283 (2001)
11. W. D. Nix, O. S. Leung, Encyc. Mater. Sci. Technol. 9262, 2001
12. Dehm, G., Wagner, T., Balk, T.J., Arzt, E., Inkson, B.J.: J. Mater. Sci. Technol. **18**, 113 (2002)
13. Pant, P., Schwarz, K.W., Baker, S.P.: Acta Mater. **51**, 3243 (2003)
14. Nix, W.D., Greer, J.R., Feng, G., Lilleodden, E.T.: Thin Sol. Films **515**, 3152 (2007)
15. Stelmashenko, N.A., Walls, M.G., Brown, L.M., Millman, Y.V.: Acta Metall. Mater. **41**, 2855 (1993)
16. De Guzman, M.S., Neubauer, G., Flinn, P., Nix, W.D.: Mater. Res. Soc. Proc. **308**, 613 (1993)
17. Ma, Q., Clarke, D.R.: J. Mat. Res. **10**(4), 853–863 (1995)
18. Poole, W.J., Ashby, M.F., Fleck, N.A.: Scripta Mat. **34**(4), 559–564 (1996)
19. Nix, W.D., Gao, H.: J. Mech. Phys. Solids **46**, 411–425 (1998)
20. H. Gao, Y. Huang, Y.W.D.W.D., Naturwissenschaftler, 86, p. 507, 1999
21. Gao, H., Huang, Y., Nix, W.D., Hutchinson, J.W.: J. Mech. Phys. Solids **47**, 1239 (1999)
22. Huang, Y., Chen, J.Y., Guo, T.F., Zhang, L., Hwang, K.C.: Int. J. Fract. **100**, 1 (1999)
23. Huang, Y., Gao, H., Nix, W.D., Hutchinson, J.W.: J. Mech. Phys. Solids **48**, 99 (2000)
24. Huang, Y., Xue, Z., Gao, H., Nix, W.D., Xia, Z.C.: J. Mater. Res. **15**, 1786 (2000)
25. Tymiak, N.I., Kramer, D.E., Bahr, D.F., Wyrobek, T.J., Gerberich, W.W.: Acta Mater. **49**, 1021 (2001)
26. Swadener, J.G., George, E.P., Pharr, G.M.: J. Mech. Phys. Solids **50**, 681 (2002)
27. Durst, K., Backes, B., Goken, M.: Scripta Mat. **52**(11), 1093–1097 (2005)
28. Durst, K., Backes, B., Franke, O., Goken, M.: Acta Mat. **54**(9), 2547–2555 (2006)
29. Uchic, M.D., Dimiduk, D.M., Florando, J.N., Nix, W.D.: Science **305**, 986 (2004)
30. J. R. Greer, W.C. Oliver and W.D. Nix, Acta Mater. 53, p. 1821, 2005; J.R. Greer, Ph.D dissertation, Stanford University, 2005

31. J. R. Greer and W.D. Nix, Phys. Rev. B 73, 245410, 2006
32. S. J. Basinski and Z.S. Basinski, "Plastic deformation and work hardening," in Dislocations of Solids, Vol. 4: Dislocations in Metallurgy, edited by F.R.N. Nabarro, North-Holland Publishing Company, Oxford, UK, p. 261, 1979
33. E. T. Lilleodden, Ph.D dissertation, Stanford University, 2001
34. Nabarro, F.R.N., Basinski, Z.S., Holt, D.B.: Adv. Phys. **13**, 193 (1964)
35. Basinski, Z.S., Basinski, S.J.: Phil. Mag. **9**, 51 (1964)
36. Basinski, Z.S.: Scripta Met. **8**, 1301 (1974)
37. G. Feng, Ph.D dissertation, Stanford University, 2005
38. M. D. Uchic, D. M. Dimiduk, J. N. Florando and W.D. Nix, Science 305, p. 986, 2004
39. J. R. Greer, W.C. Oliver and W.D. Nix, Acta Mater. 53, p. 1821, 2005; J.R. Greer, Ph.D dissertation, Stanford University, 2005
40. J. R. Greer and W.D. Nix, Phys. Rev. B 73, 245410, 2006
41. B.C. Valek, Ph.D dissertation, Stanford University, 2003
42. A. S. Budiman, N. Tamura, B. C. Valek, K. Gadre, J. Maiz, R. Spolenak, W. D. Nix and J. R. Patel. Crystal Plasticity in Cu Damascene Interconnect Lines Undergoing Electromigration As Revealed by Synchrotron X-Ray Microdiffraction. App. Phys. Lett., 88, 233515, 2006.
43. A. Maurel, J. Mercier, F. Lund, Phys. Rev. B 70, 024303, 2004
44. Tamura, N., MacDowell, A.A., Spolenak, R., Valek, B.C., Bravman, J.C., Brown, W.L., Celestre, R.S., Padmore, H.A., Batterman, B.W., Patel, J.R.: J. Synchrotron Rad. **10**, 137–143 (2003)
45. S. M. Han, Ph.D dissertation, Stanford University, 2006
46. R. Maass, D. Grolimund, S. Petegem, M. Willimann, M. Jensen, H. Swygenhoven, T. Lehnert, M. A. M. Gijs, C. A. Volkert, E. T. Lilleodden, R. Schwaiger, Appl. Phys. Lett. 89, 151905, 2006
47. MacDowell, A.A., Celestre, R.S., Tamura, N., Spolenak, R., Valek, B.C., Brown, W.L., Bravman, J.C., Padmore, H.A., Batterman, B.W., Patel, J.R.: Nucl. Inst. Meth. Phys. Res. A **936**, 467–468 (2001)
48. Cahn, R.W.: J. Inst. Met. **86**, 121 (1949)
49. Nye, J.F.: Acta Metall. **1**, 153 (1953)
50. Z. W. Shan, R. Mishra, S.A. Syed Asif, O. L. Warren and A. M. Minor, Unpublished Research, 2007
51. Budiman, A.S., Han, S.M., Wei, Q., Dickerson, P., Tamura, N., Kunz, M., Misra, A.: Plasticity in the nanoscale Cu/Nb single crystal multilayers as revealed by synchrotron Laue X-ray microdiffraction. J. Mater. Res. **27**, 599 (2012)
52. Misra, A., Hirth, J.P., Hoagland, R.G.: Length-scale-dependent deformation mechanisms in incoherent metallic multilayered composites. Acta Mater. **53**, 4817 (2005)
53. Grimes, R.W., Konings, R.J.M., Edwards, L.: Greater tolerance for nuclear materials. Nature Mater. **7**, 683 (2008)
54. M.J. Demkowicz, R.G. Hoagland, J.P. Hirth: Interface structure and radiation damage resistance in Cu-Nb multilayer nanocomposites. Phys. Rev. Lett. 100, 136102 (2008).
55. A. Misra, M.J. Demkowicz, J. Wang, R. G. Hoagland: The multiscale modeling of plastic deformation in metallic nanolayered composites. J. Mater. 60, 4, 39 (2008).
56. Hoagland, R.G., Hirth, J.P., Misra, A.: On the role of weak interfaces in blocking slip in nanoscale layered composites. Philos. Mag. **86**, 3537 (2006)
57. Kim, Y., Budiman, A.S., Baldwin, J.K., Mara, N., Misra, A., Han, S.M.: Microcompression study of Al/Nb nanoscale multilayers. J. Mater. Res. **27**, 592 (2012)
58. Anderson, P.M., Bingert, J.F., Misra, A., Hirth, J.P.: Rolling textures in nanoscale Cu/Nb multilayers. Acta Mater. **51**, 6059 (2003)
59. K. Nyilas, A. Misra, T. Ungar : Micro-strains in cold rolled Cu-Nb nanolayered composites determined by X-ray line profile analysis. Acta Mater. 54, 751 (2006).
60. C.C. Aydiner, D.W. Brown, N. Mara, J. Almer, A. Misra: In situ x-ray investigation of freestanding nanoscale Cu-Nb multilayers under tensile load. Appl. Phys. Lett. 94, 031906 (2009).

61. C.C. Aydiner, D.W. Brown, A. Misra, N. Mara, Y.-C. Wang, J.J. Wall, J. Almer: Residual strain and texture in free-standing nanoscale Cu-Nb multilayers. J. Appl. Phys. 102, 083514 (2007).
62. Misra, A., Thilly, L.: Structural metals at extremes. MRS Bull. **35**, 965–976 (2010)
63. X. Zhang, E. G. Fu, N. Li, A. Misra, Y. Q. Wang, L. Shao, and H. Wang. Design of Radiation Tolerant Nanostructured Metallic Multilayers. Journal of Engineering Materials and Technology-transactions of the ASME, 134:041010 (1–9), 2012.
64. I. J. Beyerlein, A. Caro, M. J. Demkowicz, N. A. Mara, A. Misra, andB. P. Uberuaga. Radiation damage tolerant nanomaterials. Materials Today, 16:443–449, 2013.
65. N. Li, N. A. Mara, J. Wang, P. Dickerson, J. Y. Huang, and A. Misra.Ex situ and in situ measurements of the shear strength of interfaces in metallic multilayers. Scripta Materialia, 67:479–482, 2012.
66. Koehler, J.S.: Attempt to design a strong solid. Phys. Rev. B **2**, 547–551 (1970)
67. P.M. Anderson, J. F. Bingert, A.Misra, and J. P. Hirth. Rolling textures in nanoscale Cu/Nb multilayers. Acta Materialia, 51:6059–6075, 2003.
68. Han, S.M., Philips, M.A., Nix, W.D.: Study of strain softening behavior of Al/Al3Sc multilayers using microcompression testing. Acta Mater. **57**, 4473–4490 (2009)
69. J.S. Carpenter, A. Misra, M.D. Uchic, and P.M. Anderson. Strain rate sensitivity and activation volume of Cu/Ni metallic multilayer thin films measured via micropillar compression. Applied Physics Letters, 101: 051901, 2012.
70. W Han, M. J. Demkowicz, N. A. Mara, E. Fu, S. Sinha, A. D. Rollett, Y.Q Wang, J. S. Carpenter, I. J. Beyerlein, and A Misra. Design of Radiation Tolerant Materials Via Interface Engineering. Advanced Materials, 25:6975–6979, 2013.
71. Hirth, J.P.: The Influence of Grain Boundaries on Mechanical Properties. The American Society of Metals **3**, 3047–3067 (1972)
72. Nyilas, K., Ungar, T., Misra, A.: Micro-strains in cold rolled cu-nb nanolayered composites determined by x-ray line profile analysis. Acta Mater. **54**, 751–755 (2006)
73. Yu-Zhang, K., Embury, J.D., Han, K., Misra, A.: Transmission electron microscopy investigation of the atomic structure of interfaces in nanoscale Cu-Nb multilayers. Phil. Mag. **88**, 2559–2567 (2008)
74. Budiman, A.S., Han, S.M., Li, N., Wei, Q.M., Dickerson, P., Tamura, N., Kunz, M., Misra, A.: Plasticity in the nanoscale Cu/Nb singlecrystal multilayers as revealed by synchrotron Laue x-ray microdiffraction. J. Mater. Res. **27**, 599–611 (2012)
75. Li, N., Wang, J., Misra, A., Huang, J.Y.: Direct Observations of Confined Layer Slip in Cu/Nb Multilayers. Microsc. Microanal. **18**, 1155–1162 (2012)
76. Hoagland, R.G., Hirth, J.P., Misra, A.: On the role of weak interfaces in blocking slip in nanoscale layered composites. Phil. Mag. **86**, 3537–3558 (2006)
77. Misra, A., Demkowicz, M.J., Wang, J., Hoagland, R.G.: The multiscale modeling of plastic deformation in metallic nanolayered composites. JOM **60**, 39–42 (2008)
78. Wang, J., Hoagland, R.G., Hirth, J.P., Misra, A.: Atomistic modelling of the interaction of glide dislocations with “weak” interfaces. Acta Mater. **56**, 5685–5693 (2008)
79. Wang, J., Hoagland, R.G., Hirth, J.P., Misra, A.: Atomistic simulations of the shear strength and sliding mechanisms of copper-niobium interfaces. Acta Mater. **56**, 3109–3119 (2008)
80. Wang, J., Hoagland, R.G., Misra, A.: Mechanics of nanoscale metallic multilayers: From atomic-scale to micro-scale. Scripta Mater. **60**, 1067–1072 (2009)
81. Wang, J., Misra, A.: An overview of interface-dominated deformation mechanisms in metallic multilayers. Curr. Opin. Solid State Mater. Sci. **15**, 20–28 (2011)
82. Zbib, H.M., Overman, C.T., Akasheh, F., Bahr, D.: Analysis of plastic deformation in nanoscale metallic multilayers with coherent and incoherent interfaces. Int. J. Plast **27**, 1618–1639 (2011)
83. Wang, J., Hoagland, R.G., Liu, X.Y., Misra, A.: The influence of interface shear strength on the glide dislocation-interface interactions. Acta Mater. **59**, 3164–3173 (2011)
84. Wang, J., Misra, A., Hoagland, R.G., Hirth, J.P.: Slip transmission across fcc/bcc interfaces with varying interface shear strengths. Acta Mater. **60**, 1503–1513 (2012)

85. S. Shao, H. M. Zbib, I. Mastorakos, and D. F. Bahr. Effect of Interfaces in the Work Hardening of Nanoscale Multilayer Metallic Composites During Nanoindentation: A Molecular Dynamics Investigation. Journal of Engineering Materials and Technology-transactions of the ASME, 135: 021001, 2013.
86. Wang, J., Zhang, R.F., Zhou, C.Z., Beyerlein, I.J., Misra, A.: Interface dislocation patterns and dislocation nucleation in face-centered-cubic and body-centered-cubic bicrystal interfaces. Int. J. Plast **53**, 40–55 (2014)
87. N. Abdolrahim, Hussein M. Zbib, and David F. Bahr. Multiscale modelling and simulation of deformation in nanoscale metallic multilayer systems. International Journal of Plasticity, 52:33–50, 2014.
88. Budiman, A.S., Li, N., Wei, Q., Baldwin, J.K., Xiong, J., Luo, H., Trugman, D., Jia, Q.X., Tamura, N., Kunz, M., Chen, K., Misra, A.: Growth and structural characterization of epitaxial Cu/Nb multilayers. Thin Solid Films **519**, 4137–4143 (2011)
89. Misra, A., Hirth, J.P., Hoagland, R.G., Embury, J.D., Kung, H.: Dislocation mechanisms and symmetric slip in rolled nano-scale metallic multilayers. Acta Mater. **52**, 2387–2394 (2004)
90. R. Maass, D. Grolimund, P. S. Van, M. Willimann, M. Jensen, S. H. Van, T. Lehnert, M. A. M. Gijs, C. A. Volkert, E. T. Lilleodden, and R. Schwaiger. Defect structure in micropillars using X-ray microdiffraction. Applied Physics Letters, 89:151905, 2006.
91. H. Van Swygenhoven, B. Schmitt, P. M. Derlet, S. Van Petegem, A. Cervellino, Z. Budrovic, S. Brandstetter, A. Bollhalder, and M. Schild. Following peak profiles during elastic and plastic deformation: A synchrotron-based technique. Review of Scientific Instruments, 77:013902, 2006.
92. R. Maass, P.S. Van, S. H. Van, P. M. Derlet, C. A. Volkert, and D. Grolimund. Time-resolved laue diffraction of deforming micropillars. Physical Review Letters, 99:145505, 2007.
93. Kirchlechner, C., Kiener, D., Motz, C., Labat, S., Vaxelaire, N., Perroud, O., Micha, J.S., Ulrich, O., Thomas, O., Dehm, G., Keckes, J.: Dislocation storage in single slip-oriented Cu micro-tensile samples: new insights via X-ray microdiffraction. Phil. Mag. **91**, 1256–1264 (2011)
94. Budiman, A.S., Narayanan, K.R., Berla, L.A., Li, N., Dickerson, P., Wang, J., Tamura, N., Kunz, M., Nix, W.D., Misra, A.: Plasticity Evolution in Nanoscale Cu/Nb Single Crystal Multilayers as Revealed by Synchrotron X-Ray Microdiffraction. Mater. Sci. Eng. A, **635**, 6–12 (2015)

Chapter 3
Probing Mechanics at the Extremes

Many of the new insights and updates to the classical laws of mechanics of materials as discussed in Chap. 2: **Extreme Mechanics in Advanced, Emerging Materials** were only possible through advanced experimentation techniques in the characterization of the nanoscale materials or emerging 3D architectures. This has been highlighted briefly on a few occasions as parts of the discussions to reveal how mechanics of materials have been pushed to the extreme limits in terms of elasticity, plasticity and fracture phenomena. For instance, the unique capability of the synchrotron white-beam X-ray microdiffraction was key to revealing the extreme deformability in nanolayers and the possible mechanisms behind it. In the meantime, the in-situ microbeam bending inside an SEM (Scanning Electron Microscopy) capability has enabled the innovative development and extreme design of materials and nanocomposites with enhanced mechanical properties (especially fracture) reaching or approaching the extreme limits of materials properties. These were relatively recent advances in materials characterization techniques—especially the in-situ micromechanical testing which was developed only in the past five years or so—thus the reason why this book has been written with special emphasis on particularly these advanced materials characterization and experimentation techniques.

This chapter will deal with these two advanced techniques—the synchrotron white-beam X-ray microdiffraction and the in-situ microbeam bending inside an SEM (Scanning Electron Microscopy)—in greater details. Section 3.1: **Synchrotron X-ray Microdiffraction (μSXRD)** discusses how the new approach provides new information that would be difficult to obtain with other characterization techniques. In particular, it provides a direct way to detect and measure strain gradients quantitatively. Hence, in the following section, we introduce the technique and its fundamentals that have allowed us to study the role of geometrically necessary dislocations (GNDs) and strain gradients in nanoscale plasticity, which have proved to contribute to the understanding of the size effects in crystal hardening mechanisms. These experiments provide not only scientifically useful and exciting insights, but also important practical implications, as will be discussed in greater detail later in this

A. S. Budiman, *Principles of Extreme Mechanics (XM) in Design for Reliability (DfR)*, Engineering Materials,
https://doi.org/10.1007/978-981-15-6720-9_3

book—for recent technological trends in microelectronics, energy and nanotechnology industries. These practical aspects typically involve reliability considerations (long-term durability, critical reliability assessment methodologies, etc.) for recent advances in the technological world—such as advanced Lithium ion battery (million-mile battery for electric vehicle, for instance) or the emerging Through-Silicon Via (TSV) technology that enables 3D integrated Circuit (3D IC) and the Internet of Things (IoT) applications. Both techniques are not only crucial technologies for our modern society based on data (Artificial Intelligence, next levels of social interconnectivity, etc.), but also for the fundamental understanding of the physics and materials science behind the technologies they enable. Understanding and controlling plasticity and the mechanical properties of materials on the nanoscales and with novel 3D architectures, for instance, could thus lead to new and more robust nanomechanical structures and 3D integrated devices.

The next section—Sect. 3.2: **In-Situ Micromechanical and Fracture Testing—** captures a portion of the deep discussions in the materials research world communities in the in-situ experimentation of plasticity and fracture characterization which lends us the ability to observe directly how plasticity and fracture events interact with microstructures of the materials at the nanoscales, and thus how it could affect and enable novel and extreme materials and nanocomposite design. More important perhaps is how these recent advances in the in-situ techniques have indeed led to the unraveling of novel fracture toughening mechanisms, and thus to further potential discovery/design of new materials with truly novel mechanisms and/or functionalities, such as will be shown in the subsequent chapters.

3.1 Synchrotron X-ray Microdiffraction (MSXRD)

Traditional X-ray diffraction is a technique that has been used for almost a century for elucidating the structure of materials on the macroscopic scales (0.1–10 mm). As modern electronics, photonics and even biological devices are increasingly made in smaller and smaller scales (submicron and nanometer scales), a thorough understanding of the materials structure-properties-performance relationship at such length scales (0.1–10 μm) has become critical, and thus the need for high spatial resolution X-ray diffraction. With the recent availability of bright third-generation synchrotron sources and progress in X-ray focusing optics, it is now practical to develop an X-ray microdiffraction technique and apply it to characterize materials at such small scales. In this chapter, we will describe in greater details the beamline instrumentation and performance, as well as the basic experimental procedure and data analysis involved in the implementation of this advanced experimentation technique. In addition, the special procedure to quantitatively measure the density of geometrically necessary dislocations (GNDs) from the local lattice curvature, which has been utilized in a number of our analyses as illustrated later in the following chapters, will also be discussed.

3.1.1 Diffraction Physics and Beamline Instrumentation

Bragg's law of diffraction [1] forms the fundamental principle behind the synchrotron X-ray microdiffraction technique, in which a crystalline material scatters the incident X-ray beam, so that the scattered X-rays form constructive interference when the wavelength of the incident X-rays (λ) is comparable to the interatomic spacing (d_{hkl}). This is because the spacings between adjacent atomic layers (lattice planes) in a crystalline material act as narrow slits, bending (diffraction) X-rays passing through them, leading to the focusing of X-rays, hence forming constructive interference. The necessary condition for this constructive interference is termed as Bragg's law. It was discovered and proposed by scientist Lawrence Bragg and his father William Henry Bragg.

Bragg's condition is given in Eq. 3.1

$$n\lambda = 2d_{hkl}\mathrm{Sin}\theta \tag{3.1}$$

where λ is the wavelength of the incident X-ray beam, n is the order of reflection (taking the value of any positive integer), d_{hkl} is the interatomic spacing between the diffracting crystal planes and θ is the incidence angle of the X-ray beam with the diffracting crystal planes, which is shown in Fig. 3.1. All X-ray diffraction-based instruments, from the laboratory X-ray machine to the synchrotron, function based on this law. The crystal structure of an unknown crystalline solid can be determined by analyzing the space between several crystal planes, using X-ray diffraction. Based on this law, the change in the interatomic spacing between atoms (d_{hkl}), which correlates with the strain in the given strained crystalline sample, can be determined.

The synchrotron is essentially a radiation source that generates high-energy polychromatic X-rays for diffraction and other spectroscopy experiments [2]. It consists of a particle accelerator, which accelerates the electrons to very high velocities.

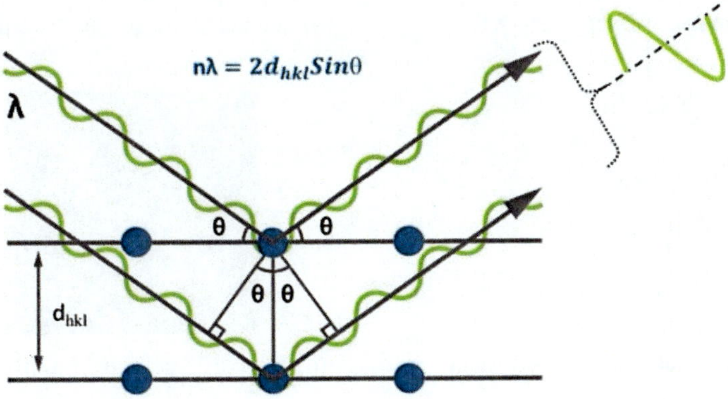

Fig. 3.1 Schematic showing Bragg's law of diffraction

Such fast-moving electrons emit high-energy electromagnetic radiation (ultraviolet light, infrared light, X-rays, neutron radiation, etc.) upon changing their direction of motion under the influence of bending magnets. Thus, it produces X-rays very high in energy. A synchrotron facility roughly spans the size of a football field, with several synchrotron sources of different capacities existing around the world [3]. Each synchrotron source contains several beam end-stations, called beamlines, which are suitable for different experimental purposes.

The scanning X-ray microdiffraction (μSXRD) of a crystalline sample can be performed at the synchrotron beamline using the highly focused and energetic polycrystalline X-ray beam of a submicron spot size. Though the principle behind μSXRD is identical to the conventional laboratory XRD, they differ greatly, and the former has proven to be superior in two very important aspects, as shown in Fig. 3.2. The first is the resolution of the X-ray beam: a laboratory X-ray beam is monochromatic, with a size generally several orders larger than the synchrotron X-ray beam; whereas, the synchrotron X-ray beam is polychromatic with a size less than 1 micron [4, 5]. Hence, it can probe very small samples and localized strains with submicron resolution.

The second difference is in the detector, which records the diffraction patterns from the sample. A laboratory X-ray generally contains only a 1D detector, which records diffraction patterns at different beam incidence angles (θ), as shown in Fig. 3.2a, c. On the other hand, the synchrotron beamline uses a 2D detector which simultaneously records several diffraction peaks of the sample subjected to a polychromatic beam,

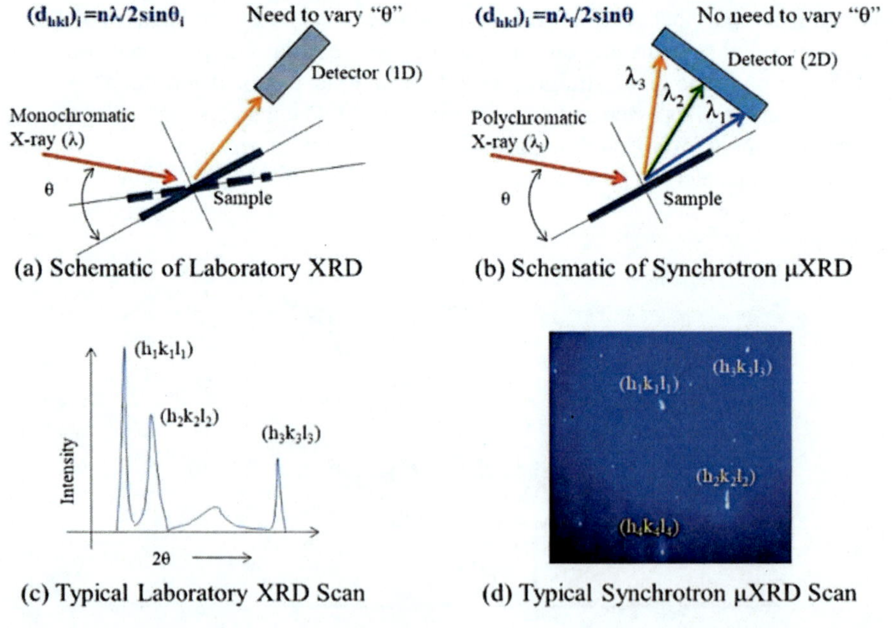

Fig. 3.2 Comparison of laboratory and synchrotron X-ray diffraction setups (With courtesy of Ref. [5] from author's copyright)

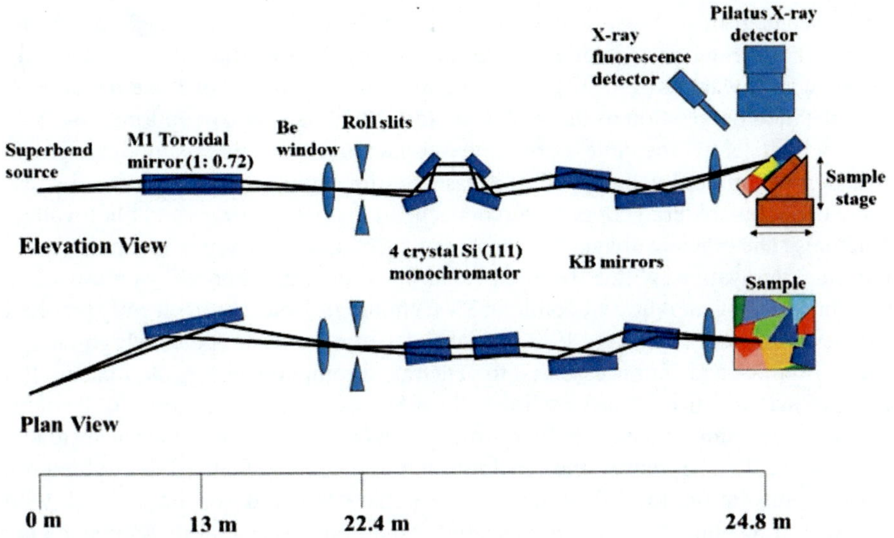

Fig. 3.3 Schematic of the synchrotron X-ray microdiffraction beamline (BL 12.3.2) at the Advanced Light Source (ALS), Lawrence Berkeley National Laboratory, Berkeley, CA, USA (With courtesy of Ref. [5] from author's copyright)

as shown in Fig. 3.2b, d. Hence, it can be used to characterize the sample in a single setup (a fixed angular position with respect to the X-ray beam). Thus, it enables fast and automatic (raster) scanning of the sample, which is the reason this technique is called scanning X-ray microdiffraction.

Our microdiffraction experiments were conducted at Beamline 12.3.2 of the Advanced Light Source (ALS) at the Ernest Orlando Lawrence Berkeley National Laboratory (LBNL), Berkeley, California [5, 6]. A schematic of the beamline layout is shown in Fig. 3.3 [5]. In this beamline, the polychromatic X-ray beam exiting from a superconductor magnet source of the synchrotron is refocused at the entrance of the experimental hutch by a platinum-coated silicon toroidal mirror (M1) operating at a grazing angle of 4.5 mrad. The final focusing is done by Kirkpatrick-Baez (KB) mirrors consisting of an orthogonal pair of 100 mm long tungsten-coated silicon substrate bent to an elliptical shape. The final polychromatic X-ray beam at the experimental stage is focused to a diameter of ~0.8 microns (μm) and consists of energy ranging from 5 to 25 keV. The experimental sample can be mounted on the sample stage, which can be moved and rotated to adjust the sample's position relative to the incident X-ray beam. The diffracted X-rays from the sample are recorded by a Pilateus 2D X-ray detector. A monochromator is available in the setup as an option, which can be engaged to get a monochromatic beam of particular energy for performing an energy scan or powder diffraction. The beamline also allows an automatic raster scanning of the sample with a single Laue pattern's acquisition time below 1 s. [4–7]. Hence, Laue patterns for points lying on the large area scan grid can be characterized within a reasonable timing (for example, a 1 cm by 1 cm area can be scanned in ~7 h using a 100 μm grid spacing). For the scanned area of the sample, the orientation and deviatoric strain maps can be constructed by the specially developed software, XMAS [8], using the measured sequence of Laue patterns. A more detailed description of the ALS beamline 12.3.2 is available in Refs. [4–8].

Very high spatial resolution stress measurements are possible using this microdiffraction beamline. Such capability is essential for highly localized residual stress evaluation near soldered copper interconnects, in microelectronics and photovoltaic structures. The efficacy of synchrotron X-ray diffraction in enabling technologically important innovations within the microelectronics and nanotechnology industries, in addition to next-generation silicon solar PV technologies, have been recently reported and they are described in Refs. [9–13]. In the case of solar PV modules, a high-energy synchrotron X-ray beam is essential to penetrate through the encapsulation and thin layers of metallization, in order to reach the c-Si solar cell. This enables for the non-destructive probing of the crystalline structure and state of strain of the encapsulated c-Si cell [14, 15]. Moreover, the synchrotron X-ray beam is focused to a submicron (0.8 μm) resolution, thus allowing for strain characterization of the c-Si cell with high spatial resolution. Such a high spatial resolution is essential to understand the distribution of residual stress in the cell, because the stress is highly localized and concentrated around the small areas near interconnects (refer to Refs. [14, 15]; also confirmed by our measurements and simulations in the present research work).

3.1.2 Synchrotron White-Beam X-ray Microdiffraction

Lab-based white-beam Laue diffraction is a standard crystallographic method used to determine crystal orientation without rotation of the sample. However, lab-based Laue diffraction is rarely used to measure strain because the precision of most Laue instruments is low compared to modern diffractometers, and because the unit cell volume cannot be determined with a standard Laue measurement. Nevertheless, with suitable instrumentation, such as used in the experimental setting described in the previous section, precise determination of crystal orientation and distortional strain is possible. Laue diffraction might even be extended by measuring the energy of one or more reflections to determine the full strain tensor in polycrystalline samples. The following description of the methodology is shown with a particular sample setup using a form of encapsulated silicon solar cell (in reality, of course, the sample could be in other forms) just for better illustration while using realistic, relevant forms of samples, such as will be used in study cases discussed in later chapters.

Experimental Procedure and Data Analysis

Figure 3.4a shows the schematic of the microdiffraction experiment setup and Fig. 3.4b shows an actual photograph of the setup. The sample needs to be mounted on the stage with its back facing the X-ray beam (Fig. 3.4b), since X-rays cannot penetrate the front glass due to absorption. To focus the X-ray beam on the cell, the back sheet was omitted in the mini PV module samples. In our experimental

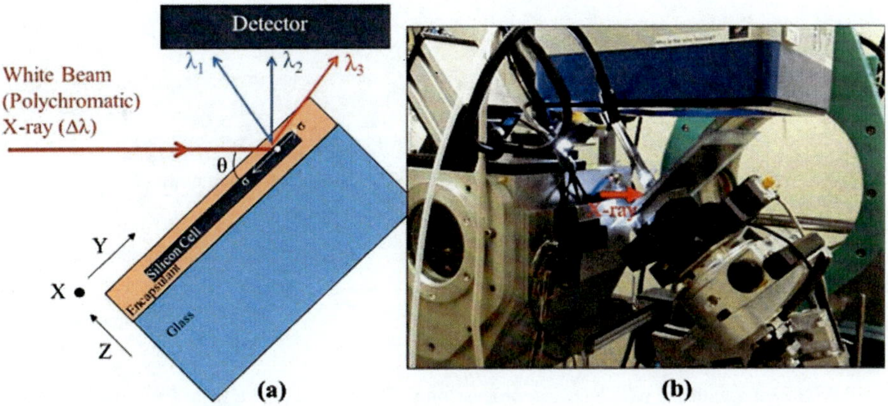

Fig. 3.4 μSXRD setup of mini PV module. **a** Schematic, **b** actual picture (With courtesy of Ref. [5] from author's copyright)

setup, the PV module samples were mounted at 45° to the X-ray beam, which is horizontal. The beamline has a hybrid pixel area detector (DECTRIS Pilatus 1 M), mounted above the sample stage, which records the 2D diffraction patterns from the sample. Hence, this setup can record multiple diffraction peaks, satisfying Bragg's condition, upon a single exposure to the polychromatic X-ray beam (white beam), also called Laue diffraction [7]. The resulting diffraction pattern is called a Laue pattern or Laue-gram. Further, the sample stage is able to move in X-, Y- and Z-directions, enabling for raster scanning. The sample was placed in the XY plane, with the Z-direction normal to the sample plane, as shown in Fig. 3.4a. The stage can be adjusted to focus the X-ray beam on the sample, as per the experimental requirement, by moving in the Z-direction, so that movement in the XY plane does not alter the focus of the X-ray beam [8]. A laser triangulation system is available for accurately (error <1 μm) focusing the X-ray beam on the sample [6]. The sample can be automatically raster-scanned point by point, using a suitable displacement step (generally 10–100 μm) along the X- and Y-directions.

The size of the PV module sample is very big, relative to the X-ray beam's resolution. Hence, scanning the entire sample is not possible in a reasonable amount of time. Moreover, the stress in the c-Si cell is highly localized and concentrated around the solder joint regions. Hence, a small region (for example, a few mm x few mm area) around the solder joint, as shown in Fig. 3.5, can be scanned with a step size of ~100 μm in a few hours. Higher resolution scans require a smaller step size and more time; therefore, the scan area and the step size should be decided based on the sample at hand and the expected stress resolution. In the automatic raster scan, the sample is scanned point by point, as shown in Fig. 3.5b. The X-ray detector records the polychromatic diffraction pattern (also called the Laue diffraction pattern or simply the Laue pattern) at each of these points for a given exposure time (~1.5 s in our experiments). Hence, for each scan, there will be $m \times n$ scan points, where m is the number of divisions of the scan region along the X-direction and n is the

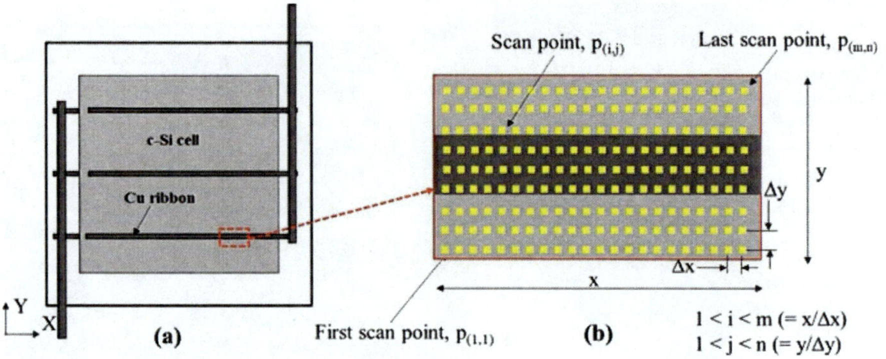

Fig. 3.5 Schematic of the μSXRD scan process of a mini PV module. **a** Mini PV module sample, **b** μSXRD scan region with the scanned points marked in yellow color (With courtesy of Ref. [5] from author's copyright)

number of divisions of the scan region along the Y-direction. The time of scan is directly proportional to the number of scanned points ($m \times n$) and the exposure time at each point.

A typical Laue diffraction pattern before indexing, from the encapsulated monocrystalline silicon solar cell sample (with surface normal to <001> and sides along <100> and <010>), is shown in Fig. 3.6a. The same pattern, after manual indexing using the XMAS program [8], is shown in Fig. 3.6b. The XMAS uses a peak-fitting routine to calculate the crystal plane orientations of the diffraction peaks, where the accuracy of fit depends on the number of peaks indexed. Hence, we use all peaks for our analysis. The indexed Laue pattern in Fig. 3.6b has 32 peaks, and

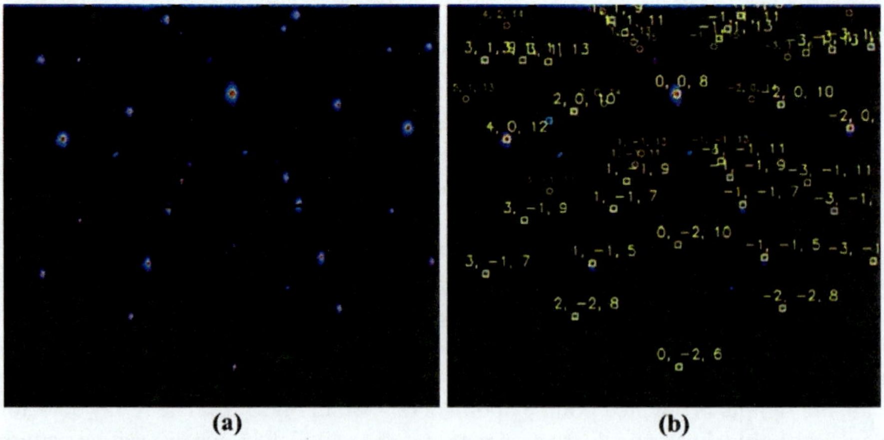

Fig. 3.6 Typical Laue pattern of single-crystal silicon cell (surface normal to <001> and sides along <100> and <010>). **a** Before indexing, **b** after indexing (With courtesy of Ref. [5] from author's copyright)

the associated fitted peak parameters are presented in Table 3.1. In general, in our analysis, there will be around 20–35 silicon peaks in a typical Laue pattern of a single-crystal silicon cell used.

Table 3.1 List of indexed silicon peaks of the Laue pattern shown in Fig. 3.6 ("h k l" are the Miller indices of the diffracting planes) (With courtesy of Ref. [5] from author's copyright)

Peak no.	h	k	l	Energy (keV)	2 theta (°)	Intensity (counts)
1	0	0	8	13.04	88.90	1.05E+06
2	0	−2	6	16.43	52.14	4.81E+05
3	4	0	12	21.98	82.16	1.05E+06
4	−2	0	6	10.75	84.39	8.82E+05
5	1	−1	5	11.14	64.33	1.05E+06
6	−1	−1	5	10.96	65.54	1.05E+06
7	2	−2	8	19.89	58.29	4.23E+05
8	−2	−2	8	19.45	59.74	5.28E+05
9	3	−1	7	16.35	64.89	7.98E+05
10	1	−1	7	13.97	71.40	1.05E+06
11	−1	−1	7	13.81	72.36	9.43E+05
12	−3	−1	7	15.81	67.36	5.10E+05
13	0	−2	10	21.27	66.36	1.26E+05
14	2	0	10	17.07	86.03	9.75E+05
15	−2	0	10	16.84	87.47	7.92E+05
16	1	−1	9	17.02	75.34	6.57E+05
17	3	−1	9	18.81	70.74	2.25E+05
18	−1	−1	9	16.88	76.09	8.41E+05
19	−3	−1	9	18.36	72.78	2.89E+05
20	−1	1	9	13.46	101.20	1.05E+06
21	1	1	9	13.55	100.29	1.05E+06
22	3	1	9	14.95	93.49	6.55E+05
23	−3	1	9	14.66	95.93	4.90E+05
24	−1	−1	11	20.02	78.48	1.20E+05
25	−3	−1	11	21.18	76.20	3.30E+04
26	−1	1	11	16.64	99.12	2.55E+05
27	1	1	11	16.73	98.39	3.29E+05
28	3	1	11	17.91	93.67	1.95E+05
29	−3	1	11	17.62	95.71	1.94E+05
30	−1	1	13	19.84	97.64	3.87E+04
31	3	1	13	20.96	93.56	1.37E+04
32	−3	1	13	20.67	95.31	1.36E+04

The number of peaks depends on the position, area and sensitivity of the detector employed in the beamline. The number of peaks decreases because of the attenuation of the X-ray beam in the presence of obstacles, such as the back metallization and solder.

In the case of IBC (Interdigitated Back-Contact) cells [5] with the Cu back metallization (20–30 μm thick), the intensity of the diffraction peaks from the underlying silicon will be very low when compared with bare silicon (where there is no back metallization). Hence, the regular peak finding routine in the XMAS cannot find these low intensity peaks. In such a scenario, the XMAS can still identify and index them using a routine called adaptive indexing [8]. In adaptive indexing, the approximate positions of the peaks corresponding to a grain (the entire cell is one grain in a single-crystal silicon) are identified by indexing the Laue pattern from bare silicon. Then, these known silicon peak positions will be used to detect low intensity peaks, to index the Laue patterns from silicon below the metallization. For example, consider the solder joint region of an IBC mono c-Si cell, shown in Fig. 3.7a. The silicon is buried under a thick Cu back metallization and the solder joint. The bending angle around the Y-axis of the silicon cell is compared with and without adaptive indexing. In this case, the routine indexing procedure cannot detect silicon, as shown in Fig. 3.7b. However, with adaptive indexing, silicon below the back metallization is indexed to show the variation of the angle around the solder joint, as shown in Fig. 3.7c. As a rule, the scan point with a Laue pattern having less than 20 indexed peaks will be

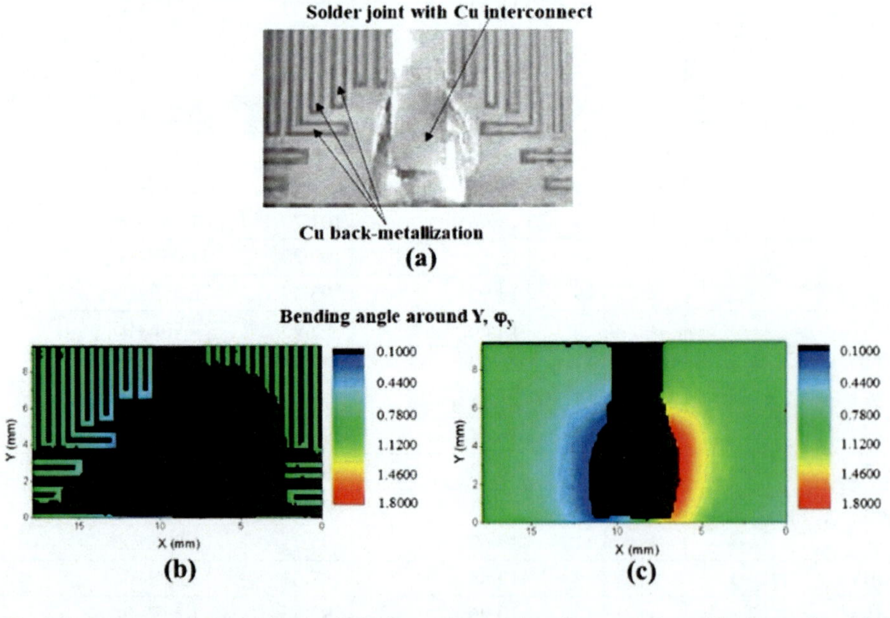

Fig. 3.7 Demonstration of adaptive indexing. **a** Scanned region around a solder joint of an IBC c-Si cell with Cu back metallization, bending angle of Si around Y-axis, **b** with general indexing, **c** with adaptive indexing (With courtesy of Ref. [5] from author's copyright)

treated as no signal (hence no result). Here, it is observed that the solder joint region is much thicker and even adaptive indexing cannot detect the silicon below it. This is because, in case of thicker obstacles such as the solder joint or interconnect, the X-ray is completely blocked from silicon, yielding no peaks at all (Fig. 3.7c). The X-ray transmission depths of the PV module materials are given in Appendix A for Ref. [16].

As it can be observed from Figs. 3.5 and 3.6, a typical μSXRD scan generates a large number of Laue patterns, based on the number of the scan points. Indexing all these Laue patterns manually would be a voluminous task. The XMAS can automatically index all the Laue patterns, if one of the patterns is correctly indexed and the associated indexing parameters are identified.

The orientation and deviatoric strain maps for the scanned area can be constructed using the measured sequence of Laue patterns, by utilizing the automatic peak indexing routine of the XMAS software [8], as shown in Fig. 3.8. The detailed procedure of the Laue pattern indexing is explained in the next section.

Laue Pattern Indexing: Evaluation of Crystal Structure and Deviatoric Strain

The polychromatic (white beam) diffraction pattern, or Laue pattern, contains very useful information about the crystal structure and strain of the sample, which can be exploited only upon proper indexing. Indexing the Laue pattern of a sample identifies its crystal structure. More importantly, it also gives the local crystal orientation

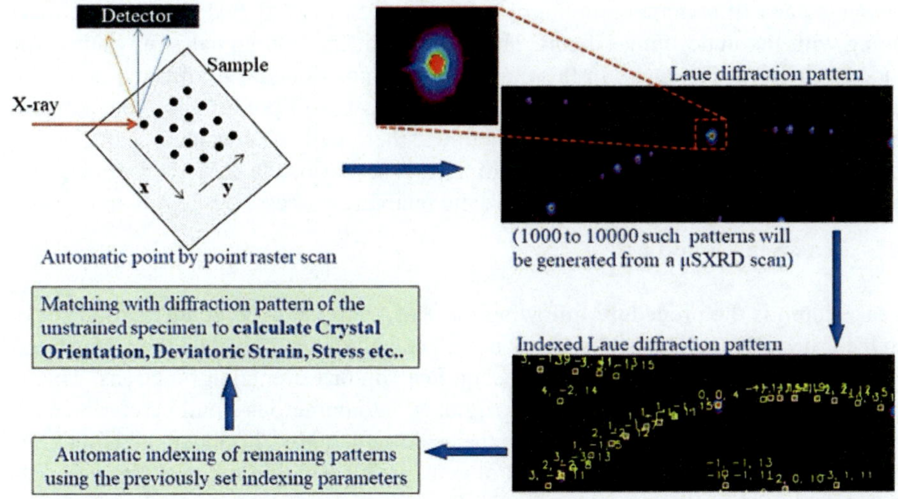

Fig. 3.8 Typical process map of automated μSXRD scan indexing using XMAS [8] (This figure was reproduced with Courtesy of Ref. [5] from author's copyright)

[8]. The diffraction peak positions in a Laue pattern slightly shift relative to their unstrained positions, due to the elastic deviatoric strains. Indexing can precisely measure such small shifts in the peak positions in a Laue pattern, and therefore, the deviatoric strains and stresses. The analysis of the Laue pattern in XMAS happens in the following 5 key steps:

A. Background noise removal.
B. Peak search and fit.
C. Experimental geometry calibration.
D. Indexing of peaks.
E. Crystal structure, deviatoric strain and stress evaluation.

As explained earlier, all these steps can be automated for indexing the Laue patterns from a scanned sample, upon the successful indexing of one of the Laue patterns in it.

A. **Background Noise Removal**

Removing noise is an important step, as it is not from the sample and affects the quality of indexing. The noise may be partially from the surroundings (the sample stage), which can be removed by several image processing techniques, such as segmentation and filtering [8]. Another reason for noise is within the sample itself. The sample may sometimes contain several crystalline materials which can cause diffraction, and we only need to index a particular material of choice. For example, the generic design solar cell contains aluminum back metallization (Al-BSF), which diffracts along with the underlying silicon. This Al-BSF diffraction signal is unwanted for indexing silicon. However, in this case, the peak positions of Al diffraction do not generally overlap with the silicon peaks. Hence, it is still possible to index silicon, even in the presence of Al peaks. Sometimes, there will be unwanted peaks, which may overlap with the silicon peaks. In such a scenario, the XMAS routine gives options to mask unwanted peaks before the indexing process.

B. **Peak Search and Fit**

Peak search is the procedure following the XMAS to identify the diffraction peaks in the Laue pattern. The peaks are identified and classified based on their integrated intensity and then fit with a 2D Gaussian, Pearson or Lorentzian functions. This is because the 2D X-ray detector area is digitized into numerous small pixels and each peak illuminates several pixels. Fitting the peaks with one of the above-said functions allows for a more accurate resolution of peak position within the peak, based on its intensity variation. We typically use the 2D Gaussian function for our peak-fitting routines, but the Lorentzian function also works well. In the case of severely distorted peaks (due to plasticity, etc.), the choice of the peak fit function affects the accuracy of the peak position significantly. The accuracy of the peak positions thus identified is key for the indexing process and the subsequent strain evaluation. Constraints on the peak intensities, peak widths and the structure factor can be used to distinguish the diffraction peaks from artefacts that resemble peaks.

C. Experimental Geometry Calibration

The next step in the analysis is the experimental geometry calibration. This is the most important step in the analysis of Laue patterns. Indexing a Laue pattern could measure very slight relative shifts in the peak positions (due to elastic strain) from their unstrained positions. However, the measurement of such small shifts needs precise information of the experimental geometry. Hence, the experimental geometry needs to be well calibrated for accurate evaluation of the crystal orientation and deviatoric strain. The vector connecting the diffraction point on the sample (diffracting volume) to the position of a peak on the area detector constitutes the lattice (scattering) vector of the respective diffracting crystal plane in the reciprocal space. Each peak in the Laue pattern is attached to such scattering vector, represented by $\boldsymbol{k}_{\mathbf{out}}$. On the other hand, $\boldsymbol{k}_{\mathbf{in}}$ is the incident X-ray beam vector. Then the normalized scattering vector ($\boldsymbol{q}$), associated with a diffraction peak, can be calculated by Eq. 3.2.

$$\vec{q} = \frac{\vec{k}_{\text{out}} - \vec{k}_{\text{in}}}{\left\| \vec{k}_{\text{out}} - \vec{k}_{\text{in}} \right\|} \tag{3.2}$$

As shown in Fig. 3.9, the spatial position of the detector is defined by the following parameters:

1. Shortest distance of the diffracting volume from the detector center channel d_c
2. Coordinates of the detector center channel x_c and y_c. Note that these are in the detector coordinate system (X_d, Y_d) and differ from the sample coordinate system.

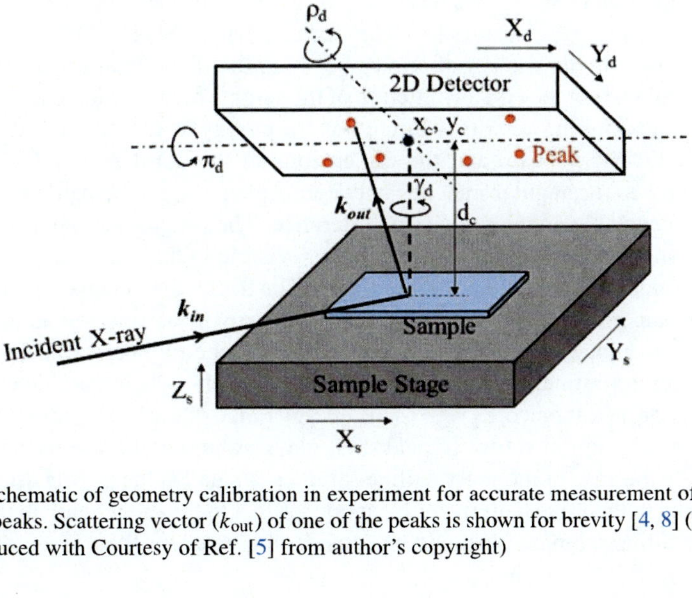

Fig. 3.9 Schematic of geometry calibration in experiment for accurate measurement of scattering vectors of peaks. Scattering vector (k_{out}) of one of the peaks is shown for brevity [4, 8] (This figure was reproduced with Courtesy of Ref. [5] from author's copyright)

3. Tilts of the detector with respect to the incident X-ray beam: Roll ρ_d, Pitch π_d, and Yaw γ_d. Out of these, the detector roll (ρ_d) will be set to zero for the reflection geometry.

Hence, the 5 independent parameters d_c, x_c, y_c, π_d and γ_d, which define the detector spatial position with respect to the diffracting volume (point of diffraction on the sample), need to be precisely calibrated for the accurate evaluation of the normalized scattering vectors (***q***) of the diffraction peaks. In general, these 5 parameters are calibrated before the start of a new experiment using a well known, unstrained reference sample. A sample Laue pattern is taken from the reference sample, mounted on the sample stage and the geometric parameters are calibrated by a non-linear least square minimization of the following function, given in Eq. 3.3.

$$\alpha_o = \frac{\sum_i w_i \left(\alpha_i^{\text{th}} - \alpha_i^{\text{exp}}\right)}{\sum_i w_i} \tag{3.3}$$

In it, $\alpha_i{}^{\text{th}}$ and $\alpha_{\text{i}}{}^{\text{exp}}$ are the theoretical and experimental angles between the pairs of scattering vectors of the diffraction peaks, and w_i are the weight functions of the least square function.

Another method of geometry calibration is possible by taking Laue patterns at different detector distances from the sample and then back tracing the origin of the scattering vectors. However, this method is complicated due to the addition of the 6th geometric parameter (roll ρ_{d}) of the detector and associated errors.

D. Indexing of Peaks

Once the normalized scattering vectors (***q***) of the peaks are calculated, the indexing routine in the XMAS selects a subset of N brightest peaks and calculates the angles between their scattering vectors [4]. This forms a list of N(N − 1)/2 angles, which will be compared with the list of angles between the theoretical scattering vectors that are calculated for the crystal structure of the sample being evaluated. The number of theoretical scattering vectors is limited by the energy band (range) of the incident polychromatic X-ray beam and the dimensions of the 2D detector. The indexing routine searches for angular matches between triplets of peaks (scattering vectors), within an acceptable angular deviation tolerance. The program computes the list of peaks that should be recorded by the detector (visible in the Laue pattern), for each triplet of peaks and their Miller indices (*h k l*). The triplet that matches the maximum number of experimental peaks on the Laue pattern will be selected as the best fit. Hence, the accuracy of indexing is proportional to the number of peaks found in the peak search routine. Indexing is an iterative process, where parameters such as angular deviation tolerance, as well as the energy range of the beam, affect the quality of indexing. The initial subset of peaks, N, plays an important role in the indexing process. For the case of monocrystalline samples, a small value of N is used because all the peaks in the Laue pattern are from one grain. On the other hand, in the case of multi-crystalline samples, N should be considerably large, as there is a possibility of overlapped peaks from multiple grains in the same Laue pattern. Each grain needs a good number of initial bright peaks for proper indexing.

E. **Crystal structure, Deviatoric Strain and Stress Evaluation**

The unstrained unit cell of a cubic crystal is characterized by 6 independent parameters, 3 edge lengths (a, b and c) of the lattice and 3 corner angles (α, β and γ). The distorted unit cell, as shown in Fig. 3.10, needs 3 additional crystal rotations (φ_1, φ_2 and φ_3) that define the distortion. Hence, 9 independent parameters can characterize a distorted unit cell of a cubic crystal. As outlined by Chung and Ice (1999) [17], in the case of a strained crystalline sample whose unit cell parameters are not known, the parameters can be identified by indexing at least 4 independent reflections (diffraction peaks). The normalized scattering vectors of the reflections in the reciprocal space ($\boldsymbol{q}_{hkl}$) can be related to the scattering vectors in the real space laboratory coordinate system ($\boldsymbol{q}_{xyz}$) by the relation in Eq. 3.4.

$$\vec{q}_{xyz} = U\vec{q}_{hkl} \tag{3.4}$$

where the 3 × 3 matrix U is the unit cell matrix. It defines the 9 parameters of the unit cell, 3 crystal rotations and 6 lattice parameters shown in Fig. 3.10. With each of the 4 scattering vectors, 4 normal directions can be observed and each normal has 2 direction angles. Hence, we require 8 equations to solve for 9 unknowns. However, these scattering vectors are normalized (ref. Eq. 3.2). Hence, the dilatational (volumetric) part of the deformation cannot be measured. This can be evaluated if the energy of one of the peaks is measured accurately through an energy scan. Ignoring the dilatational term, the deviatoric part (distortion at constant volume) of the deformation and orientation of the unit cell can still be determined by evaluating the lattice

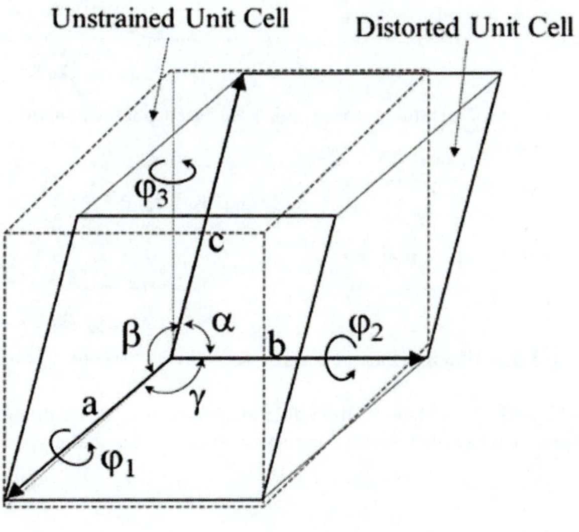

Fig. 3.10 Schematic of distorted unit cell of a cubic crystal (With courtesy of Ref. [5] from author's copyright)

parameters in relative terms, for example, "b/a" and "c/a", instead of a, b and c. With this, the lattice parameters reduce from 9 to 8 and the eight equations can be solved for the eight unknowns.

Later, the unit cell parameters can be determined, similar to the geometric calibration parameters, by solving Eq. 3.3 for the 5 lattice parameters (b/a, c/a, α, β, and γ) instead of the 5 geometric parameters (d_c, x_c, y_c, π_d and γ_d) [4].

As already stated, the Laue diffraction cannot measure the precise energy of the peaks in a Laue pattern. Hence, it cannot measure the changes in the volume of the unit cell. It can only measure the changes in the shape of the unit cell at constant volume. In other words, it can only determine the deviatoric strain (and stress) in the sample and cannot measure the dilatational term (volumetric strain or stress). This can be explained based on the simple Bragg's condition (Eq. 3.1), according to which, the changes in the shape of the unit cell at constant volume cause relative tilts in the lattice planes (diffracting planes), leading to relative shifts in the peak positions in a Laue pattern, but the energy remains the same, as shown in Fig. 3.11a. The changes in the volume of the unit cell cause a change in the interplanar distance, leading to a change in the energy of the peaks, as shown in Fig. 3.11b, but with no shift in the peak positions in the Laue pattern [4, 8, 17].

The evaluation of the deviatoric strain from the Laue diffraction is based on the homogeneity property [8], by which, the coordinate position of a point in the unstrained unit cell remains the same as that in the strained unit cell, leading to the following relation in Eq. 3.5.

$$R_o X_o = RX \tag{3.5}$$

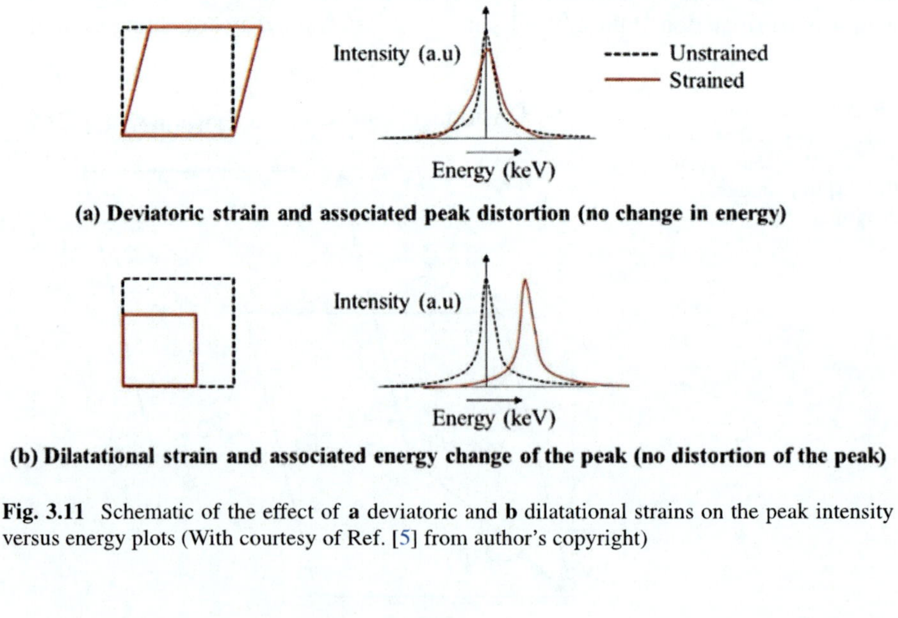

Fig. 3.11 Schematic of the effect of **a** deviatoric and **b** dilatational strains on the peak intensity versus energy plots (With courtesy of Ref. [5] from author's copyright)

X_o and X are the coordinates of a point in the unit cell, defined in terms of the Cartesian coordinates attached to the undeformed and deformed unit cells. The matrices $\boldsymbol{R_o}$ and $\boldsymbol{R}$ are the coordinate transformation matrices, mapping the Cartesian coordinates to the undeformed and deformed unit cell coordinates respectively. Equation 3.5 can therefore be rewritten, as shown in Eq. 3.6, to define the transformation matrix T, to relate the deformed state to the undeformed state.

$$X = R^{-1}R_oX_o = TX_o \tag{3.6}$$

Now, the deviatoric strain (ε') can be calculated from the transformation matrix T by removing the rotation component from it, as shown in Eq. 3.7 [17].

$$\varepsilon_{ij}^{'} = \varepsilon_{ij} - \Delta_{ij} = \frac{T_{ij} + T_{ji}}{2} - \delta_{ij} \tag{3.7}$$

where ε is the total strain, Δ is the dilatational strain and δ is the Dirac-delta function. The deviatoric stress σ can be calculated from the deviatoric strain by using Hooke's law, as shown in Eq. 3.8.

$$\sigma_{ij}^{'} = \sigma_{ij} - \sigma_H\delta_{ij} = \boldsymbol{C}_{ijkl}\varepsilon_{kl}^{'} = \boldsymbol{C}_{ijkl}(\varepsilon_{kl} - \Delta_{kl}) \tag{3.8}$$

where σ is the total stress, σ_H is the dilatational (hydrostatic) stress and $\boldsymbol{C}$ is the elasticity tensor of the sample material. The deviatoric strain and stress contain the deviatoric components only along their diagonals (directional components). The off-diagonal (shear) components are identical to the total strain and stress tensors, respectively.

3.1.3 Local Plasticity Probing Using White-Beam MSXRD

The determination of the crystallographic orientation and stress/strain (deviatoric) in each of the individual crystal has been discussed in the previous section. It is basically done using the position and change in the relative position of the Laue diffraction peaks on the CCD image. Having done extensive experiments using the technique [7, 14, 15, 18–25], we now realize another source of wealth of information that the technique provides is the shape of the Laue diffraction peak itself, as well as the change in the absolute position of the peak. Figure 3.12 shows features in Laue diffraction patterns and their associated structural and mechanical information that they contain about the material/sample.

First, the shape of the Laue peak itself provides information about plastic deformation in the crystal, especially the one involving the Geometrically Necessary Dislocations (GNDs), and then secondly, the change in the absolute position of the Laue peak gives us the rotational deformation of the crystal body. This sub-section will

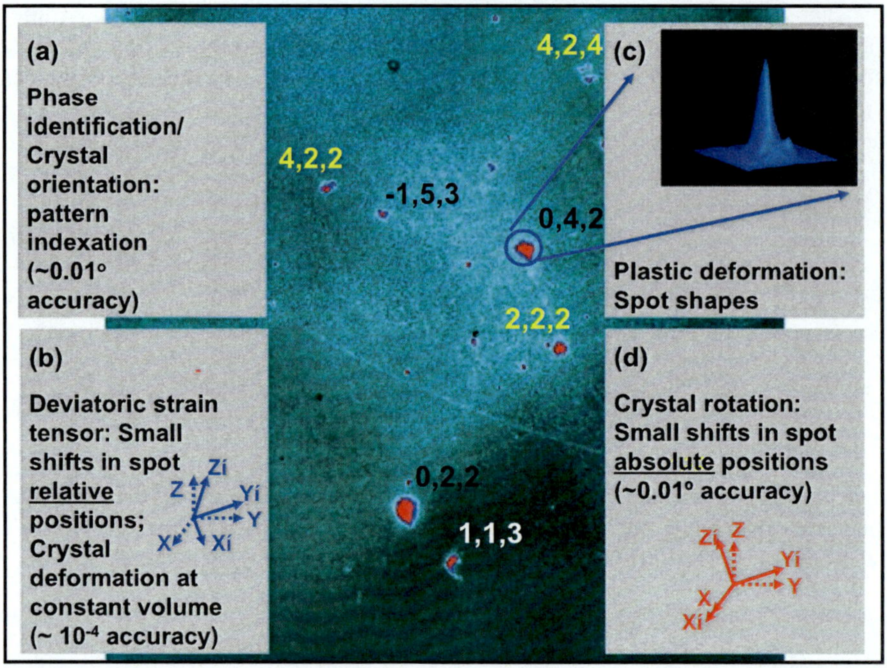

Fig. 3.12 A single white-beam CCD image consisting of multiple sets of Laue diffraction peaks (three colors: black, yellow and white) from a Cu polycrystalline sample; The determination of **a** and **b** have been described in Sect. 2.3, while the determination of **c** and **d** will be discussed in this section (With courtesy of Ref. [4] from author's copyright)

describe and discuss the data analysis procedure for the determination of plastic and rotational deformation in crystals that will be used again and again throughout the chapters in this book.

Crystal Bending, Polygonization and Rotation

In traditional X-Ray Diffraction (XRD) experiments, a peak at a certain θ angle means a particular (hkl) plane is detected, i.e. the particular θ angle, d_{hkl} interplanar distance and $\lambda_{Cu,k\alpha}$ wavelength (for instance) conspire such as to satisfy Bragg's Law:

$$n\lambda_{Cu,k\alpha} = 2d_{hkl}\sin\theta. \tag{3.9}$$

Using this methodology, only elastic deformation of crystal can be determined, where strain can simply be computed from the difference of interplanar distances between atomic planes, before and after the deformation. In contrast, plastic deformation may involve curved and polygonized crystal planes, such as shown in Fig. 3.13. In such configuration, a particular crystal plane (with a fixed d_{hkl} interplanar distance) may have a range of values of θ angle. This lattice curvature information will be lost

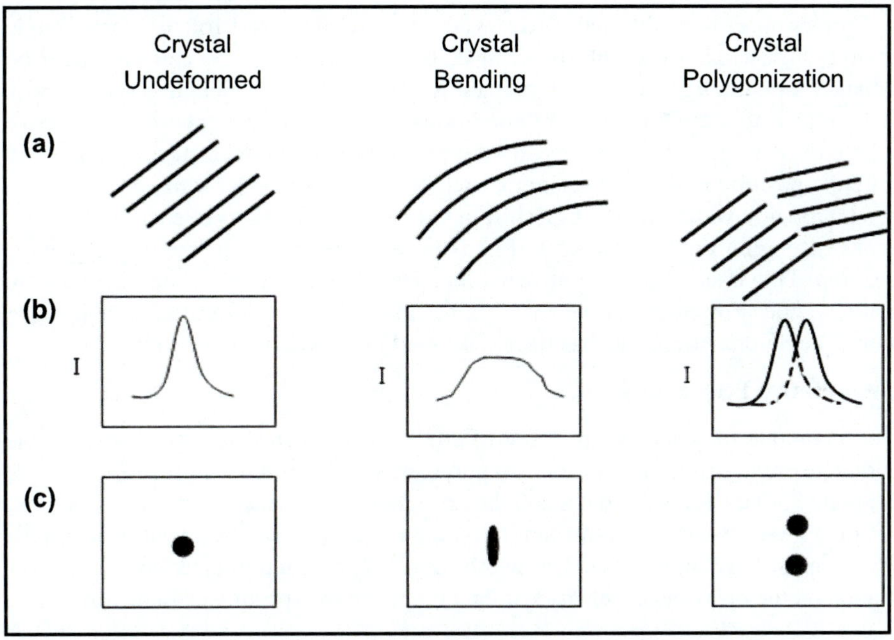

Fig. 3.13 Schematic diagrams comparing a set of crystal planes in their undeformed, bent/curved and polygonized states (**a**), while their expected Laue diffraction peaks corresponding to each of the crystal states are shown in **b**, in intensity scanning over angle consisting of the Bragg angle/s, and **c** in CCD detector space (With courtesy of Ref. [5] from author's copyright)

in a traditional XRD (using a single wavelength and no tilt). However, using a white-beam X-ray, which also has a range of wavelengths, Bragg's law can still be satisfied even though the θ angle varies. This means that with white beam a presence of lattice curvature can still be detected and even measured.

In the case of a curved crystal plane (crystal bending), a broadened Laue diffraction peak in a certain direction would be observed (Fig. 3.13) instead of a single, sharp, rounded peak typical of an undeformed crystal plane. The Laue peak broadening (also called "peak streaking") is continuous representing the continuum of θ values involved in the curved crystal. Crystal polygonization, in contrast, involves multiple, discreet peaks coming from slightly misoriented sub-crystal structures (such as shown, again, in Fig. 3.13). Such polygonization often occurs following bending of crystal where more and more dislocations are piling up, and thus become unstable with respect to glide and eventually climb to form a low-angle boundary [26–29].

This is the unique capability of our white-beam X-ray diffraction technique. Using this white-beam methodology, we can start to directly and quantitatively study the plastic deformation in crystals. This technique is especially sensitive toward the crystal bending or polygonization in their simplest geometries, where a net density of parallel same-sign Geometrically Necessary Dislocations (GND's) is formed.

Furthermore, in traditional XRD, when a body of crystal rotates, the θ angle changes and without rotating the sample, Bragg's law (Eq. 3.9) can no longer be satisfied. In contrast, using a white-beam X-ray, another wavelength would simply be available to satisfy diffraction conditions, and thus would give a constructive interference in the diffracted intensity (a Laue diffraction peak) instead on a slightly different position in the CCD camera, such as shown in Fig. 3.14. In other words, a crystal rotation would appear as a shift in the absolute positions of each of the Laue peaks belonging to the same crystal (it is to be noted that a change in the *relative* position of the Laue peaks—relative to other peaks belonging to the same crystal—in contrast, does not mean crystal rotation, but instead, shear deformation or in other words, deviatoric strains, as has been discussed in the previous section).

Quantitative Peak Study

Our technique of white-beam X-ray Microdiffraction can detect and measure the curvature of the lattice, or in the case of polygonized subgrain structure, the low-angle subgrain boundary. From how much the crystal under consideration curved, or bent, or polygonized, we can obtain quantitative data related to the plastic deformation the crystal must have undergone. The width of a Laue diffraction peak contains information on the dislocation density within a grain and the splitting of Laue diffraction

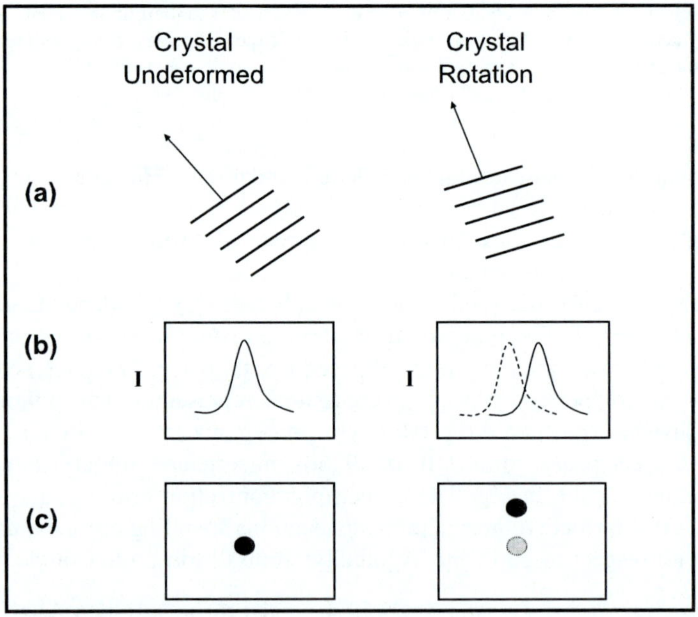

Fig. 3.14 Schematic diagrams comparing two bodies of crystal in their undeformed/unrotated and rotated states (**a**), while their expected Laue diffraction peaks corresponding to each state are shown in **b**, in intensity scanning over angle consisting of the Bragg angle (there is a shift in the absolute angular position of the peak), and **c** in CCD detector space (again, a shift in the position of the peak on the CCD detector space) (With courtesy of Ref. [5] from author's copyright)

peak into multiple individual spots can be used to measure the formation of low-angle grain boundaries.

The XMAS (X-ray Microdiffraction Analysis Software) program allows us to select an individual Laue diffraction peak from a CCD image and study it quantitatively. The peak can be plotted in 2D or 3D using the detector coordinates. More meaningfully, as far as the crystal is concerned, is the ability to convert the peak into either theta-chi (θ-χ) space or reciprocal space. In θ-χ space, the Bragg angle, θ, can be calculated using the equation:

$$\theta = \frac{\cos^{-1}[k_{\text{in}} \cdot k_{\text{out}}]}{2} \tag{3.10}$$

where k_{in} and k_{out} are the incident beam and diffracted beam. The χ angle is defined as the angle of the diffracted beam within the plane perpendicular to the incident beam. The χ angle can be found by projecting the diffracted beam into the XZ_{lab} plane, and then finding its angle relative to the Z_{lab} axis. The distortion of the θ-χ space on the CCD frame typical for our experimental geometry is shown in Fig. 3.15. The XMAS program allows conversion to remove the distortion caused by using a flat CCD to detect the Laue patterns in three-dimensional space.

Once the peaks are in θ-χ coordinates, we have the real rotation of the crystal planes associated with the plastic deformation. This angular change in the geometry of our crystals can thus be further converted into the change in the geometrically necessary dislocation structures (plasticity) that the crystal must have undergone during the deformation. For illustration purposes, we use our real experimental results (Fig. 3.16), in the form of Laue peak intensity contour on a θ-χ space, its intensity trace along a certain χ angle, and the Full Width Half Maximum (FWHM) measurements from each of the peak, to give examples of the methodology to determine the extent of plastic deformation from the peak study.

Figure 3.16 shows typical Laue diffraction peaks (in θ-χ coordinates) coming from an undeformed crystal, a plastically bent/curved crystal and a polygonized crystal into two low-angle subgrain structures. Peak broadening and split occur primarily across the theta direction in this particular experiment. The peak broadening can be quantified by defining $\Delta\theta_{FWHM}$ as the difference between the full width at half maximum (FWHM) of the peaks plotted in the θ-χ space along the θ direction at a constant value of χ, chosen to divide the peak in half. These intensity traces are plotted in Fig. 3.16 (in the column next to the Laue peaks) and are subsequently fit using a Lorentzian function as shown in the figure. Clearly from the FWHM measurements, we see peak broadening from (a) to (b) which represents plastic bending of crystal such as illustrated in (d). Similarly, the measurement of the θ angle difference between the two peaks in (c) suggests polygonization of the crystal across the theta direction as illustrated again in (d). Data can also be quantified within the reciprocal space, as an alternative to the θ-χ space.

Knowing the relevant geometrical dimensions of the crystal in the experiment, and the calibrated CCD-detector-to-sample distance, the peak broadening ($\Delta\theta_{FWHM}$)

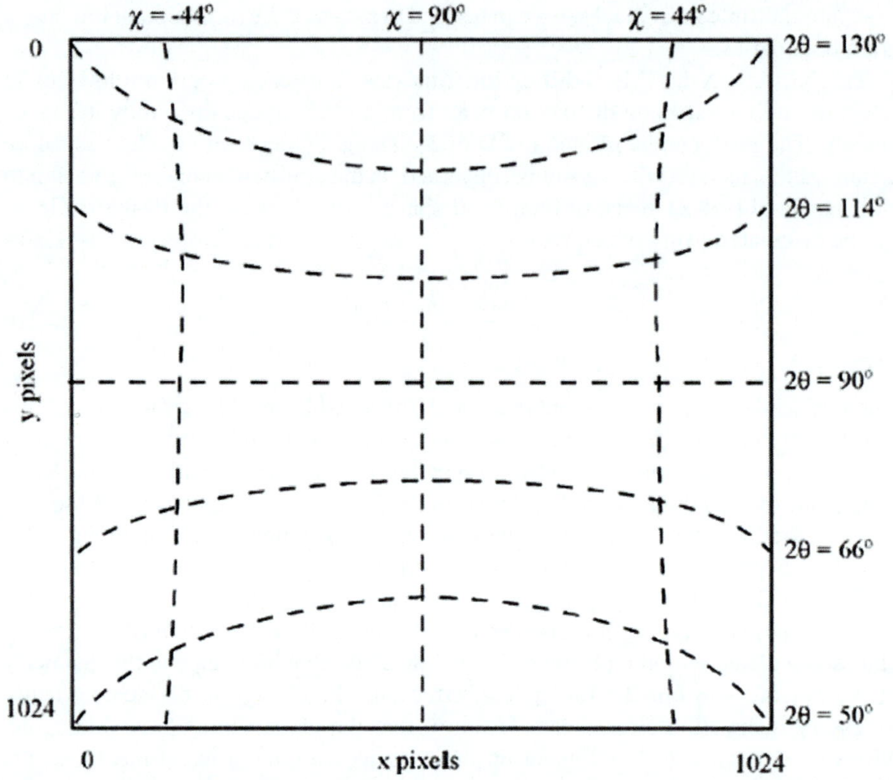

Fig. 3.15 The distortion of 2θ-χ space on the CCD detector typical for our experimental geometry; Lines of constant 2θ and constant χ are shown, as well as the orientation of the CCD pixels (Courtesy of Valek et al. [30])

can be converted into the radius of the curvature of the crystal bending, Cahn [26] and Nye [31] further suggested that from the curvature of the crystal bending, we can determine the amount of Geometrically Necessary Dislocations (GND's) that have to be in the crystal to accommodate such a change in the geometrical shape of the crystal

$$\rho = \frac{1}{Rb}, \tag{3.11}$$

where ρ is the dislocation density of GND's, R is the radius of curvature, and b is the component of the Burgers vector on the neutral axis. This is illustrated in Fig. 3.17a.

The peak split intensity profiles (Fig. 3.16) show clearly resolvable peaks indicating that the initial grain (a) breaks up into crystal segments with their own well-defined orientation (c). From the splitting of the Laue peaks, Δ_θ, the misorientation between adjacent subgrains can be measured. Using the measured misorientation,

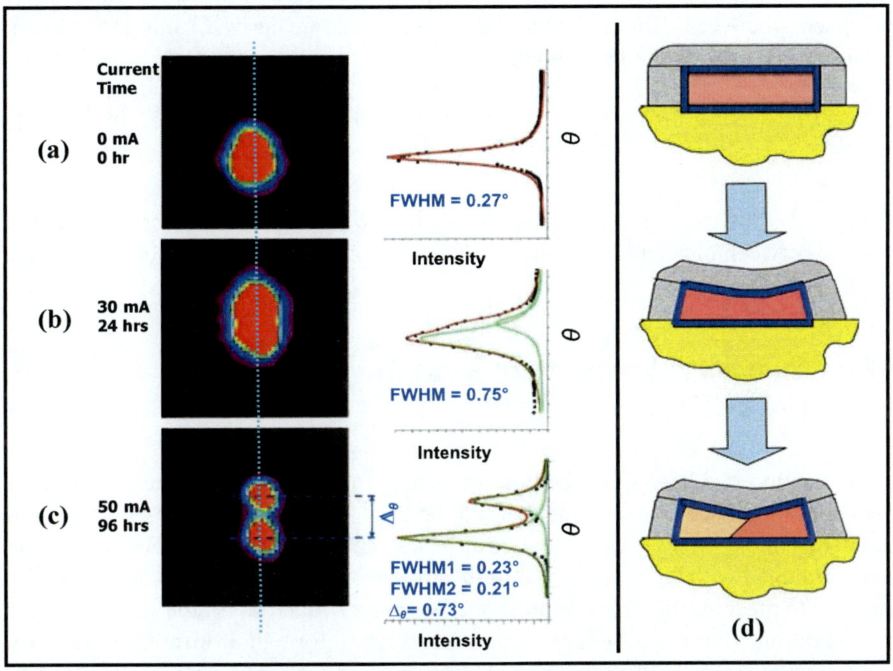

Fig. 3.16 Quantitative measurement of the evolution of a Laue diffraction peak (Peak 002) of a copper grain/crystal in an interconnect line under high current density electrical loading (electromigration) as has been reported in Ref. [18]; Broadening is shown as the peak evolves from **a** to **b**, while peak splitting, signifying the formation of low-angle grain boundaries, is evident in **c**; **d** is an illustration of the evolution of the grain/crystal that the Laue peak evolution indicates (With courtesy of Ref. [5] from author's copyright)

Fig. 3.17 Determinations of densities of GND's from geometrical shape of the crystals; **a** Crystal bending/curvature, using Cahn-Nye [26, 31] relationship, and **b** crystal polygonization, using Burgers relation for a low-angle grain boundary (With courtesy of Ref. [5] from author's copyright)

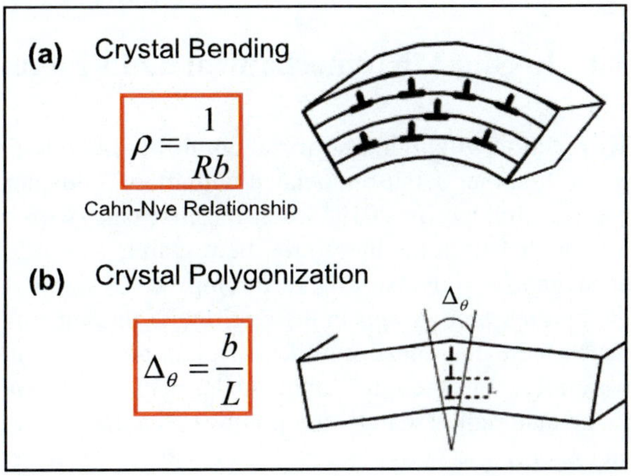

the spacing between dislocations in the boundary can be derived using the Burgers relation for a small angle grain boundary:

$$\Delta_\theta = \frac{b}{L}, \tag{3.12}$$

where Δ_θ is the measured misorientation angle, b is the Burgers vector and L the spacing between dislocations in the boundary, as illustrated in Fig. 3.17b.

3.1.4 Summary: Synchrotron X-ray Microdiffraction (μSXRD)

In this section, we have described the analytical methodologies of the synchrotron X-ray microdiffraction technique that will be used throughout the subsequent chapters in this book. We have presented how the raw data in the form of Laue diffraction peaks obtained with the scanning white-beam X-ray microdiffraction technique may translate to meaningful information about the crystals/materials of the samples under consideration. The crystallographic and stress/strain information are analyzed mostly in automated manner using the XMAS software package (Sect. 3.1.2). In contrast, the data analysis to extract plastic deformation information is mostly accomplished manually utilizing the various existing XMAS program utilities. The procedure of such plasticity analysis has been described in Sect. 3.1.3 allowing a direct and quantitative way to probe local plasticity in the crystalline materials at small (submicron and nanometer) scales.

3.2 In-Situ Micromechanical and Fracture Testing

This section will highlight first about the in-situ microfracture technique—the basic setup and other details associated with more fundamental aspects of the experimentation technique. In 2018, when the first papers on this techniques [32, 33] were first published in the literature, the research group (led by the author of this book) at Singapore University of Technology & Design (SUTD), Singapore was among the first research groups in the world with this capability to study in-situ fracture of nanoscale metal–metal multilayered composites. Many of the fundamental aspects were related to the other small-scale mechanical experimentation techniques used in the materials research communities all across the world, such as nanoindentation and micropillar compression. This discussion will be followed by the introduction of the materials synthesis techniques and sample preparations procedures to allow the in-situ microfracture tests to be conducted. We used the nanoscale metal–metal composites here as a platform to allow the study. However, in the course of developing the processes and the materials synthesis techniques, we realized we could also control further interface characteristics such as crystallographic orientation and interfacial shear strength. With the creation of these interfaces, the microscale fabrication technique of focused ion beam (FIB) is used to fabricate the test structure required to perform bending experiments to allow for a direct line of sight to catch a crack in action—interacting with microstructures of materials at the nanoscales. The MechanoElectrical contact resistance (MECR) with micropillar compression technique will be discussed in this section as well. These techniques are instrumental especially in studying interfaces and how fracture events could be influenced by them, and thus nanolayers were among the first materials systems used to demonstrate the efficacies of the techniques.

3.2.1 In-Situ Microfracture Testing

Early experimental studies done on fracture of nanoscale metal–metal multilayered composites listed in previous chapter show that the multilayered interface plays a crucial role in the resulting mechanical deformation and fracture. It would be important and crucial to study interface interactions with a propagating crack in real-time during fracture through a small-scale microfracture testing technique. The effects of the constituent layer interfaces and grain boundaries on plasticity, fracture and fatigue of the multilayers can be examined experimentally and will be further evaluated theoretically or through simulations. Current real-time observations of a crack or notch propagating across the multilayered nanocomposites in a controlled setting are limited to mainly metal/ceramic multilayered systems [34, 35].

The in-situ microfracture technique allows the simultaneous and synchronous recording of the load–displacement information, in addition to crack initiation and propagation events. The fracture load can be precisely noted, and the crack opening displacement (COD) can be determined via direct observation or digital image correlation (DIC) techniques [36]. The development of instruments [37, 38] to both apply and measure loads and displacements accurately and observe the changes and record in-situ [39] has enabled fracture testing of materials at micrometer and nanometer sizes to become a reality. Further, the availability of focused ion beam (FIB) machining, lithography, deep reactive ion etching (DRIE) and laser micromachining to prepare samples at small dimensions have encouraged the growth in this area.

Uniaxial loading in both tensile and compression modes have been done in-situ for nanoscale metal–metal composites inside a scanning electron microscope (SEM) as well as transmission electron microscope (TEM) [35, 37, 40–43]. Results from these experiments have provided many insights into the plasticity mechanisms of multilayers at the nanoscale due to interface interactions. However, these methods are not suitable for fracture studies. In the tensile mode, the maximum tensile stress is experienced throughout the entire volume of the testing sample and fracture may not happen at the weakest direction for the crack driving force. If there are sources of weak grain boundaries or voids, lower values of fracture strength may be attained for the same material through tensile mode than through bending. In compression, the crack is aligned parallel to the compression direction. When the crack propagation is the controlling mechanism of failure, a larger amount of stress is required to fail a material in compression compared to tension. However, the maximum tensile stress is conversely concentrated in a small region on the top surface above the neural axis in bending mode, making it apt for fracture. Further, ASTM standard tests to measure crack propagation and fracture toughness in bulk materials are performed in bending only [44, 45].

Consequently, the conventional methods for evaluating fracture in materials in the bulk form cannot be applied such that new experimentation techniques are required in the micron and nanoscale. Researchers have proposed non-conventional solutions of single cantilever bending [46–48], double cantilever beam (DCB) [49], pillar splitting [50, 51] and clamped beam bending [34, 52, 53] to observe fracture in-situ and allow for fracture toughness measurements for various kinds of materials. A comparison between the four in-situ fracture geometries was made to evaluate the advantages and disadvantages of using each technique to compute the fracture toughness of silicon as shown in Table 3.2 [54].

To identify key interface parameters enabling enhanced fracture and fatigue properties in nanoscale metal–metal multilayered composites, we would want to choose a method that is suitable for interfaces, enables us to view interface interactions with the notch and/or crack in-situ in the line of sight, scalable at any length scale and a stable crack growth. Two possible geometries can be used: the single cantilever bending and

Table 3.2 Comparison of the in-situ fracture geometries used to measure fracture toughness [54]

Conditions	Test methods			
	Single cantilever bending	Double cantilever bending	Pillar splitting	Clamped beam bending
Indenter type	Wedge/cone	Flat	Cube/Berkovich	Wedge/cone
In-situ required	No	Yes	No	Yes
Stable crack growth	No	Yes	No	Yes
Tolerance to misalignments	Yes	No	No	No
Free of residual stress	Yes	Yes	Yes	No
Suitable for anisotropic materials	Yes	Yes	No	Yes
Suitable for interfaces	Yes	No	No	Yes
Ease of scaling	Yes	No	No	Yes

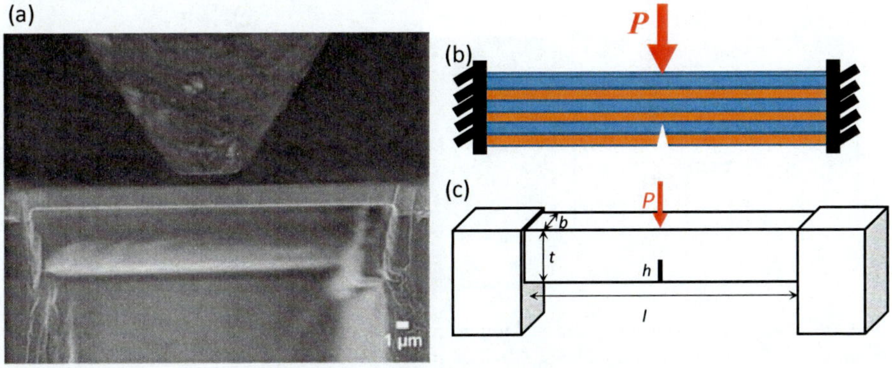

Fig. 3.18 **a** SEM image of clamped beam bending geometry for nanoscale PVD (Physical Vapor Deposition) [33] metal–metal multilayered composite where the beam is made by FIB. A conical flat tip of 5 μm diameter is used to load the beam. **b** Schematic representation of the clamped beam bending geometry with a notch across the layers and **c** 3D model representation of the clamped beam bending geometry with the beam dimensions (With courtesy of Ref. [33] from author's copyright)

the clamped beam bending. For the advantages of an in-situ requirement to observe fracture, stable crack growth, ease of making multilayered film beam structures by FIB and close resemblance to the ASTM standard geometry for fracture, the clamped beam bending is chosen to be the most suitable method to proceed with the in-situ microfracture testing of different nanoscale metal–metal multilayered composites.

Clamped beam bending was first demonstrated as a fracture test geometry at the nanoscale on PtNiAl bond coats [34]. It is based on a modification over the three-point bend test by clamping the ends instead of letting them free as shown in Fig. 3.18. Within elastic limits, simply supported beams deform four times more than rigidly clamped ones [55]. A clamped beam offers a more stable setup with crack arrest after a certain amount of propagation from the notch tip. The position of the maximum principal tensile stress changes from the notch tip to the fixed ends once the h/t ratio of the crack exceeds ~0.5 [34, 53]. This transition makes the clamped beam bend configuration transform from an unstable to a stable system for crack propagation.

The microbeams are fabricated by focused ion beam which will be elaborated in Sect. 2.3.2. The beam dimensions should adhere to an l/t ratio between 4 and 8 and an initial h/t ratio between 0.3 and 0.5 based on satisfying both plane strain and small-scale yielding criteria [54, 56]. A much shorter or longer starting crack length will not provide stable crack growth and also give rise to free surface effects. The notch sizes ($h \times w$) in all beams are varied due to the challenging process of creating these notches by FIB as a single cut across the beam as shown in Fig. 3.19. The width of the notch, w is to be as small as possible to almost a knife edge (i.e. 100–300 nm range).

The in-situ mechanical experiments are carried out inside the JEOL FESEM (JSM-7600F) using a PI-85 SEM picoindenter (Bruker Inc., USA) with a 20 mN maximum load transducer. Two beams were tested for each material system. The beams were

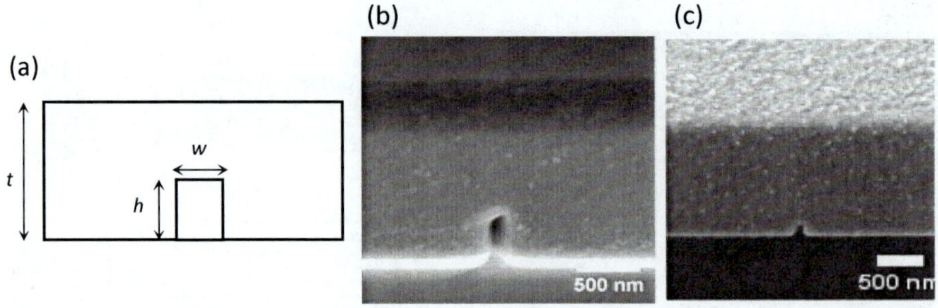

Fig. 3.19 Sharp notches made by focused ion beam (FIB): **a** schematic of a notch of width, *w* and height, *h* in the microbeam of thickness, *t*, **b** SEM image of the front notch of $h = 400$ nm and $w = 120$ nm, **c** SEM image of the back notch which is smaller than the front notch (With courtesy of Ref. [33] from author's copyright)

loaded using a conical flat tip of 5 μm diameter and carried out in displacement control mode. A constant loading displacement rate of 5 nm/s with a Q-control of 0.2, an integral gain of 0.2, a derivative gain of 1 and a specific preload was maintained for all experiments. The load–displacement data and the real-time video of the experiments were synchronized and captured during the experiment using a frame grabber with the TriboScan software (Bruker Inc., USA). Care was taken to perform drift correction before the experiment started and corrected to the final load–displacement measurements.

This geometry offers pure mode I loading as well as controlled mixed mode loading by placing the indenter at fixed offsets from the notch position. The crack can be seen to grow and arrest in stages before complete failure ensues. Such controlled crack growth is important in the measurements of R-curve or fatigue behavior of materials [53, 54] as well as in following crack trajectory and study of toughening mechanisms after the experiment.

The maximum principal stress that opens up the crack acts along the line connecting the notch tip and the indenter. Therefore, the indenter should ideally be positioned precisely across the notch for mode I loading, but misalignments up to 5% of beam length are tolerable. The indenter tip should be sufficiently blunt not to cause indentation impressions or cracking. The test is no longer valid if and when edge cracking initiates at the clamped ends. Maintaining larger l/t ratios prevents edge cracking. K_I is obtained via an extended finite element method (XFEM) simulation and has a sinusoidal dependence on h/t. Though the general form of this dependence remains constant for this geometry, each case differing in sample dimensions has to be separately simulated for quantitative determination of K_I in the absence of readily available analytical solutions [52, 53]. If there are cracks greater than the notch height present at one or both edges, the clamped beam might fail prematurely [53].

One might question why not use of single cantilever beams instead of clamped beam geometry to measure the stresses and quantify the resulting K_I. Single cantilever beams are inherently unstable loading systems subjected to maximum shear during the initial loading and the extent of shear to tensile mode drops off at later time steps. Catastrophic failure of the beams occurs, preventing the extraction of information post-mortem. The point of contact does not remain constant since the position is displaced due to the large displacement of the beam in the vertical direction. Hence, the span length becomes variable. Load offsets cannot be quantified easily with single cantilever beams too. However, there have been attempts to use notched cantilever beams to study fracture in multilayered nanocomposite beams [46, 57, 58].

There are challenges in performing fracture toughness values to determine the critical stress intensity factor, K_{IC} or J-integral. This is especially so for clamped beam bending geometry where there is a lack of analytical formulation for $K_{I.}$ Since K_I does not follow a monotonically linear relationship with increasing h/t ratio, FE modeling is performed to calculate K_I [34, 52, 53]. Standard geometries used to determine fracture toughness such as single edge notch bend (SENB) [44] cannot be scaled down easily due to limitations in specimen size and instrumentation. It is complicated to obtain a free-standing thin film specimen mounted on roller supports to replicate SENB test. That is why non-standard geometries such as single cantilever and clamped beam geometries are used.

Further, the fatigue starter cracks or notches of the microbeams are made with FIB [39]. Since the fracture toughness is sensitive to the material composition, there is a concern of the material around the crack tip to be significantly altered by ion implantation. As the dimensions of the beam specimen shrink into the micro- and nanoscale, the crack tip plastic zone becomes a more significant fraction of the specimen dimension, while also interacting with the free surface and indentation affected zone.

The changing nature of plasticity with decreasing size in nanoscale metal–metal multilayered composites implies that the yield strength is now size dependent and the fundamental fracture mechanism differs from that of the bulk. The interfaces in a multilayered system can influence the crack growth significantly [59]. Toughness measurements across multilayered interfaces are defined energetically and account for the changes in the stress intensity factor experienced by the crack tip when approaching a harder or softer layer ahead of it. Crack growth increases when the crack approaches the interface with a hard/soft transition and vice versa. This structure also brings sharp gradients in residual stresses along with the elasto-plastic property mismatch across the layers [59]. Thus, the complexity of this structure makes it challenging to calculate the fracture toughness without any form of simulation.

3.2.2 *Micron-Scale Test Structure Fabrication and Sample Preparations*

The focused ion beam (FIB) technique was used to fabricate the micron-scale pillar and beam structures in the subsequent chapters. The focused ion beam, also known as

FIB, involves the use of a finely focused beam of ions inside an SEM to mill or sputter off materials or deposit materials in a particular location for specific applications. This technique is used extensively in materials science [60–63], biology [64, 65] and the semiconductor industry [66, 67]. A liquid metal ion source, which can be gallium, helium or xenon is placed in contact with a tungsten needle, and the heated source wets the tungsten and flows to the tip of the needle. A Taylor cone is formed with the tip radius of ~2 nm. The electric field around this small tip causes ionization and field emission of the metal ions. Based on the beam current setting, the FIB beam can be used for imaging and site-specific sputtering or milling. In recent years, the FIB has been used as a machining tool to modify materials at the micro- and nanoscale [47, 54, 61, 68].

Microbeam sample fabrication for in-situ microbending experiments

It was challenging to figure out a suitable technique to make the micron beam structures of the multilayered nanocomposites without the substrate. Researchers tend to peel away the film from the substrate after deposition to obtain just the multilayered film [69, 70]. Since the multilayered nanocomposite samples were not made in-house and were limited in quantity, it was perilous to use this peeling method to get the film. This method was prone to create film samples of random sizes which were hard to control and keep for multiple uses. Further, this method was not compatible with Cu/Nb ARB (Accumulated Rolling Bonding) bulk multilayered nanocomposites too. Even though it was expensive, the FIB milling method was utilized as it saved fabrication time and provided more control over the beam and notch dimensions.

Figure 3.20 shows the schematic steps taken to make a beam from the nanoscale metal–metal multilayered composite film on top of a substrate. Special care is taken to minimize the defect introduction from the FIB process. The first step is to mill a trench in the silicon substrate from the side using 30 kV with a higher current for

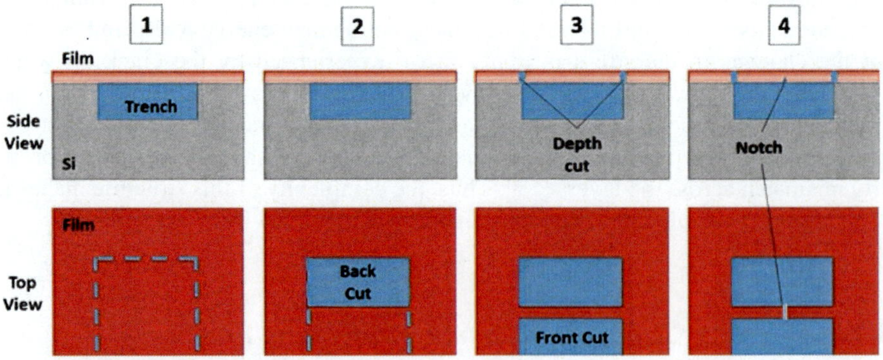

Fig. 3.20 Four-step process to make the clamped beam geometry from the edge of the sample for the case of nanoscale metal–metal multilayered composite film on top of a substrate—1. Milling of the trench occurs below the multilayered film, 2. The back cut for the removal of the film for the beam, 3. Front cut to place the beam away from the edge of the sample and 4. Notch creation across the center of the beam (With courtesy of Ref. [33] from author's copyright)

coarse cuts to create the clamped beam geometry. For the PVD samples, the silicon substrate is first milled from the edge to free the film from the substrate. For the Cu/Nb ARB samples, the top surface is polished to remove the inherent microcracks in the samples. FIB milling is done along the rolling direction of the Cu/Nb ARB sample. The length of the trench will follow that of the desired beam length, the depth of ~30 μm. After many tries on the desired width of our trench, the 100 μm width of the trench from the desired film thickness or thickness of the bulk Cu/Nb ARB multilayered nanocomposites is adopted for milling. This size is chosen based on the behavior of the microbeams especially the Cu/Nb ARB beams which plastically deform to a large extent when being bent. Overnight FIB milling is done to create these trenches.

The next step is to cut out the back trench for the width of the rectangular beam. These trenches will be around ~10 μm wide. The final beam will be a few microns away from the edge of the sample to create the final 5 μm width beam. This step prevents accidental breakage of the beam during the handling of the sample. Finer milling and polishing are performed to the final desired dimensions of the beam using a lower current. The center notch is milled at 10 pA current as one cut across the width of the beam. The notch height is about 1/3 of the thickness of beam and notch width ranges between 100 and 200 nm. This process results in the back notch being smaller in width and height than the front notch. The dimensions of the final microbeams are measured using the JEOL field emission scanning electron microscope (FESEM) (JSM-7600F). The beam dimensions are chosen based on satisfying both plane strain and small-scale yielding criteria [52, 54]. Figure 3.21 shows Cu/Nb ARB beams fabricated successfully by the above technique. This fabrication technique satisfies the requirements for the micron-size clamped beams for testing for many multilayered materials/samples discussed in this book.

Micropillar sample fabrication for in-situ microcompression experiments with the MechanoElectrical contact resistance (MECR)

Uchic et al. [60, 71] developed a simple test methodology to understand the length scale effects in metals in the submillimeter to nanometer size regime with any bulk inorganic material. It was to explore the evolution of geometrically necessary dislocations (GNDs) and statistically stored dislocations (SSDs) in a confined volume that can be easily evaluated and quantified. The pillars were fabricated using a focused ion beam (FIB) milling technique to machine cylindrical compression samples into the surface of a bulk crystal. Thereby, leaving the samples attached to the substrate at one end to eliminate the need for micromanipulation. As a result, the substrate acts as the lower compression platen during the test.

Figure 3.22 shows an example of a Cu/Nb PVD micropillar made by FIB. A high current of 3000 pA is used to mill out a ~30 μm crater in the material leaving a 1–5 μm diameter "island" in the center. This technique is known as annular milling and is used in the majority of microcompression studies [60]. The maximum height of the pillar is limited by the thickness of the film for PVD films on the substrate. An easy way was to mill till the Gallium ions encountered the substrate where the milling rate would change substantially. The pillar diameter is then adjusted depending on

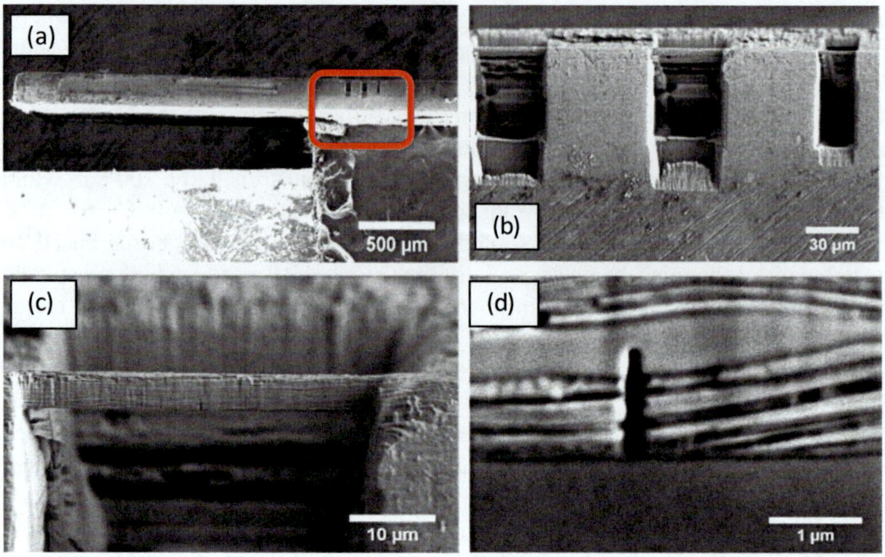

Fig. 3.21 Cu/Nb ARB beam made by FIB beam milling—**a** beams are made along the edge of the bulk sample circled with a red box. **b** 3 milled beams of various sizes in μm ($l \times b \times t$) (40 × 5 × 2), (40 × 5 × 4) and (20 × 5 × 2), **c** Close up of one of the beams to check for defects and measurements (40 × 5 × 2) and **d** the sharp notch from this beam (With courtesy of Ref. [33] from author's copyright)

the height of the pillar being milled, maintaining a ~3:1 height to diameter ratio. This ratio is chosen to prevent buckling of the pillar during compression [72]. If the ARB bulk multilayered nanocomposite sample is used, the FIB milling is more controlled to a specified height based on the desired diameter of the pillar.

The pillar is further refined with finer current in pA range and various tilt angles until a ~3:1 aspect ratio upright pillar with a flat top is created. A FIB milled location marker or platinum deposited point can be placed near the pillars to identify the location of the pillars in the bulk sample. A platinum trace can be created to enable the mechanoelectrical contact resistance (MECR) measurement between a conductive film and non-conductive substrate as shown in Fig. 3.22b. More details of this MECR technique will be discussed in the following Sect. 3.2.3.

Experimental or Instrumental Constraints

FIB is inherently a destructive technique to the specimen. When the high-energy ions bombard the sample, they will sputter atoms of the sample from the surface. The bombarding atoms (e.g. gallium) will also be implanted into a few nanometers from the surface, making the surface amorphous. For our experiments, care was taken to minimize the exposure of the milling as much as possible with a staggered milling process to make the structures and a final low voltage and current slow milling (or "polishing") in the end. However, just like in any destructive process, the remnants from the milling process have to be deposited elsewhere. We planned the milling

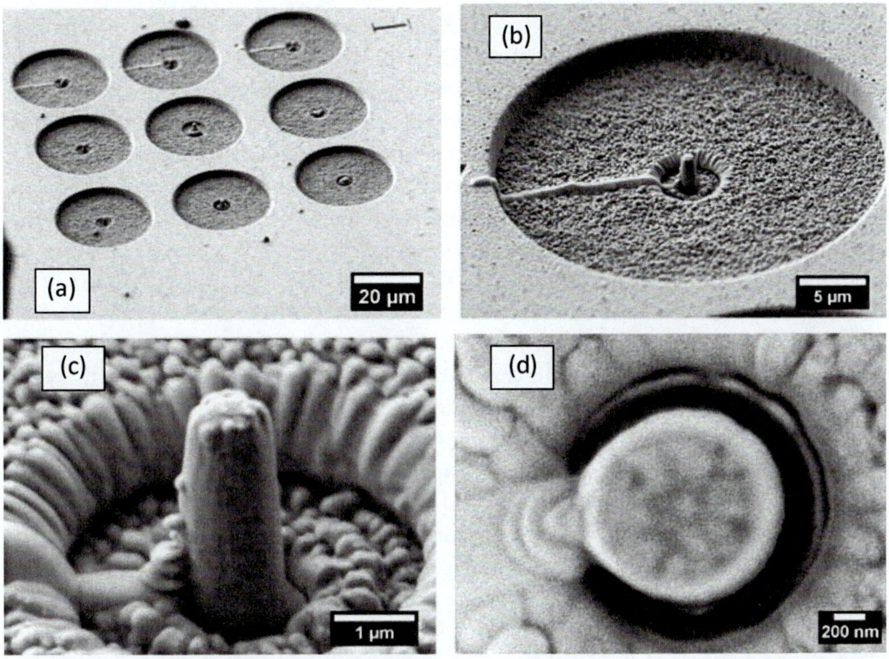

Fig. 3.22 **a** Cu/Nb PVD pillars made by FIB in a 3 × 3 grid. "I" marker to help in labeling the pillars used. 3 of the pillars have platinum lines connected between the film and the pillars, this is to enable electrical contact resistance (ECR) measurement. **b** Close up on one of the pillars in the middle of a 30 μm crater. **c** Pillar height is measured with a 50° tilt of the SEM stage and **d** diameter measurement above the pillar. Both measurements are needed for the pillar compression analysis (With courtesy of Ref. [33] from author's copyright)

steps such that the areas of interest, (where we want to understand the effect of the interface on their mechanical properties), do not get deposited with the remnants. These critical areas include the front and back sides of the beam where we view the multilayered interfaces and crack propagation from the notch across the layers and the surface area around the column of the micropillars. Any presence of an amorphous surface due to the deposited milled material on these critical areas affects the results and analysis. This was observed in our initial clamped beam bending experiments where we noticed a difference in some aspects of the mechanical behavior due to the presence of the amorphous back layer in some of the ARB multilayered beams [32].

To minimize the Gallium ion exposure on the pre-defined notch at the center of the microbeam, a very low10 pA current is used to mill as one cut across the width of the beam. However, this step results in the back notch being smaller in width and height than the front notch. If there is significant beam drift during the FIB milling, an uneven, wide notch formation is formed instead of the expected sharp, narrow notch for that particular beam.

Further, the micropillars prepared by the annular milling process have a few disadvantages. They have a noticeable vertical taper, which becomes more severe as the pillar diameter is reduced, resulting in undesirable non-uniform strain distribution within the pillar [57]. Other issues of the FIB milling technique include the rounding at the edges, different milling rates for heterogeneous materials, a non-uniform foot geometry complicating the control and assessment of the final dimensions [72]. Thus, caution needs to be taken during the FIB milling of these test structures.

There has been some literature on the effect of gallium (Ga+) induced damage on the FIB milled pillars and microbeams and their plastic deformation mechanisms. The relative depth of ion penetration and damage in materials has been observed through experimentation and simulation [73, 74]. For example, the penetration depth of 30 kV Ga+ ions in Cu is calculated to be ~24 nm [73]. Xiao et al. [75] recently investigated the effect of gallium as well as xenon ions on single crystalline and ultrafine-grained aluminum cantilevers and pillars. The minimal effect on the yield strength was observed for single crystalline aluminum samples, but gallium deposition along the grain boundaries of ultrafine-grained aluminum was observed and had a significant impact on the mechanical properties of the samples.

However, it is difficult to conclude that the effects observed in our experiments can be influenced by the existence of the surface damage layer. Burek and Greer [76] did a comparison study of gold and copper nanopillars made by FIB as well as electron beam lithography followed by electroplating and observed similar mechanical behavior from both techniques. Despite these concerns, FIB is being used extensively as the choice method in the fabrication of micron test beams and pillars for various small-scale mechanical characterization tests [39, 46, 77]. We assume that there is no significant skin effect due to the FIB milling process for all the micron test structures created for the in-situ clamped beam bending and micron pillar compression experiments.

3.2.3 Small-Scale Nanomechanical Characterization Techniques

Studies have shown that the mechanical properties of a thin film of micron thickness deviate from those of the bulk as the material dimensions are reduced to the micron or nanometer scales, and the explanations for the observed behavior are well-documented [78, 79]. This thin film can be considered as a basic fundamental block of the nanoscale metal–metal multilayered composite. The flow strength of single crystalline metallic films on Si substrates increases with decreasing film thickness regardless of grain size [80, 81]. The plastic deformation of these films is characterized by very high rates of strain hardening, caused by the storage of misfit dislocations near the film/substrate interface [82]. With more nanolayers in crystalline multilayered nanocomposites, the dislocation interactions at the metal–metal interfaces influence their mechanical properties as mentioned in Chap. 1.

Consequently, the conventional methods for evaluating the strengths of materials in the bulk form cannot be applied such that new experimentation techniques are required to measure the mechanical properties of thin films. The invention of microcompression testing by Uchic et al. [71] using focused ion beam (FIB) milling to structure micron-sized pillars of different dimensions in Ni single crystals which were subsequently deformed by using a nanoindenter, equipped with a flat punch, boosted the era of experimental micro-/nanomechanical testing. Initially focused on understanding size effects of single-crystal plasticity such as shear bands [83–86], dislocation starvation [87, 88], dislocation accumulation at the interfaces [89, 90] and crack blunting [42, 91], the field has evolved into studying the interaction between plastic deformation and microstructure components. This section covers the small-scale nanomechanical techniques of nanoindentation and MechanoElectrical contact resistance (MECR) that are used to understand the mechanics of the multilayered nanocomposites in the following chapters of this book.

MechanoElectrical contact resistance (MECR)

Almost all kinds of materials such as piezoresistive materials [92, 93], metals [94, 95] and even metallic multilayers [96, 97] experience a change of resistance or resistivity when strain is applied to the material while in tension or compression. The MechanoElectrical contact resistance (MECR) technique involves the combination of passing a current through the indenter probe to measure the resistance of the sample followed by pillar compression as shown in Fig. 3.23. This technique provides a way to detect electrical properties while performing mechanical experiments at the micro and nanoscale.

The first part of the technique involves Electrical Contact Resistance (ECR) which has commonly been used to study stress-induced phase transformations [98, 99], oxide fracture [100], MEMS contact switch resistance fatigue [101], nanoscale breakdown voltage [102] and piezoelectric material response [103, 104]. The Hysitron PI-85 SEM picoindenter by Bruker Inc., USA is a fully quantitative nanomechanical test instrument used in an SEM, with a sub-nanometer displacement noise floor and a submicro Newton force noise floor, with an additional capability of the nanoscale Electrical Contact Resistance (nanoECR) signal access module as shown in Fig. 3.24.

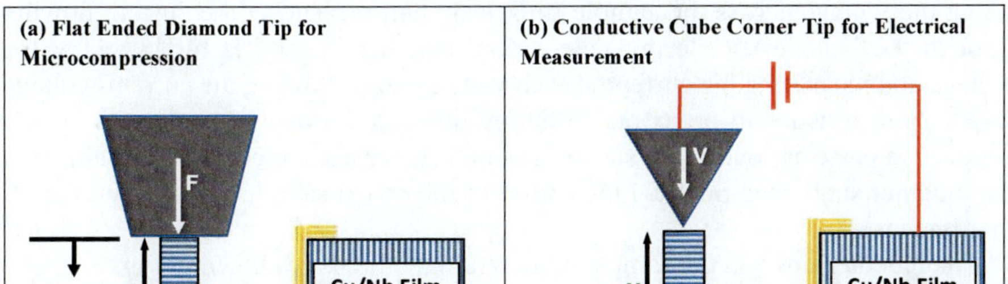

Fig. 3.23 Side view representation of **a** ex-situ uniaxial compression with the flat tip and **b** current measurement technique of the Cu/Nb micropillar before and after each compression (With courtesy of Ref. [33] from author's copyright)

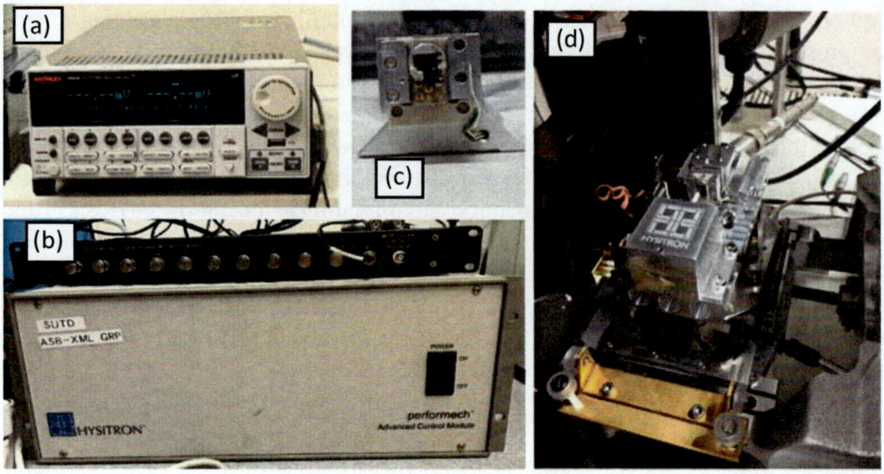

Fig. 3.24 Hysitron PI-85 picoindenter setup that consists of the main parts: **a** the Keithley 2602B source meter, **b** the nanoECR signal access module (black box) and the performance control module (white box), **c** the sample stage to enable electrical connection from sample to stage by silver paste and **d** the PI-85 picoindenter placed inside the SEM chamber before experimentation (With courtesy of Ref. [33] from author's copyright)

The nanoECR transducer provides an electrical path from the tip to signal acquisition and measurement hardware, which allows measurement of electrical contact conditions. Conductive Boron-doped diamond probes with a nominal resistivity of 3.3 Ω-cm and non-ohmic response or ceramic probes with a resistivity of 5 μΩ-cm and ohmic response are used for standard mechanical and electrical testing. A bias voltage or current is applied to the sample through a biasing stage which drives the electric flow through the tip using the Keithley 2402B voltage/current source meter (Cleveland, OH), which is remotely computer controlled through an easy-to-use flexible software interface.

There are two ways to take electrical measurements with this system. One is by fixing the voltage across the sample-tip system and measuring the current flowing through the indenter tip during indentation. Another method is by sweeping the voltage and measuring the current at each voltage step, allowing the current–voltage (I–V) characteristics to be extracted at any point in the load/unload cycle. Silver conductive paste or conductive substrates/contacts are used to connect the sample to the indenter stage electrically. Indentation of the actual sample is done away from the contacts.

The indenter tips and the sample preparation required are shown in Fig. 3.25. An example of ECR nanoindentation is a cube corner Pulsar-C conductive tip indented on the surface of the glassy carbon standard sample. A load control test is performed up to the maximum of 2000 μN and held at max load for 20 s as shown in Fig. 3.24a. A voltage sweep is performed from −0.1 to 0.1 V during this hold period and the corresponding current is measured during the whole sweep. A linear ohmic range is

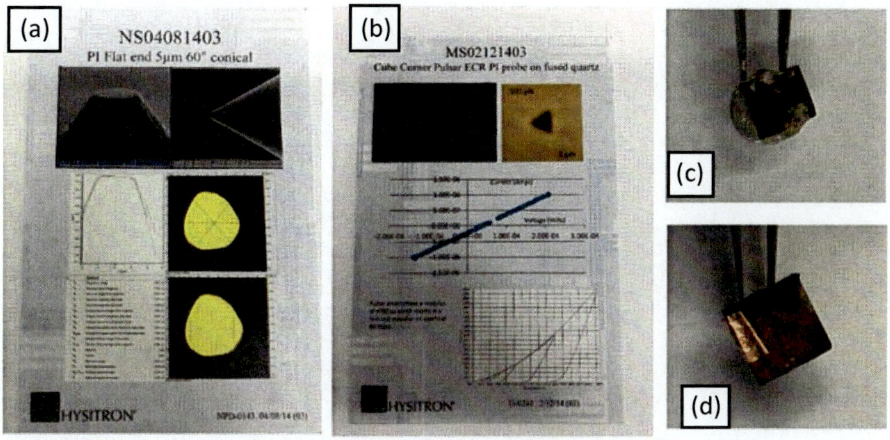

Fig. 3.25 Indenter tips used for the MechanoElectrical contact resistance (MECR) technique: **a** 5 μm conical flat diamond tip for pillar compression, **b** cube corner Pulsar-C tip for ECR measurement. Sample preparation for ECR measurement—**c** silver paste connections when adhering and close to Cu/Nb ARB bulk sample ends and **d** conductive copper tape as a connection with Cu/Nb PVD film sample (With courtesy of Ref. [33] from author's copyright)

observed as expected. The slope of the I−V curve is taken to measure the resistance of the whole setup from the tip to the sample using Ohm's law, $R = {}^{V}/_{I}$. For this particular test, the resulting resistance is 5 kΩ as shown in Fig. 3.24b.

A similar approach is used to measure the ECR of nanopillars. However, the resistivity or conductance of the material in the nanopillar is more useful for the strain sensing application. The following relationship of $R = \rho \frac{l}{A}$ is used to calculate the resistivity. The values of the pillar dimensions depend on the state of compression when the ECR measurement is made. The amount of current generated for the same amount of voltage applied and is sensitive to the load applied to the pillar or sample, with the higher amount of load with higher current. This is possible as the ECR tip is ohmic and the other resistances are much too small in comparison to the measurements made. This is an issue with the possibility of compressing the pillar too fast till the pillar is flattened. Thus, the load range for ECR measurement is between 100 and 300 μN (Fig. 3.26).

Once the initial electrical measurement of the interested multilayered nanocomposite film or pillar is completed, it is followed by the use of a nanoindenter with a flat-ended diamond tip, to perform the uniaxial compression tests externally or in-situ. For the experiments in Chap. 6, the ex-situ method is adopted. The pillar of interest is located using the scanning electron microscope, and the tip is moved to directly above the specimen. Thermal drift stabilization is then followed by compressing the pillar at a constant nominal strain rate. Load–displacement data is collected through the Hysitron Triboscan software. The resulting load–displacement curve can be easily converted to a stress–strain curve by the constant volume, homogeneous deformation assumption model [105–107].

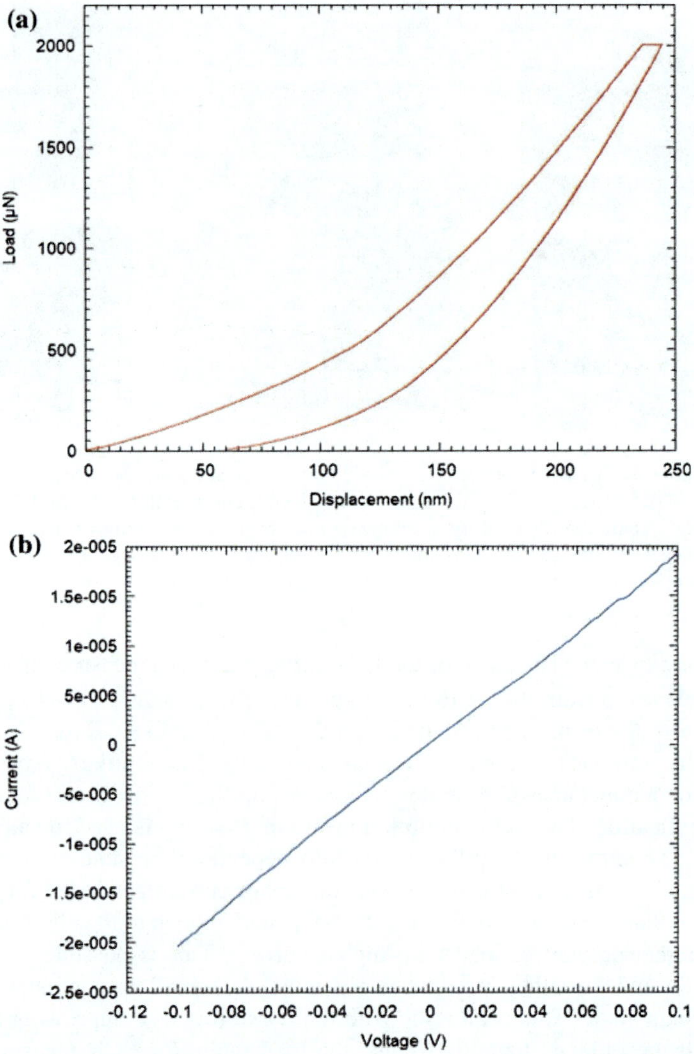

Fig. 3.26 An example of the electrical contact resistance (ECR) measurement: **a** Load–Displacement plot for a load-controlled test on a glassy carbon standard sample. **b** Current measurement during the voltage sweep from −0.1 to 0.1 V while the conductive indenter tip is in contact at the maximum load for 20 s (With courtesy of Ref. [33] from author's copyright)

A cylindrical pillar of height L_0 and cross-sectional area A_0 is assumed for stress–strain measurement. Here, the $E_{\text{multilayer}}$ refers to the modulus of the multilayered nanocomposite as determined by their volume fractions of metallic layer *M1* and *M2* (for example, Cu is *M1* and Nb is *M2*) called V_{M1} and V_{M2}. The $\nu_{\text{multilayer}}$ is the multilayered nanocomposite's Poisson's ratio which is also determined by their volume fractions.

$$E_{\text{multilayer}} = V_{M1}E_{M1} + V_{M2}E_{M2} \tag{3.13}$$

$$\nu_{\text{multilayer}} = V_{M1}\upsilon_{M1} + V_{M2}\upsilon_{M2} \tag{3.14}$$

The analysis begins with the subtraction of the Sneddon compliance, k_{Sned} [108] that arises due to the bottom cross-sectional area, A_{bottom} of the micropillar punching into the underlying material. The bottom cross-sectional area could include the influence of the taper angle of the pillar. This Sneddon displacement, u_{Sneddon} is then subtracted from the total displacement, u_{total} to calculate the displacement that corresponds to the pure deformation of the nanopillar only. Note that the total thickness of the multilayered nanocomposite is larger than the height of the pillar. This is why the multilayer elastic properties are being used for the Sneddon substrate properties.

$$u_{\text{Sneddon}} = \frac{P}{k_{\text{Sned}}} = \frac{P}{\frac{2E_{\text{multilayer}}}{1-\upsilon^2_{\text{multilayer}}} \times \sqrt{\frac{A_{\text{bottom}}}{\pi}}} \tag{3.15}$$

$$u_{\text{pillar}} = u_{\text{total}} - u_{\text{sneddon}} \tag{3.16}$$

The corrected u_{pillar} is then used to evaluate the stress versus strain behavior. Assuming that the volume of the pillar is conserved and that the pillar is deforming uniformly, then

$$A_0L_0 = A_pL_p \tag{3.17}$$

where A_0 and A_p are the initial top and the instantaneous top cross-sectional areas during deformation, and L_0, L_p are the initial and the plastically deformed lengths of the pillar. The L_p can be determined by the remainder of the initial length, L_0 of the pillar and the plastic displacement ($u_{\text{plastic}} = u_{\text{pillar}} - u_{\text{elastic}}$).

$$u_{\text{elastic}} = \frac{1}{E}\frac{P}{A_p}L_p = \frac{1}{E}\frac{P}{A_0L_0}L_p^2 \tag{3.18}$$

$$L_p = L_0 - u_{\text{plastic}} = L_0 - \left(u_{\text{pillar}} - u_{\text{elastic}}\right) \tag{3.19}$$

$$A_p = \frac{A_0L_0}{L_p} = \frac{A_0L_0}{L_0 - \left(u_{\text{pillar}} - u_{\text{elastic}}\right)} = \frac{A_0L_0}{L_0 - \left(u_{\text{pillar}} - \frac{1}{E}\frac{P}{A_0L_0}L_p^2\right)} \tag{3.20}$$

$$L_p = \frac{EA_0L_0}{2P}\left[1 - \sqrt{1 - 4\left(\frac{P}{EA_0L_0}\right)\left(L_0 - u_{\text{pillar}}\right)}\right] \tag{3.21}$$

$$\text{Calculate } A_p = \frac{A_0L_0}{L_p} \tag{3.22}$$

$$\text{True Stress} = \sigma_{\text{true}} = \frac{P}{A_p} \tag{3.23}$$

$$\text{True Strain} = \varepsilon_{\text{true}} = \varepsilon_{el} + \varepsilon_p = \frac{P}{EA_p} + \ln\left(\frac{L_0}{L_p}\right) \tag{3.24}$$

After the completion of the micropillar compression experiments, the shape of the stress–strain curve determines the elastic–plastic phenomenon in the tested metal as well as the mechanical properties, such as the elastic modulus, and yield strength of the metal at that scale. Further, easy inspection of slip or shear bands in specific orientations on the compressed pillars in correlation with the strain bursts in the stress–strain curve can help to explain the different size effects in metals [105, 106, 109].

There are some important points/pointers to take note during the MECR experiments: First, the series resistances for the whole path for the ECR part: the tip/sample contact, inherent sample behavior (i.e. dislocation behavior, phase transformation), and the resistance of the rest of the sample including the small tip/sample contact area. The interfaces between these components add either ohmic or rectifying contact resistances. Resistive heating during the experiment is negligible. Testing of the tip/sample contact resistance is essential to ensure the current/voltage values, and the resulting resistance measurements are mainly due to the sample behavior.

Second, the sample state in contact with the conductive tip influences the ECR measurements. The ECR measurement value decreases significantly if there is a native oxide layer on top of the multilayered nanocomposite film. Upon loading, the tip is not in contact with the multilayered nanocomposite film, and the current has to flow through this thin native oxide as well as over electrical barriers between the multilayered nanocomposite film, the oxide and the indenter tip. Either chemical etching to strip the oxide off or polishing of the top surface using FIB milling is performed before the ECR measurement to remove the native oxide layer. Another method is using a sharp tip like Berkovich or Cube corner to measure the resistance at the top surface of the pillar.

The MECR experiment is time consuming, as it takes 2–3 days to get five readings on one pillar sample. Furthermore, the experiment is also challenging as the SEM chamber needs to be vented, then pumped down for a few times every time the indenter tip is changed. Efforts were made to perform an in-situ MECR technique in-house using a flat punch conductive tip to perform both the ECR and compression at the same time. However, this resulted in the challenges of selecting the right conductive indenter tip to measure the ECR of the micropillars before and after compression. The current measurement was very low when the ECR flat tip was used while in compression, despite all the precautions to make a good conduction path for the sample. Oxidation on the top of the micropillars was significant enough to influence the failure of the ECR measurements in subsequent experiments. However, with more time and effort, this in-situ ECR technique may be improved further.

The last part of this technique is the micropillar compression, which is a well-established technique. However, care needs to be taken to take the pillar dimensions before and after compression as this will affect the resulting stress–strain curves.

3.2.4 Summary: In-Situ Micromechanical and Fracture Testing

The in-situ microfracture technique is described in detail in the preceding sections touching upon the rationale of this technique to see interface interactions with crack during fracture, the choice of clamped beam geometry and how to perform experiments with this technique. The micron-scale structure fabrication technique of focused ion beam (FIB) to make the microbeams and micropillars is subsequently described in detail, including some of the experimental and instrumental constraints. Concepts required to use and evaluate results from small-scale nanomechanical technique of MechanoElectrical contact resistance (MECR) are explained too. With these small-scale materials and mechanical testing methodologies, the study of the effect of interface on the in-situ fracture of multilayered nanocomposites was enabled.

References

1. Cullity, B.D.: Elements of X-ray Diffraction, 2nd edn. Addison-Wesley, New York (1978)
2. Fultz, B., Howe, J.M.: Inelastic electron scattering and spectroscopy. In: Transmission Electron Microscopy and Diffractometry of Materials, 3rd edn. Heidelberg, Berlin, Germany (2008)
3. List of Synchrotron Sources, Wikipedia: https://en.wikipedia.org/wiki/List_of_synchrotron_radiation_facilities. Viewed 30 Mar 2019
4. Budiman, A.S.: Probing crystal plasticity at the nanoscales: synchrotron X-ray microdiffraction. Ph.D. dissertation, Stanford University (2008)
5. Tippabhotla, S.K.: Ph.D. dissertation, Singapore University of Technology & Design (SUTD) (2019)
6. Tamura, N., Kunz, M., Chen, K., Celestre, R.S., MacDowell, A.A., Warwick, T.: A superbend X-ray microdiffraction beamline at the advanced light source. Mater. Sci. Eng.: A **524**(1–2), 28–32 (2009)
7. Tamura, N., Padmore, H.A., Patel, J.R.: High spatial resolution stress measurements using synchrotron based scanning X-ray microdiffraction with white or monochromatic beam. Mater. Sci. Eng.: A **399**(1–2), 92–98 (2005)
8. Tamura, N.: XMAS: a versatile tool for analyzing synchrotron X-ray microdiffraction data. In: Strain and Dislocation Gradients from Diffraction: Spatially-Resolved Local Structure and Defects, pp. 125–155 (2014)
9. Khoo, C.Y., Liu, H., Sasangka, W.A., Made, R.I., Tamura, N., Kunz, M., Budiman, A.S., Gan, C.L., Thompson, C.V.: Impact of deposition conditions on the crystallization kinetics of amorphous GeTe films. J. Mater. Sci. **51**(4), 1864–1872 (2016)
10. Budiman, A.S., Narayanan, K.R., Li, N., Wang, J., Tamura, N., Kunz, M., Misra, A.: Plasticity evolution in nanoscale Cu/Nb single-crystal multilayers as revealed by synchrotron X-ray microdiffraction. Mater. Sci. Eng.: A **635**, 6–12 (2015)

11. Budiman, A.S., Lee, G., Burek, M.J., Jang, D., Han, S.M.J., Tamura, N., Kunz, M., Greer, J.R., Tsui, T.Y.: Plasticity of indium nanostructures as revealed by synchrotron X-ray microdiffraction. Mater. Sci. Eng.: A **538**, 89–97 (2012)
12. Budiman, A.S., Han, S.M., Li, N., Wei, Q.M., Dickerson, P., Tamura, N., Kunz, M., Misra, A.: Plasticity in the nanoscale Cu/Nb single-crystal multilayers as revealed by synchrotron Laue x-ray microdiffraction. J. Mater. Res. **27**(3), 599–611 (2012)
13. Budiman, A.S., Shin, H.A.S., Kim, B.J., Hwang, S.H., Son, H.Y., Suh, M.S., Chung, Q.H., Byun, K.Y., Tamura, N., Kunz, M., Joo, Y.C.: Measurement of stresses in Cu and Si around through-silicon via by synchrotron X-ray microdiffraction for 3-dimensional integrated circuits. Microelectron. Reliab. **52**(3), 530–533 (2012)
14. Valek, B.C., Bravman, J.C., Tamura, N., MacDowell, A.A., Celestre, R.S., Padmore, H.A., Spolenak, R., Brown, W.L., Batterman, B.W., Patel, J.R.: Electromigration-induced plastic deformation in passivated metal lines. Appl. Phys. Lett. **81**(22), 4168–4170 (2002)
15. Tamura, N., Celestre, R.S., MacDowell, A.A., Padmore, H.A., Spolenak, R., Valek, B.C., Meier Chang, N., Manceau, A., Patel, J.R.: Submicron x-ray diffraction and its applications to problems in materials and environmental science. Rev. Sci. Instrum. **73**(3), 1369–1372 (2002)
16. Gullikson, E.: Center for X-ray Optics, X-Ray Interactions With Matter. http://henke.lbl.gov/optical_constants/atten2.html. Viewed 31 Mar 2019
17. Chung, J.S., Ice, G.E.: Automated indexing for texture and strain measurement with broad-bandpass x-ray microbeams. J. Appl. Phys. **86**(9), 5249–5255 (1999)
18. Valek, B.C., Tamura, N., Spolenak, R., Caldwell, W.A., MacDowell, A.A., Celestre, R.S., Padmore, H.A., Bravman, J.C., Batterman, B.W., Nix, W.D., Patel, J.R.: Early stage of plastic deformation in thin films undergoing electromigration. J. Appl. Phys. **94**(6), 3757–3761 (2003)
19. Spolenak, R., Brown, W.L., Tamura, N., MacDowell, A.A., Celestre, R.S., Padmore, H.A., Valek, B., Bravman, J.C., Marieb, T., Fujimoto, H., Batterman, B.W.: Local plasticity of Al thin films as revealed by X-ray microdiffraction. Phys. Rev. Lett. **90**(9), 096102 (2003)
20. Budiman, A.S., Tamura, N., Valek, B.C.: Materials, technology and reliability for advanced interconnects and low-k dielectrics. Mat. Res. Soc. Proc. **812**, 345–350 (2004)
21. Wu, A.T., Tu, K.N., Lloyd, J.R., Tamura, N., Valek, B.C., Kao, C.R.: Electromigration-induced microstructure evolution in tin studied by synchrotron x-ray microdiffraction. Appl. Phys. Lett. **85**(13), 2490–2492 (2004)
22. Wu, A.T., Tamura, N., Lloyd, J.R., Kao, C.R., Tu, K.N.: Synchrotron X-ray micro-diffraction analysis on microstructure evolution in Sn under electromigration. MRS Online Proc. Libr. (OPL) **863** (2005)
23. Budiman, A.S., Nix, W.D., Tamura, N., Valek, B.C., Gadre, K., Maiz, J., Spolenak, R., Patel, J.R.: Crystal plasticity in Cu damascene interconnect lines undergoing electromigration as revealed by synchrotron x-ray microdiffraction. Appl. Phys. Lett. **88**(23), 233515 (2006)
24. Budiman, A., Tamura, N., Valek, B.C., Gadre, K., Maiz, J., Spolenak, R., Patel, J.R., Nix, W.D.: Electromigration-induced plastic deformation in Cu damascene interconnect lines as revealed by synchrotron X-ray microdiffraction. MRS Online Proc. Libr. Arch. **914** (2006)
25. Budiman, A.S., Hau-Riege, C.S., Besser, P.R., Marathe, A., Joo, Y.C., Tamura, N., Patel, J.R., Nix, W.D.: Plasticity-amplified diffusivity: dislocation cores as fast diffusion paths in CU interconnects. In: 2007 IEEE International Reliability Physics Symposium Proceedings. 45th Annual, pp. 122–127. IEEE, Apr 2007
26. Cahn, R.W.: Recrystallization of single crystals after plastic bending. J. Inst. Met. **76**(2), 121–141 (1949)
27. Gilman, J.J.: Structure and polygonization of bent zinc monocrystals. Acta Metall. **3**(3), 277–288 (1955)
28. Hibbard, W.R., Jr., Dunn, C.G.: Creep and Recovery, p. 52. American Society for Metals, Park, Ohio (1956)
29. Patel, J.R.: Arrangements of dislocations in plastically bent silicon crystals. J. Appl. Phys. **29**(2), 170–176 (1958)
30. Valek, B.C.: Ph.D dissertation, Stanford University (2003)

31. Nye, J.F.: Some geometrical relations in dislocated crystals. Acta Metall. **1**(2), 153–162 (1953)
32. Radchenko, I., Anwarali, H.P., Tippabhotla, S.K., Budiman, A.S.: Effects of interface shear strength during failure of semicoherent metal–metal nanolaminates: an example of accumulative roll-bonded Cu/Nb. Acta Mater. **156**, 125–135 (2018)
33. Ali, H.P.A.: Ph.D. dissertation, Singapore University of Technology & Design (SUTD) (2019)
34. Jaya B, N., Jayaram, V., Biswas, S.K.: A new method for fracture toughness determination of graded (Pt, Ni) Al bond coats by microbeam bend tests. Philos. Mag. **92**(25–27), 3326–3345 (2012)
35. Mayer, C., Li, N., Mara, N., Chawla, N.: Micromechanical and in situ shear testing of Al–SiC nanolaminate composites in a transmission electron microscope (TEM). Mater. Sci. Eng.: A **621**, 229–235 (2015)
36. Kammers, A.D., Daly, S.: Digital image correlation under scanning electron microscopy: methodology and validation. Exp. Mech. **53**(9), 1743–1761 (2013)
37. Mara, N.A., Li, N., Misra, A., Wang, J.: Interface-driven plasticity in metal–ceramic nanolayered composites: direct validation of multiscale deformation modeling via in situ indentation in TEM. JOM **68**(1), 143–150 (2016)
38. Hay, J.: Introduction to instrumented indentation testing. Exp. Tech. **33**(6), 66–72 (2009)
39. Dehm, G., Jaya, B.N., Raghavan, R., Kirchlechner, C.: Overview on micro-and nanomechanical testing: new insights in interface plasticity and fracture at small length scales. Acta Mater. **142**, 248–282 (2018)
40. Aydıner, C.C., Brown, D.W., Mara, N.A., Almer, J., Misra, A.: In situ x-ray investigation of freestanding nanoscale Cu–Nb multilayers under tensile load. Appl. Phys. Lett. **94**(3), 031906 (2009)
41. Li, N., Wang, J., Huang, J.Y., Misra, A., Zhang, X.: In situ TEM observations of room temperature dislocation climb at interfaces in nanolayered Al/Nb composites. Scr. Mater. **63**(4), 363–366 (2010)
42. Hattar, K., Misra, A., Dosanjh, M.R.F., Dickerson, P., Robertson, I.M., Hoagland, R.G.: Direct observation of crack propagation in copper–niobium multilayers. J. Eng. Mater. Technol. **134**(2) (2012)
43. Li, N., Mara, N.A., Wang, J., Dickerson, P., Huang, J.Y., Misra, A.: Ex situ and in situ measurements of the shear strength of interfaces in metallic multilayers. Scr. Mater. **67**(5), 479–482 (2012)
44. A.S.T.M. Standard 1820.11e1: Standard Test Method for Measurement of Fracture Toughness. ASTM International, West Conshohocken, PA (2011).https://doi.org/10.1520/E1820-11E01
45. ASTM E 399-90: Standard test method for plane-strain fracture toughness of metallic materials. Annu. Book ASTM Stand **3**(01), 506–536 (1997)
46. Alfreider, M., Kozic, D., Kolednik, O., Kiener, D.: In-situ elastic-plastic fracture mechanics on the microscale by means of continuous dynamical testing. Mater. Des. **148**, 177–187 (2018)
47. Allison, P.G., Moser, R.D., Schirer, J.P., Martens, R.L., Jordon, J.B., Chandler, M.Q.: In-situ nanomechanical studies of deformation and damage mechanisms in nanocomposites monitored using scanning electron microscopy. Mater. Lett. **131**, 313–316 (2014)
48. Bohnert, C., Schmitt, N.J., Weygand, S.M., Kraft, O., Schwaiger, R.: Fracture toughness characterization of single-crystalline tungsten using notched micro-cantilever specimens. Int. J. Plast. **81**, 1–17 (2016)
49. Liu, S., Wheeler, J.M., Howie, P.R., Zeng, X.T., Michler, J., Clegg, W.J.: Measuring the fracture resistance of hard coatings. Appl. Phys. Lett. **102**(17), 171907 (2013)
50. Sebastiani, M., Johanns, K.E., Herbert, E.G., Carassiti, F., Pharr, G.M.: A novel pillar indentation splitting test for measuring fracture toughness of thin ceramic coatings. Philos. Mag. **95**(16–18), 1928–1944 (2015)
51. Sebastiani, M., Bemporad, E., Carassiti, F., Schwarzer, N.: Residual stress measurement at the micrometer scale: focused ion beam (FIB) milling and nanoindentation testing. Philos. Mag. **91**(7–9), 1121–1136 (2011)
52. Jaya, B.N., Bhowmick, S., Asif, S.S., Warren, O.L., Jayaram, V.: Optimization of clamped beam geometry for fracture toughness testing of micron-scale samples. Philos. Mag. **95**(16–18), 1945–1966 (2015)

53. Jaya, B.N., Jayaram, V.: Crack stability in edge-notched clamped beam specimens: modeling and experiments. Int. J. Fract. **188**(2), 213–228 (2014)
54. Jaya, B.N., Kirchlechner, C., Dehm, G.: Can microscale fracture tests provide reliable fracture toughness values? A case study in silicon. J. Mater. Res. **30**(5), 686–698 (2015)
55. Timoshenko, S.P., Goodier, J.N.: Theory of Elasticity. McGraW Hill Inc., New York (1934)
56. Iqbal, F., Ast, J., Göken, M., Durst, K.: In situ micro-cantilever tests to study fracture properties of NiAl single crystals. Acta Mater. **60**(3), 1193–1200 (2012)
57. Kozic, D., Gänser, H.P., Brunner, R., Kiener, D., Antretter, T., Kolednik, O.: Crack arrest in thin metallic film stacks due to material-and residual stress inhomogeneities. Thin Solid Films **668**, 14–22 (2018)
58. Treml, R., Kozic, D., Schöngrundner, R., Kolednik, O., Gänser, H.P., Brunner, R., Kiener, D.: Miniaturized fracture experiments to determine the toughness of individual films in a multilayer system. Extreme Mech. Lett. **8**, 235–244 (2016)
59. Simha, N.K., Fischer, F.D., Kolednik, O., Chen, C.R.: Inhomogeneity effects on the crack driving force in elastic and elastic–plastic materials. J. Mech. Phys. Solids **51**(1), 209–240 (2003)
60. Uchic, M.D., Shade, P.A., Dimiduk, D.M.: Plasticity of micrometer-scale single crystals in compression. Annu. Rev. Mater. Res. **39**, 361–386 (2009)
61. Greer, J.R., De Hosson, J.T.M.: Plasticity in small-sized metallic systems: intrinsic versus extrinsic size effect. Prog. Mater Sci. **56**(6), 654–724 (2011)
62. Kraft, O., Gruber, P.A., Mönig, R., Weygand, D.: Plasticity in confined dimensions. Annu. Rev. Mater. Res. **40**, 293–317 (2010)
63. Jaya, B.N., Jayaram, V.: Fracture testing at small-length scales: from plasticity in Si to brittleness in Pt. JOM **68**(1), 94–108 (2016)
64. Kizilyaprak, C., Daraspe, J., Humbel, B.M.: Focused ion beam scanning electron microscopy in biology. J. Microsc. **254**(3), 109–114 (2014)
65. Narayan, K., Danielson, C.M., Lagarec, K., Lowekamp, B.C., Coffman, P., Laquerre, A., Phaneuf, M.W., Hope, T.J., Subramaniam, S.: Multi-resolution correlative focused ion beam scanning electron microscopy: applications to cell biology. J. Struct. Biol. **185**(3), 278–284 (2014)
66. Zschech, E., Langer, E., Engelmann, H.J., Dittmar, K.: Physical failure analysis in semiconductor industry—challenges of the copper interconnect process. Mater. Sci. Semicond. Process. **5**(4–5), 457–464 (2002)
67. Giannuzzi, L.A., Drown, J.L., Brown, S.R., Irwin, R.B., Stevie, F.A.: Focused ion beam milling and micromanipulation lift-out for site specific cross-section TEM specimen preparation. MRS Online Proc. Libr. Arch. **480** (1997)
68. Di Maio, D., Roberts, S.G.: Measuring fracture toughness of coatings using focused-ion-beam-machined microbeams. J. Mater. Res. **20**(2), 299–302 (2005)
69. Mara, N.A., Bhattacharyya, D., Hoagland, R.G., Misra, A.: Tensile behavior of 40 nm Cu/Nb nanoscale multilayers. Scr. Mater. **58**(10), 874–877 (2008)
70. Huang, H., Spaepen, F.: Tensile testing of free-standing Cu, Ag and Al thin films and Ag/Cu multilayers. Acta Mater. **48**(12), 3261–3269 (2000)
71. Uchic, M.D., Dimiduk, D.M., Florando, J.N., Nix, W.D.: Sample dimensions influence strength and crystal plasticity. Science **305**(5686), 986–989 (2004)
72. Hütsch, J., Lilleodden, E.T.: The influence of focused-ion beam preparation technique on microcompression investigations: lathe vs. annular milling. Scr. Mater. **77**, 49–51 (2014)
73. Kiener, D., Motz, C., Rester, M., Jenko, M., Dehm, G.: FIB damage of Cu and possible consequences for miniaturized mechanical tests. Mater. Sci. Eng.: A **459**(1–2), 262–272 (2007)
74. Bei, H., Shim, S., Miller, M.K., Pharr, G.M., George, E.P.: Effects of focused ion beam milling on the nanomechanical behavior of a molybdenum-alloy single crystal. Appl. Phys. Lett. **91**(11), 111915 (2007)

75. Xiao, Y., Wehrs, J., Ma, H., Al-Samman, T., Korte-Kerzel, S., Göken, M., Michler, J., Spolenak, R., Wheeler, J.M.: Investigation of the deformation behavior of aluminum micropillars produced by focused ion beam machining using Ga and Xe ions. Scr. Mater. **127**, 191–194 (2017)
76. Burek, M.J., Greer, J.R.: Fabrication and microstructure control of nanoscale mechanical testing specimens via electron beam lithography and electroplating. Nano Lett. **10**(1), 69–76 (2010)
77. Lauener, C.M., Petho, L., Chen, M., Xiao, Y., Michler, J., Wheeler, J.M.: Fracture of silicon: influence of rate, positioning accuracy, FIB machining, and elevated temperatures on toughness measured by pillar indentation splitting. Mater. Des. **142**, 340–349 (2018)
78. Hirth, J.P.: The influence of grain boundaries on mechanical properties. Metall. Trans. **3**(12), 3047–3067 (1972)
79. Nix, W.D.: Mechanical properties of thin films. Metall. Trans. A **20**(11), 2217–2245 (1989)
80. Mitra, R., Chiou, A., Weertman, J.R.: In situ study of deformation mechanisms in sputtered free-standing nanocrystalline nickel films. J. Mater. Res. **19**(4), 1029–1037 (2004)
81. Venkatraman, R., Bravman, J.C.: Separation of film thickness and grain boundary strengthening effects in Al thin films on Si. J. Mater. Res. **7**(8), 2040–2048 (1992)
82. Nix, W.D.: Yielding and strain hardening of thin metal films on substrates. Scr. Mater. **39**(4–5), 545–554 (1998)
83. Paul, H.E.N.R.Y.K., Driver, J.H., Tarasek, A., Wajda, W., Miszczyk, M.M.: Mechanism of macroscopic shear band formation in plane strain compressed fine-grained aluminium. Mater. Sci. Eng.: A **642**, 167–180 (2015)
84. Bhattacharyya, D., Mara, N.A., Dickerson, P., Hoagland, R.G., Misra, A.: Transmission electron microscopy study of the deformation behavior of Cu/Nb and Cu/Ni nanoscale multilayers during nanoindentation. J. Mater. Res. **24**(3), 1291–1302 (2009)
85. Li, Y.P., Zhu, X.F., Zhang, G.P., Tan, J., Wang, W., Wu, B.: Investigation of deformation instability of Au/Cu multilayers by indentation. Philos. Mag. **90**(22), 3049–3067 (2010)
86. Xie, J.Y., Huang, P., Wang, F., Li, Y., Zhang, L.F., Xu, K.W.: Strain rate dependent shear banding deformation in Al/W nanomultilayers. Integr. Ferroelectr. **146**(1), 168–176 (2013)
87. Greer, J.R., Nix, W.D.: Nanoscale gold pillars strengthened through dislocation starvation. Phys. Rev. B **73**(24), 245410 (2006)
88. Nix, W.D., Greer, J.R., Feng, G., Lilleodden, E.T.: Deformation at the nanometer and micrometer length scales: effects of strain gradients and dislocation starvation. Thin Solid Films **515**(6), 3152–3157 (2007)
89. Beyerlein, I.J., Demkowicz, M.J., Misra, A., Uberuaga, B.P.: Defect-interface interactions. Prog. Mater. Sci. **74**, 125–210 (2015)
90. Ali, H.A., Kunz, M., Tamura, N., Budiman, A.S.: Probing plasticity mechanisms in low melting temperature metallic nanostructures using synchrotron X-ray microdiffraction. Procedia Eng. **215**, 246–262 (2017)
91. Li, Y., Zhou, Q., Zhang, S., Huang, P., Xu, K., Wang, F., Lu, T.: On the role of weak interface in crack blunting process in nanoscale layered composites. Appl. Surf. Sci. **433**, 957–962 (2018)
92. Haj-Ali, R., Zemer, H., El-Hajjar, R., Aboudi, J.: Piezoresistive fiber-reinforced composites: a coupled nonlinear micromechanical–microelectrical modeling approach. Int. J. Solids Struct. **51**(2), 491–503 (2014)
93. Meti, S., Balavald, K.B., Sheeparmatti, B.G.: MEMS piezoresistive pressure sensor: a survey. Int. J. Eng. Res. Appl. **6**(4 part 1), 23–31 (82) (2016)
94. Brown, R.A.: Electrical resistivity of dislocations in metals. J. Phys. F: Met. Phys. **7**(7), 1283 (1977)
95. Josell, D., Brongersma, S.H., Tőkei, Z.: Size-dependent resistivity in nanoscale interconnects. Annu. Rev. Mater. Res. **39**, 231–254 (2009)
96. Anwar Ali, H.P., Radchenko, I., Zhou, J., Qing, L., Budiman, A.: Designing novel multilayered nanocomposites for high-performance coating materials with online strain monitoring capability. Proc. Inst. Mech. Eng. Part L: J. Mater.: Des. Appl. 1–12 (2017)

97. Deng, L., Han, K., Hartwig, K.T., Siegrist, T.M., Dong, L., Sun, Z., Yang, X., Liu, Q.: Hardness, electrical resistivity, and modeling of in situ Cu–Nb microcomposites. J. Alloys Compd. **602**, 331–338 (2014)
98. Bradby, J.E., Williams, J.S., Swain, M.V.: In situ electrical characterization of phase transformations in Si during indentation. Phys. Rev. B **67**(8), 085205 (2003)
99. Mann, A.B., Heerden, D.V., Pethica, J.B., Bowes, P., Weihs, T.P.: Contact resistance and phase transformations during nanoindentation of silicon. Philos. Mag. A **82**(10), 1921–1929 (2002)
100. Stauffer, D.D., Major, R.C., Vodnick, D., Thomas, J.H., Parker, J., Manno, M., Leighton, C., Gerberich, W.W.: Plastic response of the native oxide on Cr and Al thin films from in situ conductive nanoindentation. J. Mater. Res. **27**(4), 685 (2012)
101. Kim, Y., Reid, J.A., Bahr, D.F.: Fabrication, structure, and performance of a microfabricated gallium electrical switch contact. J. Mater. Res. **26**(18), 2428–2437 (2011)
102. Ruffell, S., Bradby, J.E., Williams, J.S., Warren, O.L.: An in situ electrical measurement technique via a conducting diamond tip for nanoindentation in silicon. J. Mater. Res. **22**(3), 578–586 (2007)
103. Choi, Y.Y., Yun, T.G., Qaiser, N., Paik, H., Roh, H.S., Hong, J., Hong, S., Han, S.M., No, K.: Vertically aligned P (VDF-TrFE) core-shell structures on flexible pillar arrays. Sci. Rep. **5**, 10728 (2015)
104. Yun, J.M., Shin, J.H., Ryu, J., Shinde, N.M., Kim, K.H.: Piezoelectric performance of cubic-phase $BaTiO_3$ nanoparticles vertically aligned via electric field. Adv. Sustain. Syst. **2**(2), 1700133 (2018)
105. Greer, J.R., Oliver, W.C., Nix, W.D.: Size dependence of mechanical properties of gold at the micron scale in the absence of strain gradients. Acta Mater. **53**(6), 1821–1830 (2005)
106. Han, S.M., Phillips, M.A., Nix, W.D.: Study of strain softening behavior of Al–Al_3Sc multilayers using microcompression testing. Acta Mater. **57**(15), 4473–4490 (2009)
107. Kim, Y., Budiman, A.S., Baldwin, J.K., Mara, N.A., Misra, A., Han, S.M.: Microcompression study of Al–Nb nanoscale multilayers. J. Mater. Res. **27**(3), 592–598 (2012)
108. Harding, J.W., Sneddon, I.N.: The elastic stresses produced by the indentation of the plane surface of a semi-infinite elastic solid by a rigid punch. In: Mathematical Proceedings of the Cambridge Philosophical Society, vol. 41, no. 1, pp. 16–26. Cambridge University Press, June 1945
109. Mara, N.A., Bhattacharyya, D., Dickerson, P., Hoagland, R.G., Misra, A.: Deformability of ultrahigh strength 5 nm Cu/Nb nanolayered composites. Appl. Phys. Lett. **92**(23), 231901 (2008)

Chapter 4
Recent Cases in Advanced Micro/Nanoelectronics, Microsystems and MEMS Devices and Technologies

The development of the modern integrated circuits (microelectronics) and related nanotechnology device structures has brought about the need to understand material behavior, including mechanical properties, at the submicron and nanometer scale. In addition, nanomechanical devices will play an ever more important role in future technologies. Already advanced 3D interconnects at the submicron and nanometer scales are common in today's memory and microprocessors. New devices based on micro-/nano-electromechanical systems (M/NEMS) and nanotechnology are increasingly becoming a reality in the marketplace. The creation of such small components requires a thorough understanding of the mechanical properties of materials at these small length scales. Here we propose to directly examine some of the effects that arise when crystalline materials are plastically deformed in small volumes. The principles of extreme mechanics (XM) in elasticity and plasticity were used in the following recent cases to enable Design for Reliability (DfR) for advanced microelectronics, MEMS and nanotechnology applications.

One type of such small-scale structure is represented by the metal wires used in microelectronic chips as vertical interconnects (3D Through-Silicon Via). Metal thin films patterned into micron–scale conductor lines (horizontal and vertical) comprise the communication network of all integrated circuits. When the electrical current density running through these increasingly smaller and smaller wires becomes large enough (in the order of MA/cm^2), atoms start to migrate, inducing voids and hillocks to form under certain circumstances and eventually resulting in the final catastrophic failure of the device. The unique and powerful capability of synchrotron X-ray microdiffraction was used here to reveal the unique stress states in these vertical interconnect structures as a result of typical annealing processes involved in the making of such 3D Integrated Circuits (IC). Utilizing submicron-focused polychromatic synchrotron beam developed in the Beamline 12.3.2 at the Advanced Light Source (ALS), Berkeley Lab, the technique proved advantageous as a local probe of mechanical behavior (stress, elastic deformation) and plastic deformation

A. S. Budiman, *Principles of Extreme Mechanics (XM) in Design for Reliability (DfR)*, Engineering Materials,
https://doi.org/10.1007/978-981-15-6720-9_4

in small-scale devices. Furthermore, with this facility, in-situ annealing experiment was enabled, which is a rare opportunity that is not always practical with other characterization techniques. This capability enables us to investigate the evolution of the structure of the crystals as they deform due to the enormous thermal, mechanical or electronic loading (wind force of electrons moving from one end of the interconnect line to the other, for instance) associated with the operations of the devices. This is an important piece of information for the fundamental understanding of reliability degradation processes in 3D interconnects, thus it has important implications and great applications (product design, product warranty, etc.) for the microelectronics/semiconductor chip industries (microprocessing chips, memory chips, etc.). This will be discussed in the first section—Sect. 4.1: **Through-Silicon Vias (TSV) for Enabling 3D Integrated Circuits (IC)**.

Three more similar recent cases will be highlighted further following this section. Section 4.2: **Nanoplasticity in Low-Melting Temperature Metals** will discuss the unique fabrication of these nanostructures and their distinctive mechanical behaviors especially with respect to the potential applications in advanced microsystems packaging technology. Further, unique mechanoelectrical behaviors of Cu/Nb nanolayered structures were revealed using the MECR technique (discussed in previous chapter) with potential applications in Structural Health Monitoring (SHM) in Sect. 4.3: **Extended Linear Sensitivity Regime in Cu/Nb Nanolayers for Strain Sensing Applications**. Finally, Sect. 4.4: **Stress-Induced Voiding in Advanced Submicron Interconnects** will discuss yet another application in enabling next technology nodes in the semiconductor industry roadmap.

4.1 Through-Silicon Vias (TSV) for Enabling 3D Integrated Circuits (IC)

One type of small-scale structure is represented by the metal wires used in microelectronic chips as vertical interconnects (3D Through-Silicon Via). TSV (Through-Silicon Via) has been used for 3-dimensional integrated circuits. Mechanical stresses in Cu and Si around the TSV were measured using synchrotron X-ray microdiffraction. The hydrostatic stress in Cu TSV went from high tensile of 234 MPa in the as-fabricated state to −196 MPa (compressive) during thermal annealing (in-situ measurement), to 167 MPa in the post-annealed state. Due to this stress, the keep-away distance in Si was determined to be about 17 μm. Our results suggest that Cu stress may lead to reliability as well as integration issues, while Si stress may lead to device performance concerns.

4.1.1 Background and Overview

Through-silicon via (TSV) has recently gathered much interest for 3D interconnects in the microelectronics industry due to their potentials for vast improvements in the performance of microelectronics devices as well as their promises in enabling advanced multi-level chips integrating diverse CMOS technologies with each other as well as with emerging technologies such as MEMS and bio-chips [1, 2]. One of the key enablers for the successful implementation of 3D interconnects using TSV is the control of the mechanical stresses in the Cu TSV itself as well as in the surrounding silicon substrate [3]. Stresses induced intrinsically by thermal treatment in the TSV interconnect schemes, as well as stresses induced extrinsically, for example, due to silicon wafer thinning or chip-package interactions, introduce new reliability challenges for through-silicon via (TSV) technology, while stresses in the silicon substrate surrounding the copper-filled TSV can affect electron mobility in the transistors and thus determine the keep-away zone for Front-End Of Line (FEOL)/chip design [4].

In an effort to shed lights on these topics, the synchrotron-based X-ray microdiffraction technique [5] is the most suitable to measure crystal orientations and stresses in Cu TSV samples. As the high-brilliance synchrotron-sourced X-rays can penetrate and easily distinguish signals of the metallic structures from that of other materials in their surroundings, strain measurements can be done in our samples while the copper-filled TSV is still buried inside the silicon substrate. This allows strain measurements that are as close as possible to conditions in the real operations of the device. Furthermore, this unique capability of the technique also enables us to monitor strains in-situ during the course of the service of the device.

Past efforts to measure stresses in the 3D interconnect schemes using TSV technology involved mainly micro-Raman technique [6, 7] to measure stresses in the silicon substrate surrounding the Cu TSV and laboratory X-ray diffraction (XRD) technique [7] to obtain average stresses in an array of Cu TSV's. With the synchrotron-based X-ray microdiffraction technique, both stresses in Cu TSV as well as in the Si surrounding it can be measured simultaneously and can thus be compared and their correlation can further be investigated. It is thus another unique advantage of utilizing this technique.

The synchrotron technique of scanning X-ray microdiffraction has been described thoroughly elsewhere [5]. The power of this technique as a local stress probe for micro- and nanoscale devices stems from its two most important features. First, the submicron-sized focused X-ray beam that enables measurement of stresses at submicron resolution. Secondly, the continuous range of wavelengths in the white X-ray beam allows measurements of the deviatoric components of the stresses in addition to the standard ability of an X-ray diffraction technique to measure hydrostatic components of the stress tensors. This technique has proven to be useful in the study of mechanical stresses in advanced microdevices [8] as well as novel micro- and nanoscale structures [9, 10]. Impacts of such findings on reliability and robust integration of the devices were also reported [11, 12].

4.1.2 Experimental Methodologies—Probing Extreme Stress Generation

The TSV test structure used in this study was electrodeposited after metallization (i.e. no high-temperature treatment in the course of the TSV fabrication). The TSV test structures are of 20 μm diameter and 90 μm height and assembled in an array with 90 μm pitch as shown in Fig. 4.1.

The synchrotron X-ray microdiffraction experiment was conducted at the Beamline 12.3.2 at the Advanced Light Source (ALS), Lawrence Berkeley National Laboratory (LBNL). Both monochromatic (at 10 keV energy) as well as polychromatic (white beam) X-rays were used to enable measurements of hydrostatic and deviatoric components, respectively, of the stress tensor. The deviatoric components of the stress tensor in the Cu TSV were studied at the post-annealed state as well as during annealing at 200 °C using two different samples. The sample was mounted on a precision XY MICOS stage and the TSV of interest was raster scanned under the focused X-ray beam. The focused X-ray beam size was 1.0×1.0 μm (full width at half-max intensity). The sample was scanned in 1-μm steps.

The hydrostatic component of the stress in the Cu TSV was measured in the as-received state as well as during the annealing at 200 °C for 1 h and finally in the post-annealed state. All the hydrostatic stress measurements were performed on the center of the Cu TSV using monochromatic X-ray beam with 10 keV energy. Cu hydrostatic stress corresponding to the post-annealed state is +167 MPa (tensile), while during the annealing it is −196 MPa (compressive). The hydrostatic stress

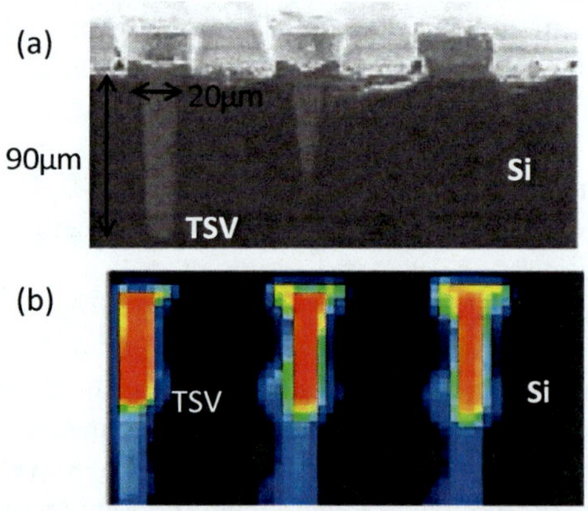

Fig. 4.1 The TSV test structure: **a** the SEM image that shows the cross-sectioning (by polishing) of an array of TSV's at an angle so that some of the TSV's will be optically visible as shown in the figure and some will still be buried under the silicon substrate thus allowing stress measurements in Cu TSV to be as close as possible to the conditions in the real operations of the device, and **b** the synchrotron X-Ray microFluorescence (XRF) mapping that shows all the TSV's and thus enable stress measurements and crystal orientation mapping by X-Ray microDiffraction (XRD) (Reproduced from Ref. [5] with permission from Elsevier)

measurement performed on the same Cu TSV before annealing (ie. in the as-received state) resulted in a high tensile stress of +234 MPa.

4.1.3 Extreme 3D Stress: Enabling Advanced Interconnects to Extend to the Third Dimension

The white-beam (polychromatic) X-ray measurement results were shown in Fig. 4.2a, b for Cu TSV at the post-annealed condition (after annealing at 200 °C for 1 h). This is presented first as this is the most relevant stress state for practical considerations in terms of reliability and robust integration of a TSV scheme in industry.

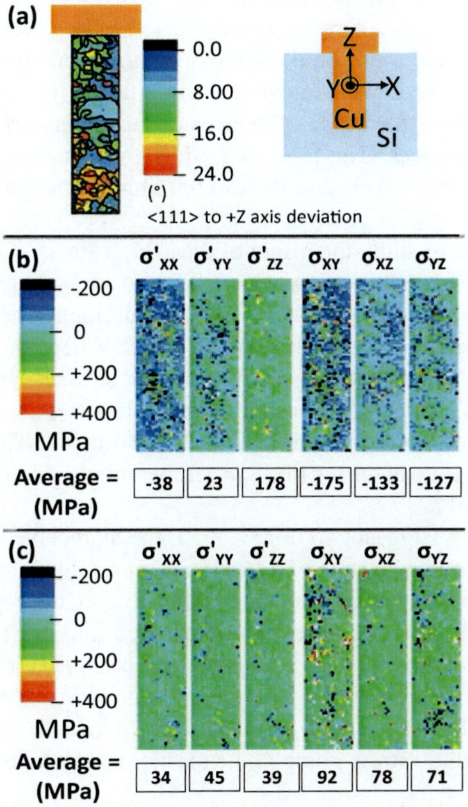

Fig. 4.2 White-beam X-ray measurement results showing **a** the grain orientation mapping as well as **b** the deviatoric stress mapping of the Cu TSV sample in the post-annealed state (after annealing at 200 °C for 1 h), and **c** the deviatoric stress mapping during the annealing at 200 °C. The XYZ coordinates used here are as described in (a) with respect to the laboratory coordinates (Reproduced from Ref. [5] with permission from Elsevier)

Figure 4.2a shows the grain orientation mapping of the Cu TSV indicating relatively large grains with wide range of orientations. Figure 4.2b shows the stress mapping of all the components of deviatoric stresses (normal as well as shear) and their distribution across the TSV domain. The normal deviatoric stresses σ'_{XX}, σ'_{YY} and σ'_{ZZ} are basically the full normal stresses (σ_{XX}, σ_{YY} and σ_{ZZ}) without the hydrostatic components (the hydrostatic stresses were measured separately). The shear deviatoric stresses σ_{XY}, σ_{YZ} and σ_{YZ} hold their usual meaning. These stress maps suggest the stress state of the Cu TSV is dominated by high shear deviatoric stresses (averaging 145 MPa). The overall stress level corresponding to this post-annealed state is provided by the hydrostatic stress which was already measured to be relatively high tensile of about 167 MPa.

Both the high shear (deviatoric) stresses as well as the relatively high tensile hydrostatic stresses at this post-annealed state could have important implications in terms of reliability and robust integration of the TSV scheme. The high tensile hydrostatic stress of 167 MPa here is comparable to the stress level in a conventional Cu interconnect lines and thus could lead to debonding, cracking as well as stress-induced voiding as has been well known in traditional Cu lines, which in turn could lead to degradation of performance of devices and eventually the final catastrophic events of failure of devices. The high shear deviatoric stresses could also lead to plastic damages which could eventually lead to failure of the interconnect device as well.

It is thus important to study the origin of these high stresses and understand how the stresses evolve. Figure 4.2c shows the deviatoric stress state of Cu TSV during the high-temperature annealing at 200 °C (the measurement was in-situ and took 6–7 h to complete). The deviatoric stress state of Cu TSV here shows a much-relaxed state both in terms of the normal as well as the shear stresses compared to the post-annealed state. This is expected as the high-temperature treatment allows relaxation of stresses both by plastic deformation as well as diffusion. The hydrostatic stress corresponding to this state of during the high-temperature annealing was compressive and was about 196 MPa.

These findings thus suggest two things. First, the as-received TSV appears to be already in a very high tensile state of 234 MPa. This could be due to the significant grain growth during room temperature annealing which Cu has been widely known to exhibit [13], that is from the time it was deposited into the TSV structure to the time the stress was measured in this experiment (this period was estimated to be about two months here). Similar room temperature grain growth in Cu TSV has also been previously reported by Okoro et al. [14]. Such grain growth in a confined volume would lead to hydrostatic stress going to the tensile direction due to grain-boundary elimination [13].

Furthermore, Fig. 4.3a shows the cross-sectional SEM image of a Cu TSV in the as-received state and it is evident that some of the grains here (especially in the middle of the Cu TSV structure) are larger than grains in the edges of the Cu TSV structure. This indicates there could have been room temperature growth in the middle part of the Cu TSV structure since the time Cu was deposited into the TSV structure (presumably with uniformly fine grain structure similar to those on

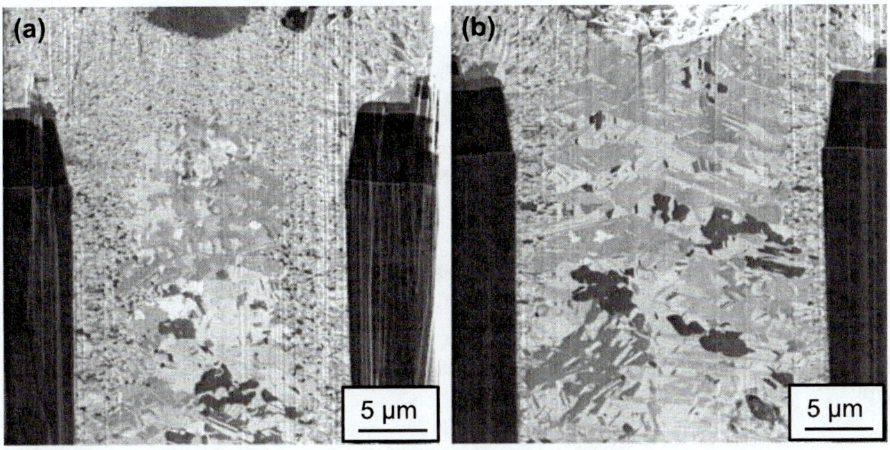

Fig. 4.3 SEM images showing **a** cross section of a Cu TSV in the as-received state (approximately two months after the Cu deposition into the TSV structures) and **b** cross section of another Cu TSV in the post-annealed state (after annealing of 200 °C for 1 h). These images indicate significant grain growth during the annealing phase of 200 °C for 1 h (Reproduced from Ref. [5] with permission from Elsevier)

the edges of the Cu TSV structure) to the time the sample was received by our group and cross-sectioning was performed. This is consistent with our above hypothesis.

Secondly, the hydrostatic stress state of Cu seems to largely follow the expansion mismatch with the Si substrate during the heating as well as the cooling of the annealing phase. The $\Delta\alpha\Delta T$ for Cu and Si elements and for the temperature difference associated with the thermal annealing at 200 °C correspond to a stress difference of about 294 MPa which is reasonably close to the actual measured relaxation/increase of hydrostatic stresses in Cu TSV during heating/cooling, respectively (430 MPa relaxation and 363 MPa increase). The slight discrepancies could be due to the combination between further grain growth during annealing which has indeed been observed in this experiment as shown in Fig. 4.3a, b which would strain the crystal further to the tensile direction and plastic deformation during high-temperature annealing that would relax the crystal.

The measurement of stresses in Si surrounding the TSV was done from the top of the TSV rather than from the side as was done for Cu TSV as shown in Fig. 4.2a, b. The measurement was done only in the post-annealed state with the TSV cap which consists of Cu as well as tin solder materials first removed (by polishing) leaving behind only the Cu TSV with the Si surrounding for the polychromatic X-ray beam scanning. The result was shown in Fig. 4.4a. The result is comparable with the simulation result by other groups [15], which was based on an analytical solution called Lame solution [16] for a tube with the outer diameter taken here to be infinite and with an inner pressure (or in this case a hydrostatic stress).

The measured stress (σ_{XX}) map here bears a rather close resemblance to the simulation result based on the Lame solution/model [15, 16]. The measured σ_{XX} on the

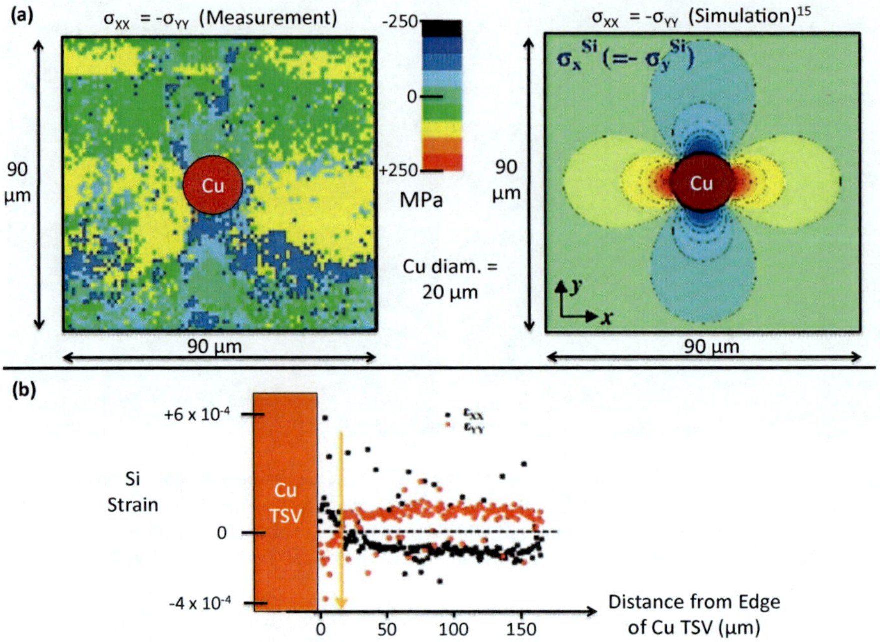

Fig. 4.4 Si Stress State Surrounding the Cu TSV showing **a** the measured stress map from the top of the TSV sample as compared to the simulation result performed by other research groups [14] and **b** the strain/stress profile from the side of the TSV sample showing the "keep-away" zone from the edge of the Cu TSV of about 17 μm (Reproduced from Ref. [5] with permission from Elsevier)

right- and left-hand sides of the TSV is mostly tensile whereas on the above and below of the TSV it is mostly compressive—a trend that is largely and qualitatively consistent with the simulation result [15]. This close resemblance to the Lame solution/model strongly indicates that the stress state in the Si surrounding the TSV is largely controlled by the hydrostatic component of the Cu stress state (i.e. the "inner pressure" of the "tube" in the Lame solution) in the TSV.

Figure 4.4b shows the measurement of the ε_{XX} and ε_{YY} which was done from the side—similar to the Cu TSV stress measurement in Fig. 4.2a, b. This measurement in particular was intended to show the strain/stress profile near the edge of the Cu TSV thus providing insights into the "keep-away" zone information. The measurement error in strain due to uncertainty in instrumental calibration here is 1×10^{-4}. Both strain profiles (ε_{XX} and ε_{YY}) show high strains very near to the edge of the Cu TSV (maximum $\varepsilon_{XX} = 6 \times 10^{-4}$ (tensile) and maximum $\varepsilon_{YY} = -4 \times 10^{-4}$ (compressive). The high strains then reduce to a level which is practically zero ($\pm 1 \times 10^{-4}$) after a distance of about 17 μm for both ε_{XX} and ε_{YY} consistently (shown in Fig. 4.4b with the orange arrows). This thus indicates a "keep-away" zone of about 17 μm from the edge of the Cu TSV within which the stresses in Si might change the electron mobility of the transistor device significantly such that electrical performance of the device might be out of control [4].

Our findings in the Cu as well as Si stress states in the TSV 3D interconnect scheme thus allow us to make some prediction that the key of controlling the reliability as well as the performance issues of the TSV for the reliable and robust integration of the 3D interconnect scheme might be in the high (tensile) hydrostatic stress in the Cu TSV in the as-received state. As such high tensile hydrostatic stress in Cu TSV could be originated from the room temperature grain growth in Cu during the shelf time, one may speculate that any method to minimize the room temperature grain growth in Cu TSV significantly could lead to lower stresses first in the Cu TSV itself and then in turns in the Si surrounding the TSV thus minimizing the reliability concerns as well as the device performance issues.

In summary, the stress state in the Cu TSV is dominated by both high hydrostatic stress as well as high shear stress. Both could lead to reliability as well as integration issues in the Cu TSV and lead to high stresses in the Si surrounding the Cu TSV which could in turn lead to device performance concerns. Measuring and controlling the stresses and especially how they lead to degradation of the device are thus important in the technology development as well as reliability improvement of the 3D interconnect schemes.

4.2 Nanoplasticity in Low-Melting Temperature Metals

Low-T_M (T_M = Melting Temperature) metals such as tin (Sn) and indium (In) have played key roles in semiconductor microelectronics packaging industries, including more recently in nanostructures. In this section, we present unique fabrication of these nanostructures and their distinctive mechanical behaviors especially with respect to the potential applications in advanced microsystems packaging technology. Mechanical behaviors and especially the strain-rate responses at the nanoscales of low-melting temperature metals, such as indium have not been much studied. Indium is one of the key materials or alloy-components in advanced microelectronics and nanotechnology industry, and understanding their mechanical behaviors at nanoscales becomes increasingly important to ensure lifetime reliability of their applications in novel nanoscale devices or advanced systems (for packaging at the nanoscales, for instance). Synchrotron X-ray microdiffraction has been utilized to examine defect structures of nanoscale materials as well as their strain-rate responses. Nanoscale or advanced microelectronics packaging, for instance, require acceptable levels of drop test results. For these low-melting temperature materials especially, this technique offers a unique advantage as conventional methods such as TEM and EBSD will expose the structure to high-energy electron beams that may significantly alter the microstructure and defect structure during analysis. Using this approach, we found interesting differences in terms of X-ray peak broadening after deformation with different strain rates, which could indicate differences in plasticity mechanisms in the microstructures of indium, which could be important for its applications in nanodevices. Understanding these differences could lead to better control of mechanical properties of low-melting temperature metals at the nanoscales and thus have important implications for nanodevice reliability.

4.2.1 Background and Overview

Mechanical behaviors and especially the strain-rate responses at the nanoscales of low-melting temperature metals, such as indium have not been much studied. Understanding the importance of the mechanical behavior of small-scale structures at the micron and nanoscale has become increasingly important to ensure reliability and lifetime of microelectronic devices. With the push toward lead-free solder products since its prohibition in 2006 [17], new interconnection materials of soft solders such as indium [18], tin [19] and their alloys [20, 21] are being implemented as bonding materials used in electronic packaging for low-temperature applications. Indium is a promising replacement due to its low-melting temperature, ductility, excellent wetting properties, high electrical conductivity and electrical reliability [22, 23]. Some of the examples are solder bumps for hybrid packaging in microelectronics [24, 25] and thermal interface materials (TIM) [26, 27]. However, a reflection of such a low-melting temperature means that indium solders are mostly operated at a very high homologous temperature even when their in-service condition is at room temperature. This could induce microstructural change due to relatively fast solid-state diffusion, creep and fatigue resulting in potential changes in response to the service environment conditions [18]. Hence, one of the concerns about these indium solders is the potential environmental effects that can cause microstructural changes. We do not wish these lead-free solders to be fragile and cracked easily because these damages may cause the malfunction of the portable electronic equipment under normal or excessive usage.

Broad-level drop impact tests such as the free drop test [28, 29] and pulse-controlled drop test [30, 31] have been developed to evaluate solder joint reliability of surface mounted electronic components for mobile applications in an accelerated test environment. However, to correlate mechanical characterization between board-level drop tests and product system level drop tests remain unclear and need further investigation. Further, research into the reliability of indium as interconnection materials [16, 21, 27] implies that thermal fatigue cracking of indium during thermal cycling can be a critical factor affecting the reliability of microelectronic devices using indium solders.

Instead, to attain the desired functionality and lifetime of small-scale components, a thorough understanding of nanoscale indium deformation mechanisms is required as a start. At room temperature, indium is a face-centered tetragonal (FCT) and space group of I4 / mmm. The melting temperature of indium is ~156 °C, corresponding to a homologous temperature of 0.7 in ambient conditions. For this material, dislocation motion even at room temperature may become non-conservative via thermally activated processes like cross-slip and dislocation climb. Its mean free path will be significantly shorter while the overall distance is sufficiently long. The enhanced interaction of this dislocation with other dislocations will lead to dislocation density increase via multiplication in the indium microstructures.

Indium exhibits strain rate sensitivity where the flow stress varies with the deformation rate. Strain rate and nanoindentation creep mechanical behaviors have been studied in detail [32–35]. Lee et al. [36] performed indium nanopillar compressions of different diameter sizes (350–920 nm) and different strain rates through ex-situ pillar compression experiments. They observed a noticeable increase in indium nanopillar strength from ~17 to ~37 MPa with the strain rate increasing from 0.0001 to 0.01 s^{-1} with a strain rate exponent of 6.12 but no effect of the size of nanopillars.

Budiman et al. [37] further evaluated the defect density evolution of indium micropillars of 920 nm in diameter through synchrotron Laue X-ray microdiffraction (μSLXRD) before and after pillar compression. This is a non-destructive ex-situ technique that can be performed before and after uniaxial compression tests to understand the defect density evolution of metallic micropillars. The dislocation defects such as statistically stored dislocations (SSD) or geometric necessary dislocations (GND) can be detected by this technique [9, 38, 39]. Results from these μSLXRD measurements show the increase in dislocation density due to mechanical deformation and is a strong function of strain rate.

The primary objective of this work is to examine the effect of strain rate before and after the uniaxial compression of ~1 μm diameter indium micropillars through ex-situ μSLXRD characterizations of the defect density evolution in the microstructures. Indium has a low-melting temperature of ~156 °C and a face-centered tetragonal (FCT) crystalline structure with a c/a = 1.075 [40]. Thus, thermally activated processes for vacancy diffusion and dislocation cross-slip or climb are expected to have a non-trivial contribution in the mechanical deformation of these structures. Our μSLXRD results reveal that the Laue diffraction peaks broaden after the mechanical compression tests, suggesting defect accumulation during the deformation with a larger extent of accumulation with a higher strain rate. A variation of the Williamson-Hall (W–H) plot called the Ungár's method [41, 42] is implemented to quantify the net dislocation density increase due to the deformation process. This analysis reveals an estimated ~1.5 times net increase of defects in the pillar with the increase of strain rate from 0.001 to 0.01 s^{-1}.

4.2.2 *Experimental Methodologies—Probing Extreme Plasticity and Strain-Rate Effects*

Synchrotron Laue X-ray Microdiffraction

To examine the defect structure of submicron scale specimens of low-melting temperature metals such as indium, the Synchrotron Laue X-ray Microdiffraction technique (μSLXRD) was utilized. This technique has been widely used in measuring densities of geometrically necessary dislocations (GNDs) [37, 38, 43], electromigration effects in copper interconnects [8, 44], stresses of silicon solar cell modules [45–47] and silicon anodes in lithium ion batteries [48]. The X-ray beam comes from a powerful synchrotron source, which can be focused into a submicron spot size, close to the grain size of most single crystalline materials. X-rays allow deeper penetration depths, sample preparation, and measurements under various conditions. Bragg's Law is satisfied with a white X-ray beam of a continuous range of wavelengths, allowing a huge number of reflections to be obtained simultaneously even when the lattice is locally rotated or bent. It is this sensitivity to local lattice curvatures that makes the μSLXRD technique is suitable for probing plasticity at small scales since strain gradients and geometrically necessary dislocations (GNDs) are directly related to the local lattice curvatures [9, 38, 39].

This technique involves scanning the sample with the focused X-ray beam at submicron resolution, thus gaining structural information about the crystal and its defects in the diffracted volume through the shapes of the Laue diffraction peaks. Using this approach, we can monitor the change in the Laue diffraction peaks before and after deformation. A quantitative analysis of the Laue peak widths then allows us to estimate the density of geometrically necessary dislocations (GNDs) in the sample. The absolute number of GNDs in the crystal can be determined using the relevant dimensions of the sample. A comparison of the numbers of GNDs before and after, or even during, the deformation provides information about the change in microstructure associated with plastic deformation. Conventional structural characterization methods such as transmission electron microscope (TEM) [49] and Resonant Ultrasound Spectroscopy (RUS) [50, 51] will expose the structure to high-energy electron beams and sounds that might alter the internal defect structure during analysis.

The microstructure of 920 nm diameter indium micropillars was characterized with Beamline 12.3.2 at the Advanced Light Source synchrotron facility at Lawrence Berkeley National Laboratory. Each indium micropillar sample was mounted on a MICOS high precision XY-positioning stage of resolution of 0.05 μm. The location of the micropillar was identified through white X-ray beam raster scanning. A second X-ray scan square size of $10 \times 10\ \mu m^2$ with 0.5 μm step sizes was conducted to obtain the detailed diffraction data of the micropillar. The μSLXRD patterns were collected using a MAR 133 X-ray charge coupled device (CCD) detector of resolution of 100 μm pixel size. 3–4 h of data collection was required to obtain about 400 CCD frames with an exposure time of 5 s for each frame for every 10 s of electronic readout time. A customized XMAS software [39] was used to analyze the resulting μSLXRD patterns. Further details have been discussed in previous publications [37, 38, 52].

Laboratory XRD techniques that separate the effects of strain, crystal size and non-uniform strain

The X-ray diffraction peak broadening is observed caused by both crystallite size and lattice strain due to the large volume of grain boundaries in typical polycrystalline materials [53]. Therefore, the peak broadening must be considered to examine the structural properties of materials. Accordingly, some XRD parameters such as the peak width, peak intensity, and XRD peak shift are originated by the crystallite size and lattice strain. The crystallite size and lattice strain are related to Bragg's angles as $1/\cos\theta$ and $\tan\theta$, respectively [54].

The generation of strain in the crystalline lattices may be caused by crystal imperfections, lattice dislocations, stacking faults, coherency stresses and many more [55]. The traditional Debye–Scherrer's (D-S) method offers only lower bound on the crystallite size [53]. Based on the results of an XRD pattern, several methods have been used to estimate the crystallite size and lattice strain of materials. In this paper, we applied Ungár's method which is an analytical method used to estimate the crystallite size and lattice strain contributions to the line broadening by modifying existing methods of Williamson-Hall (W–H) analysis and Warren–Averbach (W–A) analysis.

Strain broadening is the broadening of fundamental Bragg reflections within the adjacent vicinity of the Bragg reflections. This scattering effect is intimately related to the scattering caused by the long-range strain fields of linear-type defects called dislocations. However, strain anisotropy is related to the strain broadening of diffraction peaks. The size broadening of the diffraction peaks is uniform but the strain broadening increases in reciprocal space [56]. The separation of the size and strain effects have been established by the Williamson-Hall (W–H) analysis and Warren–Averbach (W–A) analyses. However, a special logarithmic series expansion of the real part of the Fourier coefficients of a peak profile has to be used to consider strain broadening by dislocations in large crystals as shown in Eq. (4.1) [57, 58]:

$$\ln|A(n)| = -\rho^* n^2 \ln\left(\frac{R_e}{n}\right) + Q^* n^4 \ln\left(\frac{R_2}{n}\right) \ln\left(\frac{R_3}{n}\right) \pm O\left(n^6\right) \tag{4.1}$$

where n is the Fourier parameter, ρ^* is the formal dislocation density, Q^* is the correlation factor representing the fluctuation of the dislocation density $Q^* = \rho^{*2} - \rho^*$ [57, 58]. R_e is the outer cutoff radius of dislocations, R_2 and R_3 are auxiliary constants [56]. The formal and true values of dislocation density ρ^* and correlation factor Q^* are related to each other as [57, 58]

$$\rho^* = \rho\left(\pi g^2 b^2 \overline{C}\right)/2, \quad Q^* = Q\left(\pi g^2 b^2 \overline{C}\right)^2/4, \tag{4.2}$$

where $\overline{C}$ is the average contrast factor of dislocations in the case of a particular *hkl* reflection and b is the Burgers vector of dislocations. The values of $\overline{C}$ can be calculated numerically on the basis of the crystal structure the elastic properties and the slip systems of dislocations in the crystal.

Williamson-Hall method is based on the combination of two terms including size broadening and strain broadening [59]. It was suggested that the FWHM of peak profiles could be written as the sum of these two broadening effects [60]:

$$\Delta K = 0.9/D + \Delta K^D, \tag{4.3}$$

where the first term is related to the size broadening from Scherrer's formula [61], D is the average grain size and ΔK^D is the strain contribution to peak broadening. Here, $\Delta K = 2\cos\theta(\Delta\theta)/\lambda$ where $\Delta\theta = \frac{FWHM}{2}$ in angle, θ is the diffraction angle and λ is the wavelength of X-rays respectively. However, dislocations give different contrasts depending on the relative positions of the Burgers and line vectors of the dislocations and the diffraction vector. Not assuming ΔK^D is linearly or quadratically related to K, ΔK^D can be written in the following form [57–59]:

$$\Delta K^D = A(\rho^*)^{1/2} + A'(Q^*)^{1/2} \tag{4.4}$$

where A and A' are parameters determined by the effective outer cutoff radius of dislocations, R_e and the auxiliary parameters R_1 and R_2, respectively. If dislocations are the source of strain in a crystal, the proper scaling factor of the breadths of peak profiles is $\left(K\overline{C}^{1/2}\right)$ instead of nearly K. Using Eqs. (4.2) and (4.4), Eq. (4.3) will be [59]:

$$\Delta K = 0.9/D + \left(\pi A b^2/2\right)^{1/2}\rho^{1/2}\left(K\overline{C}^{1/2}\right) + \left(\pi A' b^2/2\right)Q^{1/2}\left(K^2\overline{C}\right) \tag{4.5}$$

Equation (4.5) will be known as the Modified W–H plot.

If the size and strain effects are present in the crystal at the same time, one of the fundamental equations of the Fourier coefficients in the Warren–Averbach analysis is [56, 62]:

$$\ln A(n) = \ln A^s(n) + \ln A^D(n), \tag{4.6}$$

where the s = size and D = distortion. Inserting Eqs. (4.1) and (4.2) into (4.6), the following is obtained for the real part of the Fourier coefficients [59]:

$$\ln A(n) \cong \ln A^s(n) - \rho B n^2 \ln(R_e/n)\left(K^2\overline{C}\right) + Q B^2 n^4 \ln(R_2/n)\ln(R_3/n)\left(K^2\overline{C}\right)^2, \tag{4.7}$$

where $B = \pi b^2/2$. In the classical W-A analysis, the scaling parameter is K^2. In the case of dislocated crystals, the proper scaling parameter is $K^2\overline{C}$ instead of K^2. This will be known as modified W-A analysis in agreement with the notation used in [59]. From Ungár's previous works [41, 42, 63], it was observed that the breadths or the Fourier coefficients are changing anisotropically in terms of g or the *hkl* indices. However, the global increase or decrease indicates the presence of strain.

Materials—fabrication of micropillars

The indium pillars were fabricated by electron beam lithography and metal electroplating technique developed by Burek and Greer [64]. The specific procedure and methodology to create these micropillars are described in detail here [10, 36, 37]. The

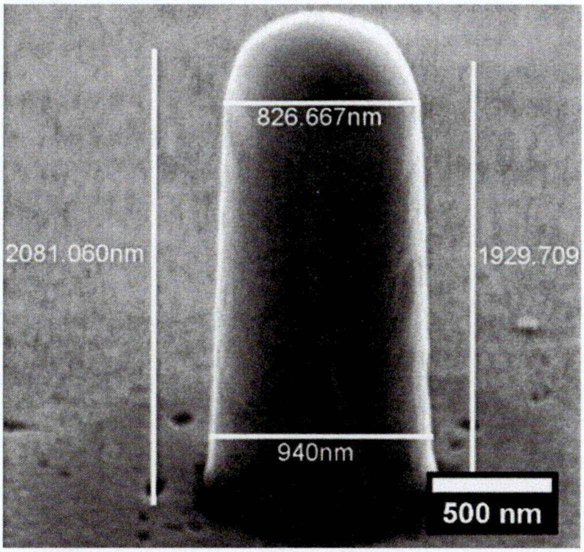

Fig. 4.5 SEM image of an as-fabricated indium micropillar (Reproduced from Ref. [38] with permission from Elsevier)

resulting indium micropillars were approximately 920 nm in diameter. Figure 4.5 shows an example of one of the fabricated micropillar with an aspect ratio of ~3:1. A rest period of fifteen days was maintained for all the samples to ensure that thermally activated processes reached equilibrium at room temperature prior to testing.

Micropillar compression experiments

The micropillar compression tests were conducted at Stanford University using an Agilent NanoXP (Agilent/MTS, Knoxville, TN) nanoindenter outfitted with and a custom fabricated flat-ended diamond tip. The test was operated in a continuous stiffness mode. Two indium pillars were selected and compressed at two different engineering strain rates of 0.01 and 0.001 s^{-1}. After the compression and before the μSLXRD characterization, the deformed indium micropillars were stored in the dry ice sublimation environment (−78.5 °C) to reduce the room temperature annealing effects. The Laue X-ray diffraction peak widths measured before and after uniaxial compression were then compared.

4.2.3 New Insights on Plasticity of Low-Melting Temperature Metals at the Nanoscales

A series of tests were performed to understand the effect of plasticity and creep on indium micropillars through uniaxial pillar compression, synchrotron X-ray microdiffraction, and further theoretical calculation through analytical methods.

Pillar compression results (Stress–Strain)

Figure 4.6 shows the uniaxial compression stress–strain data of strain rate of 0.01 s^{-1} to a final plastic strain of near 28%. The engineering stress–strain curves reveal that this sample was quite soft with flow stress of ~13 MPa which was measured at 5% engineering strain. Similar results were observed at the strain rate of 0.001 s^{-1}. Post-compression SEM images of both micropillars shown in Fig. 4.7 show that the compression was accommodated by sidewall wrinkling, bulging and folding. There were no discrete slip bands observed in this deformed sample.

Synchrotron X-ray Laue Microdiffraction (μSLXRD) Results

μSLXRD measurements were done on both pre- and post-compressed indium micropillars at different strain rates. From the initial μSLXRD measurements, the

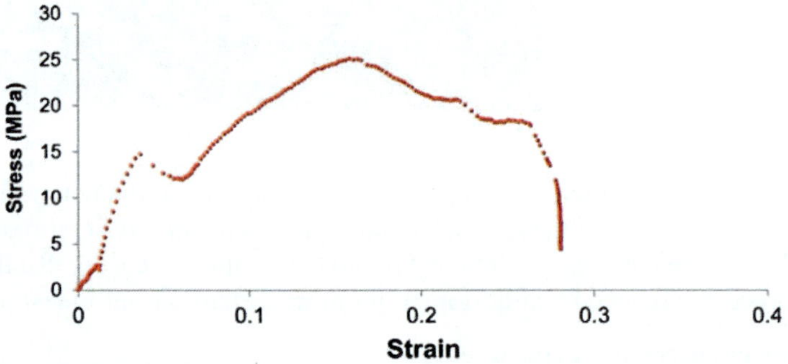

Fig. 4.6 Typical engineering stress–strain curve of the 920 nm indium micropillar (Reproduced from Ref. [38] with permission from Elsevier)

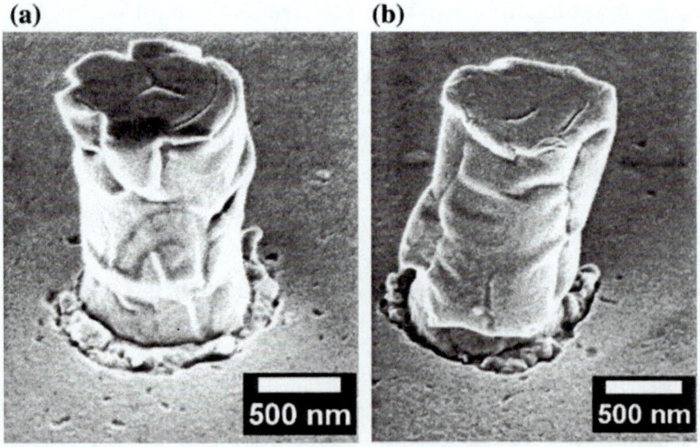

Fig. 4.7 SEM micrographs of post-compressed indium micropillars at a strain rate of 0.001 and 0.01 s^{-1} (Modified from Ref. [38] with permission from Elsevier)

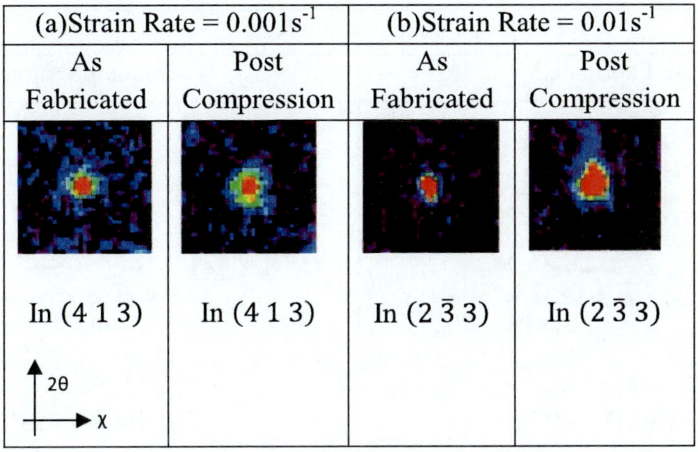

Fig. 4.8 Indium **a** (4 1 3) Laue diffraction peak at a strain rate of 0.001 s^{-1} and **b** (2 −3 3) Laue diffraction peak at a strain rate of 0.01 s^{-1} from the as-fabricated and deformed micropillar states (Modified from Ref. [38] with permission from Elsevier)

micropillars were indexed as a tetragonal single-crystal structure with out-of-plane orientation near (1 1 1) with a misalignment of 7°. For the 0.001 s^{-1} strain rate, diffraction spot from the (4 1 3) crystal plane was shown in Fig. 4.8a. Similarly, diffraction spot from the $\left(2\,\overline{3}\,3\right)$ crystal plane was shown in Fig. 4.8b. These figures illustrate that the shape of all the indium diffraction peaks remained nearly isotropic before and after the uniaxial compression with no significant asymmetric broadening. However, the size of the post-compression Laue diffraction spots was significantly larger than those before compression.

Figure 4.9 shows the intensity plots of the Laue diffraction peaks: (4 1 3) for strain rate of 0.001 s^{-1} and $\left(2\,\overline{3}\,3\right)$ for strain rate of 0.01 s^{-1}. The intensity traces were carried out along a particular χ-axis. The maximum intensity for both peaks was about 5×10^4 counts. After ~28% compressive strain, there is a lack of crystal rotation as the center points remain stationary at 96.3° (at a strain rate of 0.001 s^{-1}) and 80.8° (at a strain rate of 0.01 s^{-1}). A large number of active systems were available during the pillar compression.

However, the amount of diffraction peak broadening was quite different with the full-width-half-max (FWHM) values increased by 0.255° (for strain rate of 0.001 s) and 0.466° (for strain rate of 0.01 s^{-1}). These increases are likely a result of defect accumulation due to dislocations. It is important to note that these increases in Laue diffraction peak broadening represent the lower limit of defect accumulation due to the room temperature annealing during the post-compression μSLXRD measurements. The extent of the diffraction peak broadening at 0.001 s^{-1} is approximately 44% less than the one compressed at a faster rate of 0.01 s^{-1}.

(a) In (4 1 3) peak, (Strain Rate = $0.001 s^{-1}$)	
As Fabricated	Post Compression
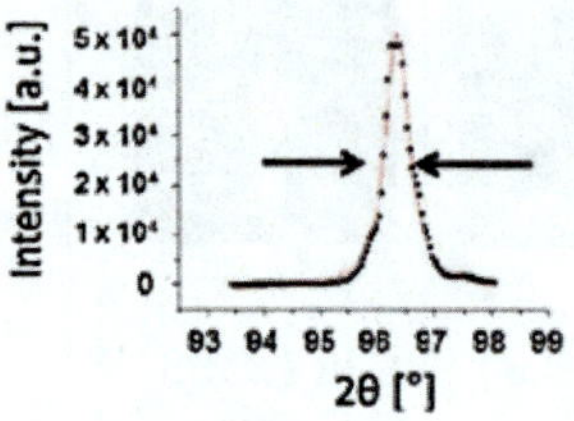	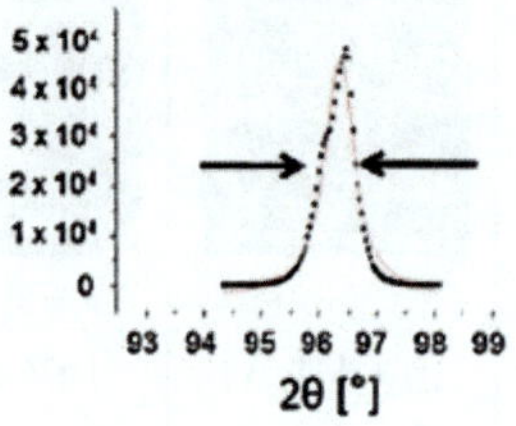
$X_c = 96.3°$ FWHM = 0.298°	$X_c = 96.3°$ FWHM = 0.553°
(b) In $(2\,\bar{3}\,3)$ peak, (Strain Rate = $0.01 s^{-1}$)	
As Fabricated	Post Compression
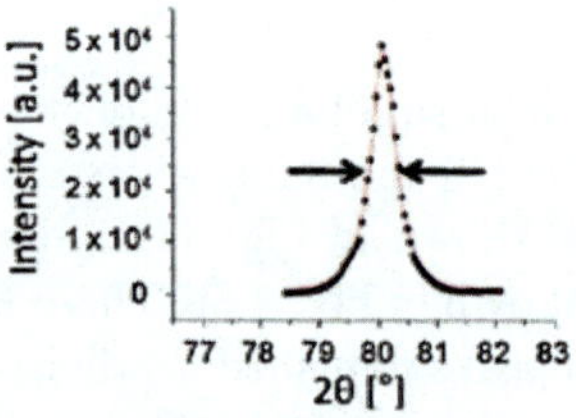	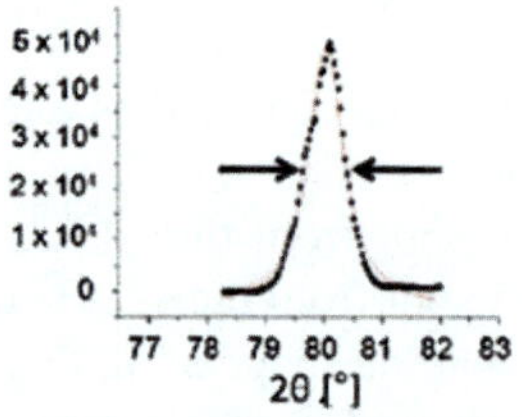
$X_c = 80.8°$ FWHM = 0.312°	$X_c = 80.9°$ FWHM = 0.778°

Fig. 4.9 The intensity distribution of indium Laue peaks before and after the uniaxial compression of Laue diffraction peaks of **a** (4 1 3) at a strain rate of 0.001 s^{-1} and **b** $(2\,\bar{3}\,3)$ at a strain rate of 0.01 s^{-1} (Modified from Ref. [38] with permission from Elsevier)

Analytical calculation using Ungár's method

Previously in Budiman et al. [38]'s paper, the increase in dislocation density was quantified approximately using Cahn-Nye equation [65, 66], $\rho = \frac{|\kappa|}{b}$ where $|\kappa|$ is the local lattice curvature and b is the magnitude of Burger's vector. The initial dislocation densities of both micropillars were similar of ~4 × 10^{13} / m^2. Assuming the only factor in the diffraction peak broadening is the dislocation defect accumulation within the indium micropillars during deformation, the equation was applied using $|\kappa| = FWHM$ similar to previous publications [8, 67–69]. Using the FWHM values in Fig. 4.9, the dislocation defect densities were reported in [37] showing nearly 2.5 times increase in dislocation density for strain rate of 0.01 s^{-1} as compared to 0.001 s^{-1}.

The diffraction peak broadening of all the indium Laue diffraction peaks was plotted according to the modified Williamson-Hall plot as described earlier. A line fitting was done to obtain a smooth curve is obtained as shown in Fig. 4.10. The C factors were averaged by assuming that all slip systems permitted in an FCT crystal are equally populated and that edge and screw dislocations are present with equal probability.

The intercept of the curve at $K = 0$ yields $D_{WH} \cong 900$ nm for the particle size that matches the theory that the micropillar is single crystalline. From the modified W–H plot, the exact dislocation density cannot be determined since the value of constant A is defined by the effective outer cutoff radius of dislocations which on the other hand, depends on the tail of the profiles. However, we can take the magnitude of the fitting parameter of $K\overline{C}^{1/2}$ abbreviated as M to describe the extent of dislocation density.

It was found that the M value after compression of the micropillar was always larger than before compression. The value of M at a strain rate of 0.001 s^{-1} was

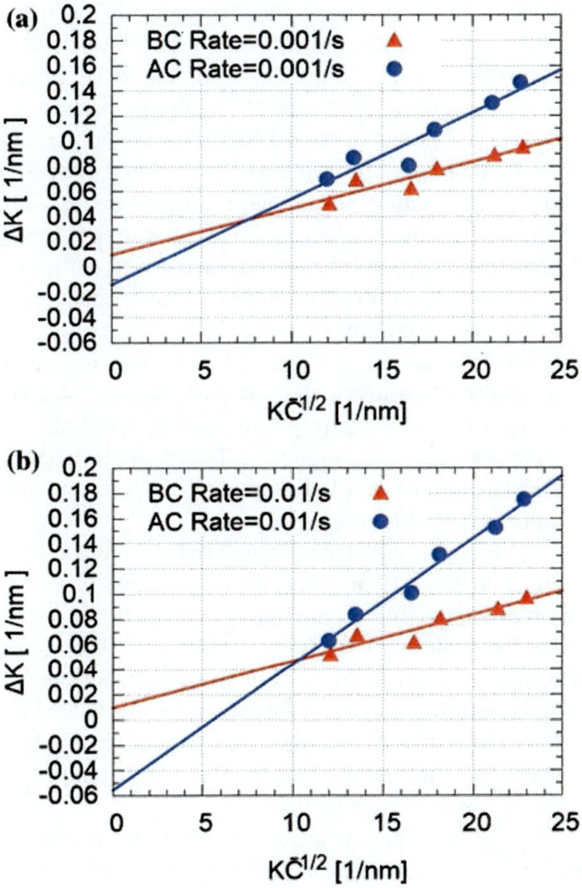

Fig. 4.10 Modified Williamson-Hall (W–H) plots of Indium micropillars before compression (BC) and after compression (AC) at a strain rate of **a** 0.001 s^{-1} and **b** 0.01 s^{-1} using Ungár's method [59] (Reproduced from Ref. [37] with permission from IEEE)

increased by 1.85 times. The value of M at a strain rate of 0.01 s^{-1} was increased by 2.69. The dislocation density increased by 1.46 times at a faster strain rate of 0.01 s^{-1}. This is close to the results observed previously by Lee et al. [36] as well with a dependence of strain rate sensitivity for the majority of its indium pillars.

4.2.4 Extreme Plasticity: Enabling Advanced Nanoscale Packaging at Low Temperatures

The strain-rate response of indium microstructures

There is a substantial strain rate effect in the indium deformation with the micropillar deformation at 0.01 s^{-1} strain rate exhibiting larger peak broadening and steeper slope through the modified W–H method over 0.001 s^{-1} strain rate. At room temperature, indium micropillar experiences substantial thermal contribution to dislocation motion. The indium dislocations change slip direction through dislocation cross-slip and climb. This continuous switching in dislocation travel direction increases the likelihood of dislocation–dislocation interactions, cross-slip, and the generation of new dislocation via new source formation [37]. Indium undergoes recovery and recrystallization at the ambient conditions because of its low-melting temperature of ~156 °C. Some of the dislocations created through uniaxial compression can be annihilated by these thermally activated processes. However, the synchrotron and Ungár's method results demonstrated the indium pillars exhibit a net increase of defect density. This suggests that the defect multiplication rate is considerably higher than the combined annihilation rate of the recovery, recrystallization, and surface evaporation processes [37]. This mechanism has been reported in the bulk deformation of indium as well [36].

The uniaxial compression of indium micropillars together with the synchrotron Laue microdiffraction technique is an attempt to estimate the stress and plastic strain as a function of time during a constant temperature. The mechanical response of the testing machine is analogous to that of an electronic package. The assembly stores the elastic strain energy during loading then converts the elastic strains into plastic strains in the solder during relaxation.

The increase in the dislocation density by about ~1.5 with increasing strain rate from 0.001 to 0.01 s^{-1} shows the increase in the work hardening of the indium micropillars at room temperature. This work hardening data can be used to determine the indium solder joint stresses and strains during a thermal transient in an electronic assembly. In the package, the displacement rate is determined by the rate of temperature change and the expansion mismatch between components. Strain rates will be higher for rapid temperature changes and large assemblies. Thus, more data is needed over the operating temperature of the microelectronic package to characterize stresses during a transient fully.

In summary, the evolution of defect densities in nanometer-sized indium pillars before and after uniaxial compression was characterized by using the non-destructive μSLXRD technique. We observe Laue diffraction peak broadening, suggesting that the defect density likely increased in the course of deformation using the μSLXRD technique where the FWHM values were found to vary strongly anisotropically as a function of diffraction order. The results were confirmed analytically by Ungár's method of separating size and strain broadening in the XRD profiles and found interesting differences in terms of X-ray peak broadening after deformation with different strain rates. We estimate that the defect density increase by approximately 1.46 and 2.69 times for compressive strains of 28% with a strain rate of 0.001 s^{-1} and 0.01 s^{-1}, respectively. This increase of defect concentration is consistent with indium's low-temperature behavior where it is hypothesized that this bulk-like effect in indium microstructures is due to thermally activated processes like dislocation climb and enhanced cross-slip. Such processes are activated in the deformation of indium at room temperature due to its low-melting temperature.

These results help confirm the previously reported extent of increase in the dislocation density of indium microstructures fabricated by the same technique at different strain rates. A thorough understanding of indium microstructure deformation generated in the course of this work will further enrich the current small-scale plasticity state of the art, especially in the area of low-melting temperature materials and contribute significantly in commercial applications and reliability of indium and indium-based alloys in the microelectronics packaging industry.

4.3 Extended Linear Sensitivity Regime in Cu/Nb Nanolayers for Strain Sensing Applications

Multilayered nanocomposites, such as Cu/Nb, may have unique mechanoelectrical behaviors as revealed and presented in this section using the MECR technique (discussed in the previous chapter) with potential applications in Structural Health Monitoring (SHM). Cu/Nb nanolayers have been known for their extraordinary mechanical properties of very high flow strength, ultra-light weight and stable plastic flow to large strains, represent a class of novel composite nanomaterials in which there arises rare opportunities to design new materials from the ground up and to tailor their properties to suit exactly their performance requirements. These materials can withstand very high strains in the elastic regime without any inelastic relaxation due to plasticity or fracture compared to its bulk counterparts. This extended elastic regime opens up new possibilities for tuning the physical and chemical properties of materials as well as bringing novel functionalities, such as high-performance coating materials with online strain monitoring capability. Our findings in this article suggests basic feasibility study of a Cu/Nb multilayered nanocomposite with 20 nm layer thickness having a novel functionality for online strain monitoring capability, in addition to its more known application as a high-performance coating materials due to its extraordinary strength and deformability. A linear trend of resistivity with respect to engineering strain for strains in excess of 3.5% was observed and suggests a significant regime for use for strain sensor/detection/monitoring capability.

4.3.1 Background and Overview

Materials have been designed mostly through "trial and error" ever since the Iron Ages, long before much was understood about them [70, 71]. By chance, our ancestors in the sixthcentury found out by heating iron above a certain temperature and quickly quenching it into water, they could create much harder and stronger metallic weapons [70]. Later, it was understood that it was the presence of carbon in iron and its martensitic transformation that was responsible for its enhanced hardness and strength. Edisonian (i.e. "trial and error") approaches to new materials discoveries continue through the years as the main focus of research went into the invention and improvement of materials and their fabrication processes [72].

Now, we have the abilities to construct the microstructural building blocks of materials from our detailed understanding of defects and dislocations, grain boundaries and interfaces and the tools to manipulate and actually see them through nanomaterials fabrication and characterization techniques [73]. With these skills, we are increasingly ready for the building of the fundamental knowledge base needed to progress toward designing and making a material with specific and desired function or property. Thus, a materials by design approach can contribute significantly toward accelerating the pace of discovery to deployment of advanced materials systems.

President Barack Obama's White House in the United States started the Materials Genome Initiative in 2011 which recognized the importance of accelerated materials development to the well-being and advancement of society [74–77]. With this initiative, new computational materials design methodologies of first principle calculations and molecular dynamics have been incorporated to speed up the discovery processes, creating new materials in a systematic way [78–80]. However, little effort has been looked into identifying key material properties in existing established materials and directing the research into discovering new functionalities.

Such a notion of materials design or "directed discovery" where the systematic iterations of materials synthesis, characterization and simulation could lead to new materials discovery, development and deployment ***faster*** than the traditional methods of trial and error was proposed. Such accelerated materials discovery is illustrated in Fig. 4.11 with an analogy to a process whereby each iteration will get us closer to the landing zone, i.e. the materials system/solution found for a particular application with the red arrows there represent the "directed discovery" design methodology. The iterations could be accelerated through the fundamental understanding and physical modeling aided by the computer modeling and simulation, and thus could lead to completely new and unexpected materials system that will not have been gained through the traditional "trial and error" methodologies.

In this section, we utilize multilayered nanocomposite materials, in particular metal–metal nanocomposites as a model material system platform, where we

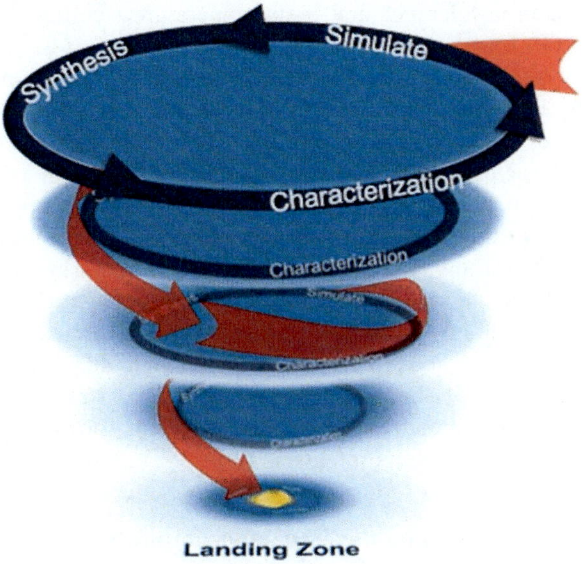

Fig. 4.11 Materials design by "directed discovery" could lead to new materials discovery/design [72] faster and to completely new materials system that will not have been gained through trial and error methods (Reproduced from Ref. [85] with permission from SAGE)

could easily create and modulate interfaces, and study them and their interactions with other microstructural features such as grain boundaries, dislocations and defects in nanoscale confined volumes (thus mimicking nanostructured materials or devices/systems) and simultaneously develop systematic design methodologies for new nanomaterials for targeted applications [81–85]. This interface material system could be built from the ground up—one microstructural building block, i.e. one nanolayer, at a time and thus we have full control of the overall/bulk resulting properties of the material. The key physics of these basic building blocks have also recently been identified (which are the interface mechanics and energetics between the constituent nanolayers) and currently actively being investigated making this material system particularly attractive for the demonstration of such design methodology for accelerated materials discovery and development.

Metallic Multilayered Nanocomposites

Metallic multilayered nanocomposites (i.e. metal nanolayers) have attracted much attention in recent studies due to their unique and sometimes unexpected combination of properties of ultra-high strength and deformability [86]. They exist either as metal–metal [86] or even metal-graphene [87] nanolayers. Their remarkable properties are mainly due to interfaces [81, 88–91]. The plot shows the Hall–Petch relation of hardness proportional to $\frac{1}{\sqrt{h}}$ giving rise to "smaller is stronger" [92–95]. Further, nanopillars are made from these nanolayers system to understand more about the effects of plasticity in small volume and dislocations in the nanolayer system [96–99]. For this paper, we will concentrate on the Cu/Nb multilayer system that has been well-established in simulation and experimentally for its high flow strength, good radiation tolerance and high-temperature stability due to its incoherent interface [100–106].

In addition to these extraordinary mechanical properties which would be useful for high-performance coating materials, the Cu/Nb nanolayers have also been found to exhibit a behavior during its deformation that could prove useful for strain sensing/detection/monitoring capability. Budiman et al. performed successive ex-situ compressions of Cu/Nb micropillars up to accumulative strain of 35% [85]. The layers exhibited a maximum flow stress of 1.6 GPa at 24% accumulative compressive strain as shown in Fig. 4.12a, b. Monochromatic synchrotron X-ray Microdiffraction experiments were conducted for each successive strain. By comparing the X-ray ring width and how it broadens as a function of strains, a relationship between dislocation density and strain in both the Cu and Nb nanolayers was established where a linear relationship was observed up to 3.5% strain followed by the saturation of the ring width broadening at higher strains as shown in Fig. 4.13a, b. This represents an extended regime of linear response in terms of defect density with respect to deformation.

Resistivity of multilayered nanocomposites

Electron–defect scattering effects in materials occur when electrons encounter defects in a material, resulting in resistivity [107–109]. In the bulk regime, resistivity measurements have been done on Cu/Nb filament drawn alloy wires to understand its enhanced electrical performance over traditional Cu wires [110, 111]. Early work for resistivity in metal nanolayers focused more with antiferromagnetic coupling effect of nanolayers for electromagnetic application [112, 113]. Deviations from bulk resistivity calculations occur due to the layer thickness of nanolayers can reach equal to or smaller than the typical mean free path of its bulk components [114]. The interfaces between multilayers can act as scattering centers for electron–electron and electron–phonon interactions as the electrons are likely to cross different layers between two scattering events, influencing the overall resistivity of the multilayered composite

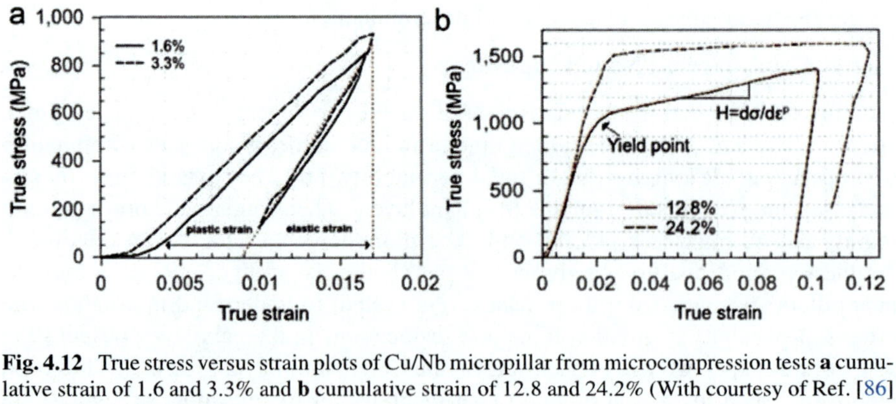

Fig. 4.12 True stress versus strain plots of Cu/Nb micropillar from microcompression tests **a** cumulative strain of 1.6 and 3.3% and **b** cumulative strain of 12.8 and 24.2% (With courtesy of Ref. [86] from author's copyright)

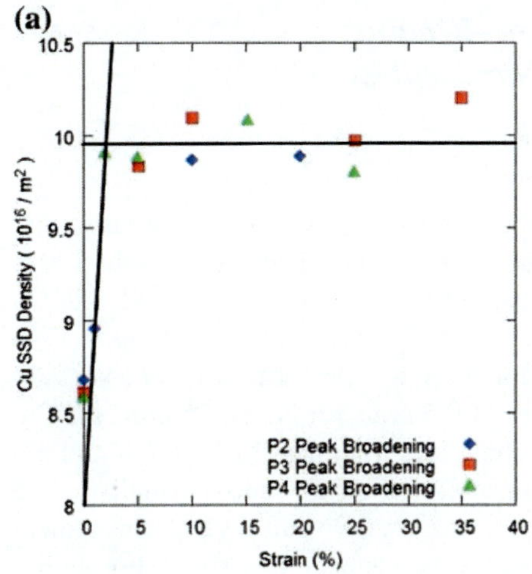

Fig. 4.13 Statistically stored dislocation (SSD) density with respect to various macroscopic compressive strains of **a** copper (3 1 1) and **b** niobium (3 1 0) (With courtesy of Ref. [86] from author's copyright)

[113]. Skomski et al. established a model on the relationship between resistivity and the layer periodicity of the Co/Pd multilayers where resistivity is increasing with reducing layer periodicity til it reaches a saturation plateau at very low periodicities [113]. Griffin et al. looked into resistivity measurements of magnetron sputtered Cu/Nb thin film multilayers with thicknesses ranging from 10 to 1000 nm using standard four-probe measurement system down to 4K and found electrical resistivity increasing with layer periodicity for the same total thickness due to greater number of interfaces [115, 116]. It emanates from the coupling between enhanced elasticity and interface-defect interaction creating atomic disorder that influences the effective electrical resistivity of nanolayers. This opens up new possibilities for the development of novel functionalities, such as strain sensing.

As defect density in the interfaces of Cu/Nb nanolayers increases linearly with increasing strains and defects are sites where electrons would be scattered, it can thus be expected that Cu/Nb nanolayers would exhibit linearity of resistivity also with strain. Such behavior would prove useful for strain sensing/detection/monitoring capability. It is thus our aim in the present study to investigate if the Cu/Nb nanolayers do exhibit such linearity of resistivity with strain during deformation. A high-performance coating that has structural as well as functional capabilities, such as online strain monitoring capability, is essential for many important and strategic applications, from use in remote oil and gas pipes, to use as both radiation damage protection as well as sensing materials simultaneously for satellites and space technologies [52, 118–122].

4.3.2 Experimental Methodologies—Probing Mechano-Resistivity at the Nanoscales

Materials and Micropillar Fabrication

The Cu/Nb metallic multilayered nanocomposites were grown by the "bottom-up" fabrication process of Physical Vapor Deposition (PVD). The PVD method was done by electron beam evaporation of alternatively depositing layers of Nb and Cu on single-crystal a-cut sapphire ($11\bar{2}0$) substrates (MTI Corporation, Richmond, CA) at an elevated temperature of 200 °C [83]. Deposition was performed under a vacuum environment of at least 6.7×10^{-6} Pa at rates of 0.5 nm/s for Nb and 5 nm/s for Cu. Total thickness was about 3 μm and individual layer thickness of both Cu and Nb was 20 nm. Special attention was made to ensure good quality and thermally stable (i.e. continuous layers of Cu/Nb nanolayers) single-crystal multilayers. The growth of these nanoscale Cu/Nb single-crystal multilayers has been described thoroughly elsewhere [86, 123, 124]. With this technique, the Cu/Nb nanoscale multilayers have a strongly textured {111} FCC Cu on {110} BCC Nb out-of-plane Kurdjumov–Sachs orientation relationship.

FEI Helios NanoFab Dual Beam Focus Ion Beam (FIB) at the Microelectronics Reliability & Characterization Laboratory at Nanyang Technological University (NTU), Singapore was utilized to fabricate the micropillars. Special care was taken to minimize the Ga+ ion beam exposure of the top surface of pillar structure to minimize the defect introduction from the FIB process. A pedestal of 30 μm in outer diameter and 10 μm in inner diameter was first made with coarse current and the pedestal was reduced down to the final pillar dimensions using finer ion beam currents. Dimensions of the micropillars and tapers were measured by Scanning Electron Microscope (SEM). The machined pillars have a nominal diameter (d_o) of about 1 μm and a pillar height of about 3 μm. The nanolayer direction is perpendicular to the cylinder axis. The 3:1 length-to-diameter ratio was used to avoid possible buckling of the sample during compression. The taper angles of the pillars were measured to be about 2°. To ensure a conductive path from the pillar to the sample stage, a thin strip of platinum was deposited from the base of the pillar to edge of the crater consisting of the Cu/Nb multilayer film as shown in Fig. 4.14. Further, silver paste was applied from the film edge to the stage holder base across the non-conductive sapphire base.

Nanomechano-electrical Experimentation

Uniaxial compressions for micropillars using flat diamond tips have been done, both in-situ and ex-situ to understand plasticity induced deformation mechanisms of materials at the nanoscale [52, 119–121]. Ex-situ electrical measurements of conductive materials have been done mainly by probe or nanoindentation methods [122–124]. To determine the relationship of linearity of resistivity with compressive strain, a tool that had the capability of performing both mechanical and electrical testing is desired.

A successive ex-situ uniaxial compression followed by electrical current measurement technique was adapted in the present study. The Hysitron PI-85 nanoindenter

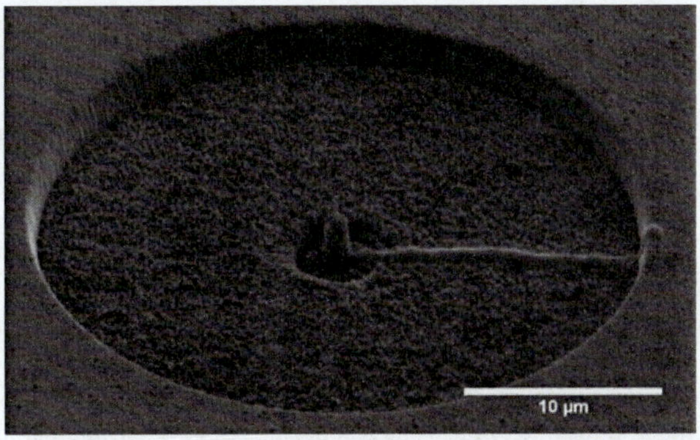

Fig. 4.14 SEM image of one of the fabricated micropillars (Pillar 1) of diameter 1016.518 nm and height (after tilt correction) of 3375 nm giving an approximate height to diameter ratio of 3.3:1. A thin platinum strip was added from the pillar to the edge of the crater to ensure a conductive path from the pillar to the sample stage (With courtesy of Ref. [86] from author's copyright)

(Minneapolis, MN) with a Performech controller inside JEOL 7600F Scanning Electron Microscope (SEM) with Electrical Contact Resistance (ECR) capability, as discussed in the previous chapter, was used to perform the uniaxial compression testing and electrical measurement of these micropillars was conducted at the Magnetic and Mechanical Materials Characterization Design Laboratory at Singapore University of Technology and Design (SUTD). The maximum load achievable for this instrument was 10,000 μN.

Two types of nanoindenter tips were used for this experiment. A flat-ended cylindrical non-conductive diamond tip with an inscribed circle of 5 μm in diameter was used for uniaxial compression of the micropillars. A conductive ceramic cube corner tip of nominal resistivity of 5 μΩ-cm was used to measure the electrical current before and after each compression. This type of testing has been done for in-situ electrical measurement for nanoindentation of silicon by Ruffell et al. using conductive boron-doped diamond Berkovich tip [125]. In the present study, however, an ex-situ electrical characterization of micropillars during deformation was conducted (not an in-situ experiment) due to our current lack of conductive tip with an inscribed circle typically used for both uniaxial compression as well as electrical resistance measurement simultaneously.

The initial resistance of the film and the top of the pillar was measured with the conductive cube corner tip during a voltage sweep and measures the current at each voltage step. This was done by applying a voltage to the sample stage to which samples were electrically connected and current flows through the sample and tip with the latter held at ground potential. An external Keithley 2602B Sourcemeter provided the voltage, measured the circuit current (sensitivity ~10pA) and transferred this to a personal computer with Hysitron data logging software. The tip indented the top surface of the pillar at a nominal constant load rate of $\dot{l} = 10$ μN/s, holds for 10 s

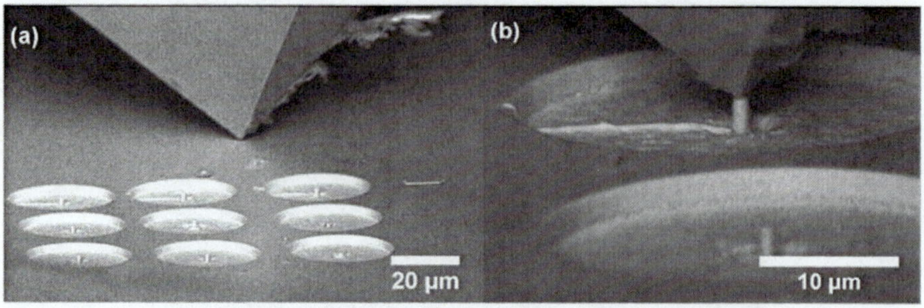

Fig. 4.15 SEM micrographs of the initial resistance measurements with the Cube corner tip—**a** at the surface of the film near the micropillars and **b** at the top of one of the micropillar (With courtesy of Ref. [86] from author's copyright)

allowing the sweeping of the voltage to measure the current, allowing current–voltage (I–V) characteristics as shown in Fig. 4.15.

During measurement, the current flows through the following series resistances of the tip, tip/sample contact, sample holder, Cu/Nb nanolayer and the sapphire. Contributions by the tip and sample holder were found to be much lower compared to the actual resistance of material and are taken to be negligible. Care was taken to compress less than 0.5% of the pillar during each current measurement and as centered as possible.

The indenter tip was then switched to flat-ended cylindrical non-conductive diamond tip. The PI-85 nanoindenter was programmed to perform a nominally load-controlled test. The experiment started by collecting the load–displacement data in the continuous stiffness measurement mode of the instrument. All the micropillars were compressed at a nominal constant load rate of $\dot{l} = 10\ \mu\text{N/s}$ which provided a rough approximate constant displacement rate of $\dot{h} = 2\,\text{nm/s}$. This translates to nominal engineering strain rate of $\dot{\varepsilon} = \frac{\dot{h}}{l_o} = 0.0006\,\text{s}^{-1}$ for a micropillar with a length of 3 μm as shown in Fig. 4.16.

Fig. 4.16 SEM image of Cu/Nb 20 nm micropillar being compressed with the flat cylindrical tip (With courtesy of Ref. [86] from author's copyright)

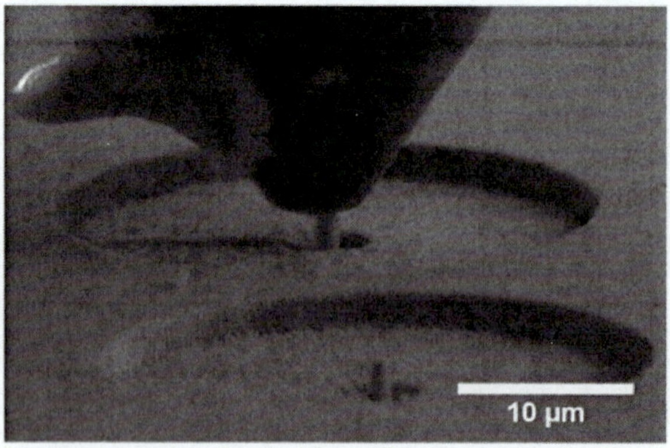

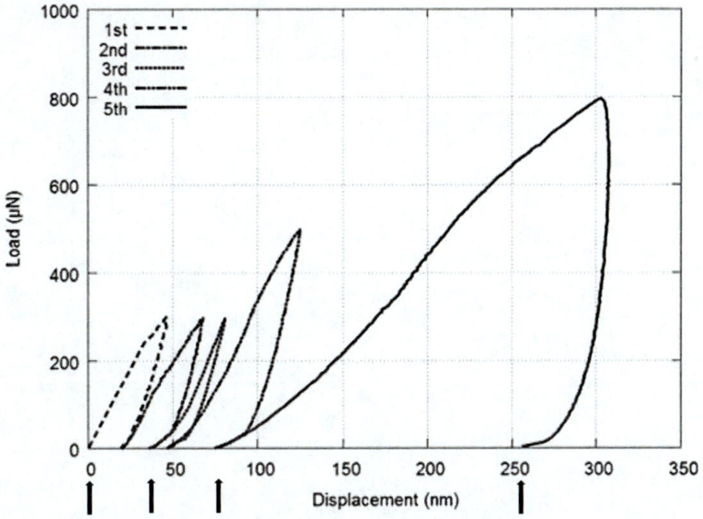

Fig. 4.17 Load–displacement curves of 5 successive compressions of 20 nm Cu/ 20 nm Nb layer thickness micropillar (Pillar 1). The arrows along the Displacement axis represent the current measurements were made during the compression of Pillar 1 (With courtesy of Ref. [86] from author's copyright)

Load–displacement data in the continuous stiffness measurement mode of the instrument was collected. At each successive compression strains, the tip was switched to conductive Cube corner tip to measure the current through a voltage sweep from −0.1 to 0.1 V. This enabled us to measure the resistance of the micropillar and how it evolved during micropillar compression/deformation. The ex-situ micro-compression of the micropillar followed by current measurement process was repeated five times successively with the respective current measurements indicated by the green arrows as shown in Fig. 4.17.

4.3.3 Extreme Strain-Resistivity: Extending Linear Strain Sensitivity for Strain Monitoring

New Insights on Extreme Strain-Resistivity in Advanced Metallic Nanolayers

Two Cu/Nb nanolayered micropillar samples (each deformed with a different strain schedule up to 10%) of similar dimensions were compressed using a flat punch followed by ex-situ current measurement. The measured load–displacement and resistance measurements show variation within the nominal experimental scatter. The results from one of these pillars (Pillar 1) is discussed throughout this section based on this confirmation and the SEM images before and after compression are shown in Fig. 4.18.

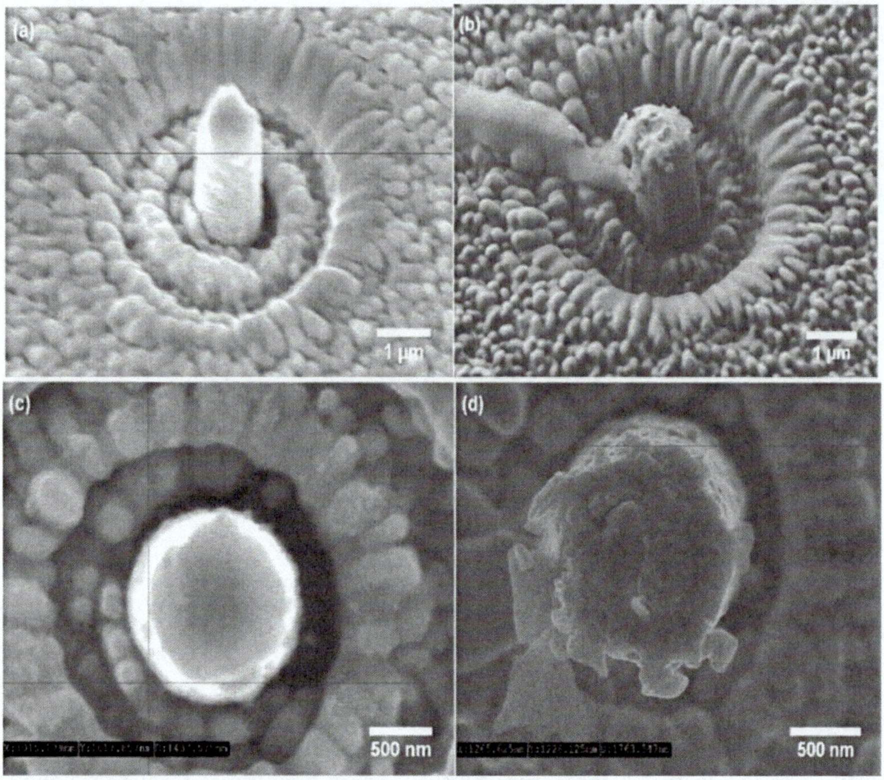

Fig. 4.18 SEM micrographs of 20 nm Cu/ 20 nm Nb layer thickness micropillar (Pillar 1) **a** before the 1st compression and **b** after the final 5th compression by 30° tilt of the sample, and their corresponding topviews **c** before the 1st compression and **d** after the final 5th compression (With courtesy of Ref. [86] from author's copyright)

Figures 4.17 and 4.18 show the load–displacement curves of the successive compression of the micropillar and the corresponding SEM images before the start of the experiment and at the end of the final microcompression of the micropillar, respectively. The micrographs of the micropillar before and after the successive compressive strains as shown in Fig. 4.18 indicate largely uniform/homogeneous and uniaxial deformation of the micropillar. The load–displacement data obtained during the compression were then converted to uniaxial true stresses and strains using the assumption that the plastic volume is conserved throughout the deformation using a constant volume and homogeneous deformation model [97, 126, 127]. The true stress–strain mechanical response of the pillar sample from the compression tests is shown in Fig. 4.19. In this case, the volume measured from initial cross-sectional area A_o (based on d_o) and length L_o of the undeformed pillar is equivalent to the final volume of the deformed sample, therefore $A_oL_o = A_pL_p$ where A_p and L_p denote final pillar area and length, respectively. The subtraction of Sneddon compliance was

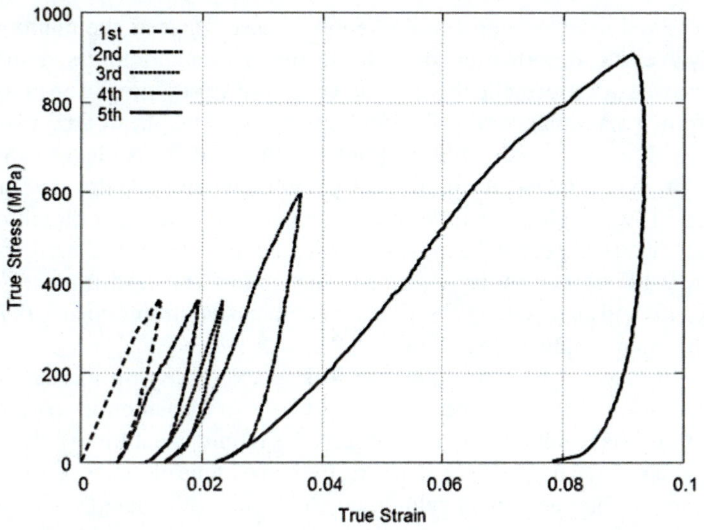

Fig. 4.19 The true stress versus true strain plot of the micropillar synthesized from the Cu/Nb multilayers with 20 nm Cu/20 nm Nb layer thicknesses (With courtesy of Ref. [86] from author's copyright)

considered to only calculate the displacement that corresponds to the pure deformation of the micropillar only. The modulus and Poisson's ratio of Cu is 120GPa and 0.34 and that of Nb is 105GPa and 0.4.

The first stress–strain curve in Fig. 4.19 (the red curve) indicates initial contact issues with the rough surface of the pillar was influencing the load–displacement curve and thus the true stress versus true strain relationship, especially on the loading part of the curve. But from the subsequent stress–strain curves, it shows the unloading curves of all the stress–strain curves indicated largely self-consistent slopes, hence Young's modulus of the Cu/Nb nanolayered pillar sample (of about 66–116 GPa which is on similar range as reported in previous studies [86, 103, 128].

It is also seen from the stress–strain response (Fig. 4.19 and Table 4.1) that pillar subjected to the total true strain ε^{T} ($\varepsilon^{elastic} + \varepsilon^{plastic}$) of 1.298, 1.943, 2.365, 3.613%, the single-crystal Cu/Nb nanolayers deform mostly linearly throughout the loading

Table 4.1 Calculated values of the true stress versus strain values of compressed micropillar (With courtesy of Ref. [86] from author's copyright)

Compression	1st	2nd	3rd	4th	5th
Total true strain (%)	1.298	1.943	2.365	3.613	9.147
Plastic strain (%)	0.777	0.888	0.749	1.514	6.604
Elastic strain (%)	0.521	1.055	1.616	2.099	2.543
Max stress (MPa)	366.287	363.171	361.699	598.637	910.005
Composite Young's modulus (GPa)	63.980	69.385	69.826	78.241	116.977

regime and show little to almost no hysteresis. These suggests the micropillar was still largely in elastic deformation mode during the first four compressive strains, and only went to plastic deformation in the fifth compressive straining. The unloading of the pillar from a maximum strain of 1.298% results in a true plastic strain of 0.777%. Similarly, analyzing the unloading response of the Cu/Nb nanolayer subjected to maximum strain of 3.683%, it can be clearly observed that it results in a true plastic strain of ~1.514%. Slight increase of maximum true stress at higher total true strain from 1 to 3.6% is expected in mostly elastic regime. When the total true strain increases to 9.147%, the nanolayers seem to transition from elastic to plastic regime. No shear bands and bulges were observed for compressive strains up to 9.147% based on the SEM micrographs in Fig. 4.18.

At each instance of Electrical Contact Resistance (ECR) tip measurement, the voltage was swept across between −0.1 and 0.1 V to measure the desired current reading in μA. The resultant I-V curve observed is ohmic in nature. Applying Ohm's law, ECR is calculated by ECR = V/I, taking the slope of the plotted I versus V. Taking the Cu/Nb micropillar as a small cylindrical conductor with a length of l and a cross section of A, resistivity ρ of the Cu/Nb nanolayer is calculated by the traditional definition of $\rho = \mathrm{ECR} \cdot \frac{A}{l}$. Attempts were made to reduce the resistivity error by the accurate measurement of current with the ECR tip and of size of the micropillars before and after each measurement. For this study, resistivity was only measured at room temperature of 298 K due to the technique involved.

Using measured current data obtained from the cube corner tip experiments and the dimensions of the micropillars, resistivity was calculated and plotted in Fig. 4.20 in relation to total percentage strain. The described micropillar (labeled Pillar 1)

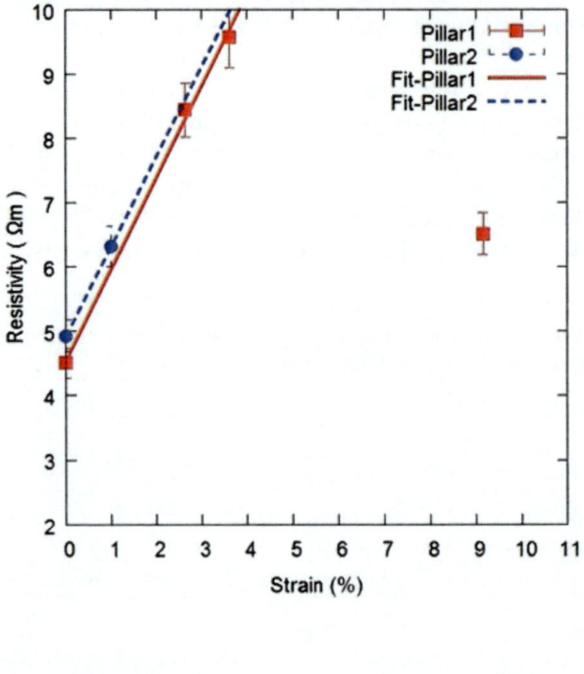

Fig. 4.20 Resistivity versus percentage strain of the compressed 20 nm Cu/20 nm Nb layer thickness micropillars (Pillar 1 and Pillar 2) (With courtesy of Ref. [86] from author's copyright)

showed a linear trend of resistivity up to ~3.6% strain followed by a sudden drop at 9.1% strain. Another set of measurement results from Pillar 2 is included to show confirmation of this trend.

The Cu/Nb multilayer micropillar system can be modeled as a stack individual alternating resistors of copper and niobium arranged in series to a voltage supply. The typical bulk resistivity values of copper and niobium are 1.67×10^{-6} Ωcm and 15.2×10^{-8} Ωcm at room temperature, respectively [129]. For a 3 μm film, there will be ~150 layers of copper and niobium in 50:50 portion. The bulk resistivity of this system is found to be about 1.37 μΩm. However, the initial measured resistivity of micropillar is observed to be much higher at 4.51 μΩm. This is due to the size effect of the micropillar with nanoscale 20 nm multilayers resulting increased electron scattering on the external and internal interfaces of copper and niobium conductors even at room temperature [130].

Extending Limits of Linear Strain Sensitivity for Advanced Strain Monitoring Applications

Figure 4.20 strongly suggests linear resistivity with increasing strains, up to total accumulative compressive strain of 3.613%. This is consistent with previous measurement of defect density using synchrotron X-ray microdiffraction [86] which also reported linear increase of defect density with increasing strains up to 3.3%. We propose here that the resistivity increase with increasing strains up to about 3.3–3.6% is related to increase of defect density (mostly dislocations) in the many interfaces between the Cu and Nb nanolayers in the micropillar sample as shown in Fig. 4.20. This is exactly as has been proposed in the literature [88, 89] about deformation mechanisms in Cu/Nb nanolayers especially in the range of individual layers of tens of nanometers. In this range, compression of the sample leads to emission of dislocations from the bulk of the nanolayers, but as there are high probabilities that there are interfaces just a few atomic sites away (due to the fact that the individual layers are only a few of tens of nanometers), dislocations are mostly just going to go to the interfaces and thus mostly just accumulate in the interfaces. The more dislocations accumulate in the interfaces, the higher defect density that the micropillar sample contains, the higher electrons will be scattered during an electrical measurement, leading to higher resistivity of the material.

It is to be noted here that up until this point of strains about 3.3–3.6%, the deformation in the micropillar sample is still mostly elastic (the bulk of nanolayers of Cu and Nb are both mostly elastically deformed, only with a few local stress concentrations which might lead to dislocation emission which would almost right away being absorbed by the interfaces, thus leaving behind the bulk of the nanolayers still mostly in elastic regime. This is consistent with the stress–strain curves in Fig. 4.19, which show little to zero plastic hysteresis. These nanoscale effects in the multilayered geometry of the sample obviously thus lead to an extended elastic regime. This is again consistent with Budiman et al. [86]. This linearity in the increase of resistivity with increasing strains up to 3.6% thus confirms our initial hypothesis that this class of novel nanolayered material could have a strain sensing/detection/monitoring capability, especially with such an extended elastic regime as shown in Fig. 4.20. For instance, in the application as high-performance coating materials with online strain monitoring capability for say remote gas pipes, increased level of measured resistivity detected by the coating material of the pipes suggests automatically increased level of strains possibly due to increased gas pressure from within the pipes. This could then indicate some issues in the pipes that need repair or fixing immediately before possibilities of major outbreak of the system. Or at the very least, such novel function of the coating material would then allow online strain or health monitoring of the pressured gas pipes and the overall grid transportation system.

Beyond 3.6% of strains, we observed significant decrease in resistivity, however we currently only have 1 data point in this regime of deformation. Based on Budiman et al. [86], the defect density seems to reach plateau beyond 3.3% (after linear increase up to that strain level). It was the current understanding that such defect density could be due to opposite dislocations neutralize each other in the interfaces (i.e. crystal recovery) and thus it leads to maturation of the defect density increase in the interfaces, and thus in the overall micropillar sample. As some of these dislocations could neutralize each other and thus recovered the crystal leading to smoother interfaces in the nanolayered pillar sample, it could then be expected that the electron scattering would be less and thus the resistivity of the pillar sample might become lower, as observed in the Fig. 4.20. This conjecture is however rather preliminary at this stage and certainly needs further investigations.

It has thus become clear here that such a novel function of the nanolayered materials stemmed from the atomic design of their interfaces. In the case of our nanoscale Cu/Nb nanolayered composite system here, it provides us with weak, incoherent interfaces between the Cu and Nb nanolayers which then could easily absorb and accumulate dislocations, such that the defect density increases largely linearly with increasing strains up to strains of 3.3% [85], which also, as we proposed in the present study, leads to resistivity increases linearly with increasing strain up to 3.6% as shown in Fig. 4.21. This thus opens up the possibilities for designing nanoscale metallic composite materials with modulated strain sensing/detection/monitoring sensitivities, depending on the needs of the applications, based on atomic engineering of the interfaces in the nanolayered materials.

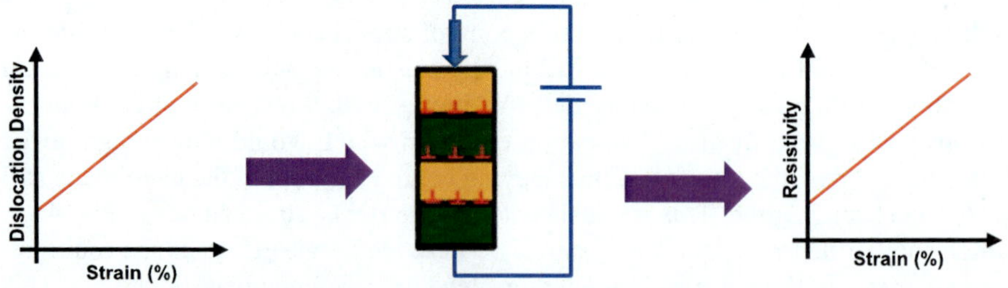

Fig. 4.21 With the understanding of the relationship between Dislocation Density versus Strain, a resistivity versus strain relationship is established through the application of voltage to the micropillar during microcompression (With courtesy of Ref. [86] from author's copyright)

In this study, micropillars of nanoscale Cu/Nb multilayers were electrically characterized during ex-situ successive uniaxial compressions to a total strain ramp-up of (revised value of 11.7%). After each compressive straining cycle, the samples were subjected to resistivity measurements. The results showed resistivity increases linearly with increasing strains up to 3.6%, which confirms the early promise of the materials for novel functionalities for online strain sensing/detection/monitoring capability. This feature, we propose, is related to the specific atomic design of the interfaces in this particular material system. This study thus suggests possibilities and perhaps platforms for more systematic ways to a materials by design approach by a directed discovery methodology based on atomic engineering of interface. Through this platforms to achieve designer interface, we propose this as a new, and perhaps more systematic, way to accelerate materials design and for achieving the objectives of the materials genome initiative.

4.4 Stress-Induced Voiding in Advanced Submicron Interconnects

In this section, we present evidence of a shift in the Stress Migration (SM) peak profile temperature for smaller interconnect linewidths typically associated with the nanoscale technology nodes. With other parameters (fabrication, materials, line thickness and via diameter being kept nominal among all these samples), this clear shift toward the lower temperatures for smaller linewidths appear to indicate a size effect in the Stress Migration in advanced Cu interconnect scheme. The synchrotron X-ray microdiffraction experiment is used to show that plasticity is involved in the stress relaxation process at about 200 °C, but not at higher temperature nor at room temperature. Such plasticity-assisted strain relaxation in interconnects especially at lower temperature range could explain the critical temperature shift observed in the present study, in addition to the typical diffusion-assisted mechanism. Further, the synchrotron X-ray microdiffraction experiments also suggests indications of plasticity-assisted voiding. Numerical finite element analyses were also conducted in conjunction with the experimental study, to provide greater insight. The modeling result demonstrates the importance of creep plasticity in causing thermal stress relaxation in Cu interconnects.

4.4.1 Background and Overview

Advanced interconnects are built in increasingly very small structures, represented by the nanoscale technology nodes. One of the key challenges for advanced Integrated Circuit (IC) generations in these nanoscale technology nodes is the Stress-Induced Voiding (SIV) or Stress Migration (SM) in interconnects [131, 132]. The margin for reliability of copper-based interconnect systems is ever shrinking. Manufacturing is impacted by material selection, design rules and test structure design. The key for designing robust components for advanced technology nodes is indeed in the understanding of the factors that trigger the onset of SIV.

In this study the typical SM daisy chain structures (M1/M2) which is a variation of a BEOL process for a 32 nm CMOS technology fabricated by Vitesse Semiconductor Corp was used as the interconnect test structure. Samples of 4 different linewidths were used (4.5, 2.0, 0.9, 0.4 μm) with all having the same via diameter, overlay as well line length and thickness. Typical SM reliability testing (thermal loading) were conducted at six different temperatures (110, 135, 170, 195, 225 and 255 °C) with electrical resistance measurements. The resistance increase after about 400 h of SM baking based on 10 DUT's (Device Under Testing) for each test temperature for each linewidth is shown in the Fig. 4.22. To study plasticity and mechanical stresses in the device and to understand how the stress evolution during the device's real and accelerated service conditions in-situ synchrotron-based white-beam (Laue) X-ray microdiffraction technique (ALS, Berkeley Lab) was used. The synchrotron technique of scanning white-beam X-ray microdiffraction has been described in a complete manner in a previous publication [133].

The synchrotron-based X-ray microdiffraction technique [8, 133–135] is a non-destructive micron resolution method to measure crystal orientations and stresses in the samples. As the high-brilliance synchrotron-sourced X-rays can penetrate and easily distinguish signals of the metallic structures from that of other materials in their surroundings, strain measurements can be done in the samples easily [45]. This allows strain measurements that are as close as possible to conditions in the real operations of the device. The focused X-ray beam allows stress measurements in the submicron regime. Further, the white X-ray beam provides a white range of wavelengths that allows deviatoric stress component measurements in addition to the traditional X-ray diffraction technique to measure hydrostatic components of the stress tensors. This synchrotron-based X-ray microdiffraction technique has also been used to investigate microstructure-controlled or stress-controlled reliability as well as performance issues in advanced microelectronics [11, 12, 134, 136], nanotechnologies [9, 10, 38, 52, 137, 138] along with next-generation energy systems [43, 45, 99, 117, 118].

4.4.2 *Critical Temperature Shift for Stress-Induced Voiding in Advanced Cu Interconnects for 32 nm and Beyond*

Stress migration temperature profile has been widely recognized and first reported by McPherson and Dunn [139].

$$\frac{\Delta R}{R_o} = C(T_o - T)^N exp\left(\frac{E_A}{KT}\right) \tag{4.8}$$

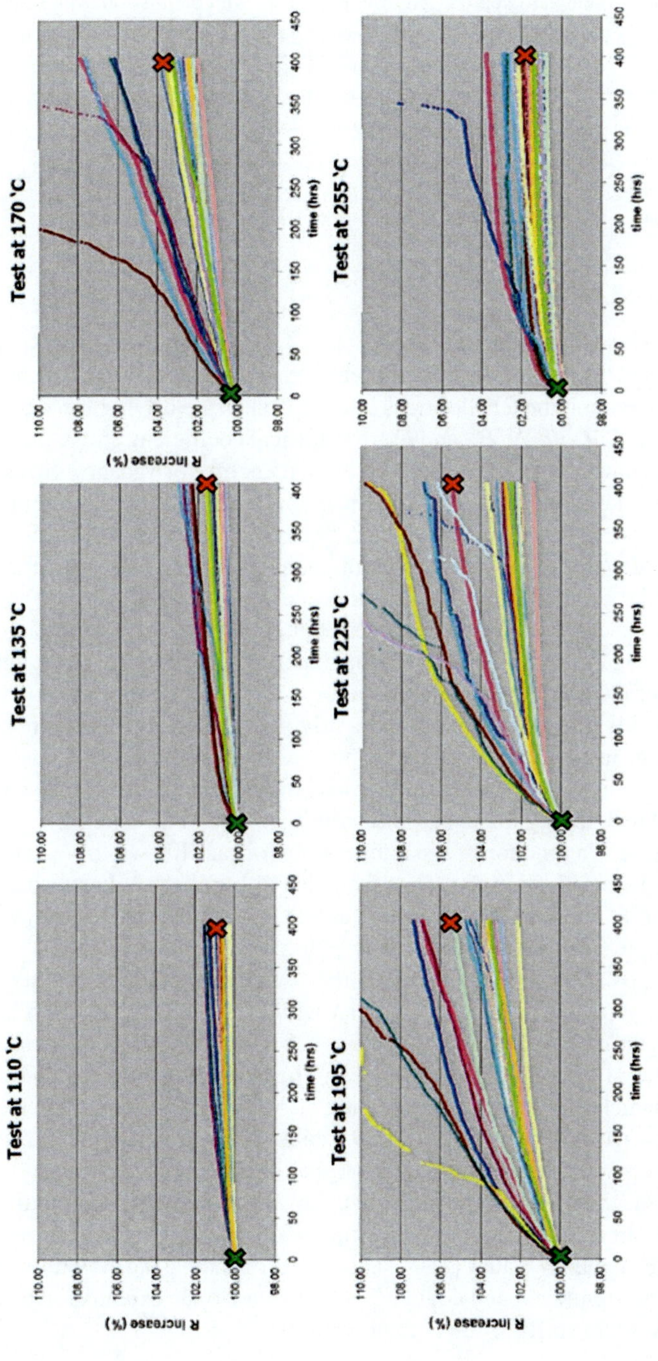

Fig. 4.22 Typical SM test measurements: ΔR/R versus time (Reproduced from Ref. [136] with permission from Elsevier)

Table 4.2 McPherson-Dunn parameters experimentally obtained by fitting the electrical resistance data for each linewidth samples (Reproduced from Ref. [136] with permission from Elsevier)

	0.4 μm linewidth	0.9 μm linewidth	2.0 μm linewidth	4.5 lμm linewidth
T_0 (°C)	350	350	350	350
E_a (eV)	0.71	0.77	0.78	0.79
N	5.89	3.96	4.11	2.84
C	8.75E−08	5.92E−07	6.06E−07	6.78E−09
R^2 stdev	0.04	0.11	0.08	0.09
T_{peak} (°C)	153	173	176	191

The McPherson-Dunn equation above (Eq. 4.8) describes how SM resistance increase ($\boldsymbol{\Delta R}$ and $\boldsymbol{R_o}$ are resistance increase and initial resistance respectively) depends on two factors: the tensile stress in the metal as the driving force of voiding and the diffusion ($\boldsymbol{C}, \boldsymbol{T_o}, \boldsymbol{T}, \boldsymbol{N}, \boldsymbol{E_A}$ and $\boldsymbol{K}$ are empirical coefficient, stress-free temperature, test temperature, creep exponent, activation energy for vacancy diffusion and universal gas constant, respectively). The stress component decreases as temperature approaches stress-free temperature (which is typically about 300–400 °C for Cu thin films, for example) while the diffusion component always increases as temperature increases thus the profile with a peak in the intermediate temperature. Voiding is minimum at low temperatures because even though stress is high, diffusion is low, meanwhile voiding is also minimum at high temperatures because even though diffusion is high, stress which is the driving force for voiding is low. At some intermediate temperatures, voiding is worst because both driving force and diffusion are enabled.

There appears in the present study to be a clear shift in the SM peak toward lower temperature as we go toward smaller and smaller linewidths (representing the more advanced technology nodes) as shown in Fig. 4.22. The experimental data of resistance increase as a function of test temperature for each linewidth could be fit into the McPherson-Dunn equation above and from that fitting all the unknown parameters ($\boldsymbol{C}$, $\boldsymbol{T_0}$, $\boldsymbol{N}$ and $\boldsymbol{E_A}$) could be determined. Table 4.2 provides all the parameters for each of the sample linewidths with the best fitting.

As the SM peak temperature shifts to the lower temperatures for the smaller linewidths, it is apparent from Table 4.1 that while $\boldsymbol{E_A}$ is relatively constant (from the best fitting), the creep exponent, $\boldsymbol{N}$, increases as linewidth gets smaller and smaller (representing the advanced technology nodes). The largest linewidth of 4.5 μm has $\boldsymbol{N}$ equals to 2.84 which is a typical value for bulk ductile metals, whereas the narrowest linewidth of 0.4 μm has $\boldsymbol{N}$ equals to 5.89 which is from metals creep references usually represents work-hardened metals. Higher creep exponent, $\boldsymbol{N}$, usually means there are more mechanisms for the metal to creep, such as dislocation motion, dislocation multiplication, etc. Insertion of any kind of obstruction to dislocation motions usually leads to higher $\boldsymbol{N}$ value [140]. This trend of increasing $\boldsymbol{N}$ with narrowing linewidth may be significant and could potentially lead to the explanation of why the SM peak temperature shifts to lower temperatures.

As fabrication process is nominal, this shift toward lower temperature could thus be a size effect in which dislocation motion and processes may play a more important role as compared to in the more typical Stress Migration events. It appears that while SM voiding in the larger linewidths is mostly governed by the stress driving force and diffusion as in the McPherson-Dunn equation, the SM voiding in the narrow linewidths could be controlled in addition by plasticity (dislocation motion, multiplication, etc.). This could make sense as in the lower temperatures (such as 153 °C for the 0.4 μm linewidth samples) diffusion is likely to play smaller role, whereas plasticity could still be activated as it is mostly driven by the high stress in the metals. More importantly, the stress state of narrow lines is more likely to be triaxial rather than biaxial in the wider lines (i.e. closer to stress state of thin films) thus providing higher shear stresses which are driving force for dislocation motion and activities. This lower temperature mode is more dangerous as it is closer to real operational condition of microelectronics devices. This is especially a going concern for the upcoming advanced technology nodes (32 nm and beyond) with smaller and smaller linewidths and via dimensions.

Probing Local Plasticity

We have some preliminary evidence that plasticity could play a more prominent role in the lower temperature range. Our synchrotron results show plasticity in advanced Cu interconnect lines undergoing thermal stressing, not at the highest of stress temperatures, but instead at a rather low temperature of 200 °C (Fig. 4.23).

The plasticity here manifests in the form of asymmetric broadening of Cu Laue peaks observed in both the 2 μm- (Fig. 4.24) and 4.5 μm-wide interconnect lines. In-situ Laue synchrotron X-ray microdiffraction study could not be performed here for lines with 0.9 and 0.4 μm linewidths due to lack of resolution (Focused X-ray beam

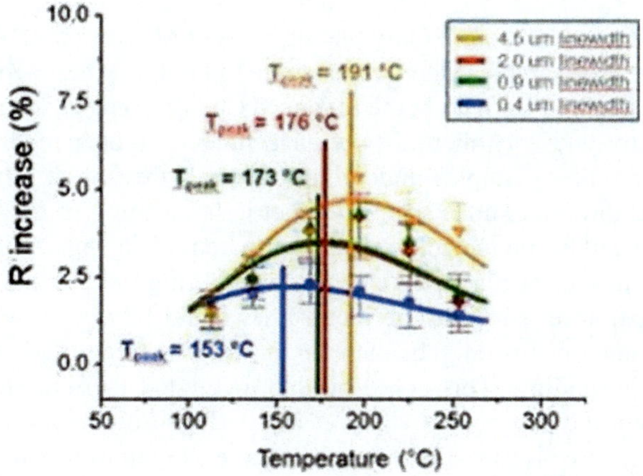

Fig. 4.23 Stress Migration temperature profiles for Cu samples with different linewidths. The data point represents the median value while the error bar represents the range of all data (10 DUT's per test temperature per linewidth) (Reproduced from Ref. [136] with permission from Elsevier)

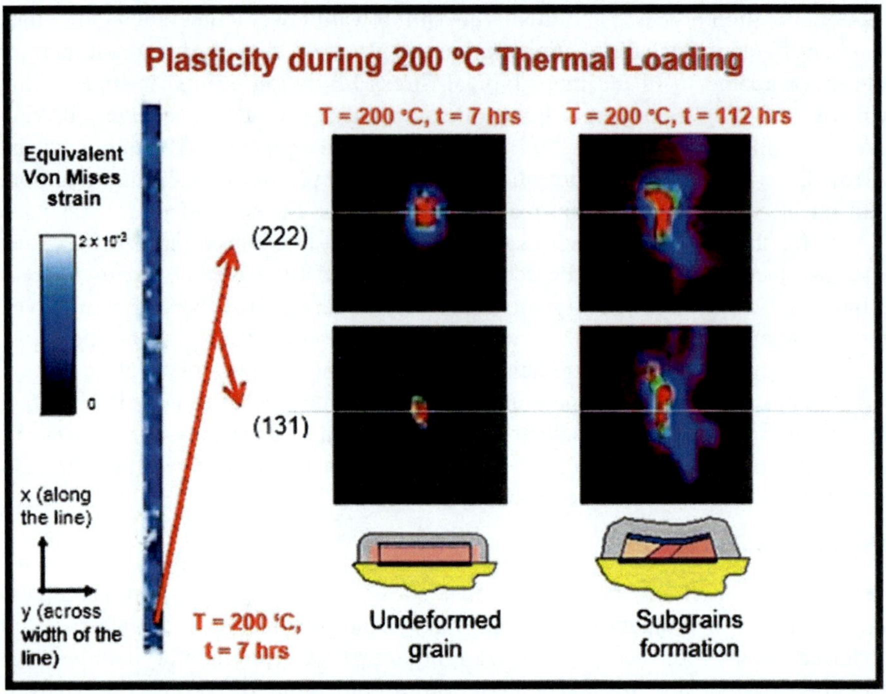

Fig. 4.24 Plasticity during 200 °C thermal loading in 2 μm-wide interconnect line. Streaked (asymmetrically broadened) Cu Laue peaks were observed in one end of the M2 line segment (length = 20 μm) just above the via as shown above (red arrow) after about 100 h of thermal loading. Similar observation was found with the 4.5 μm-wide sample (Reproduced from Ref. [136] with permission from Elsevier)

size is about 0.8 μm × 0.8 μm). Only one segment of M2 lines each from the 2 μm and the 4.5 μm-wide samples were scanned due to limited experimental time in the ALS Beamline 12.3.2. Plasticity in the advanced Cu interconnect and its observation using the synchrotron X-ray microdiffraction technique has been reported earlier [8].

Similar interconnect samples undergoing thermal stressing at a higher temperature of 350 °C did not exhibit such asymmetric broadening of Cu Laue peaks in the post-mortem study conducted by synchrotron Laue X-ray microdiffraction, indicating that no significant plasticity was involved during the thermal treatment. At such high temperature, stress in the metal lines could be mostly relaxed through voiding by vacancy diffusion. The samples in the present study however did not show observable voiding. This could be due to limited thermal stressing/baking time in the beamline experiment (only ~100 h). These observations thus suggest that plasticity was involved in the thermal stress relaxation during low-temperature loading (in addition to vacancy diffusion). We also have evidence (Fig. 4.25) that the strain mapping of the interconnect line after some period of thermal loading does show a general stress level that is substantially lower than that of the beginning

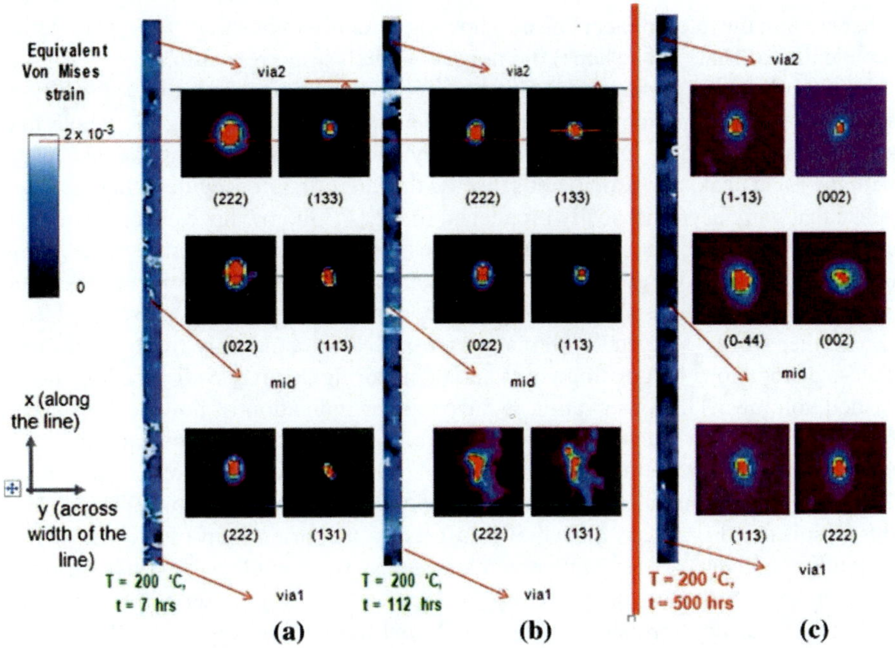

Fig. 4.25 Complete evolution of a Cu interconnect line (2 μm width, 20 μm length), which is the same interconnect line in Fig. 4.3, in terms of von Mises strains and Cu Laue reflection peak shapes during a 500-h thermal loading/baking experiment; **a** at time equals to 7 h, **b** at time equals to 112 h and finally **c** at time equals to 500 h of the thermal loading/backing experiment (Reproduced from Ref. [136] with permission from Elsevier)

of the thermal loading. This plastic relaxation during thermal loading is again only at low-intermediate temperatures (not high temperatures, not room temperature). Figure 4.25 shows basically an evolution of the same interconnect line through a thermal loading/baking experiment (for 500 h) as observed by synchrotron X-ray microdiffraction with Laue (white beam) technique. Each of the figures (a), (b) and (c) shows the von Mises strain map of the interconnect line and some samples of Cu Laue reflection peaks that were observed during the baking from three different locations in the line (via 1, via 2 and the mid of the line). The evolution of what happens inside the interconnect line was recorded at (a) at time equals to 7 h, at (b) at time equals to 112 h and finally at (c) at time equals to 500 h of the thermal loading/backing experiment. The first thing to note is the evidence of plasticity at via 1 as has been discussed in Fig. 4.24 and in the paragraphs above. However in Fig. 4.25b, it also became clear how the plasticity led directly into the strain release right in the area where we took the Laue portraits from (i.e. in via 1). This is evident from the strain mapping. As we can see, the strain mapping of Fig. 4.25b shows much darker color in via 1 area, which indicates much lower von Mises stress. This is a strong indication that the metal stress has been relaxed by plasticity, in addition to by vacancy diffusion, especially at low-intermediate temperatures. It is also evident that other areas in the interconnect line not showing evidence of plasticity (i.e. asymmetric broadening of Laue peak shape) did not show much strain relaxation.

Yet, further and perhaps more interestingly, Fig. 4.25c shows two things: first, the strain map of the line is now in general much darker indicating the whole line has become much more relaxed, and secondly there is yet another significant change with the Laue peaks via 1 area. Let's discuss the latter first. It is evident that the Laue peaks that were asymmetrically broadened in Fig. 4.25b now have become rounded again in shape, indicating the absence of excess geometrically necessary dislocations (GND's). While we don't have yet the microstructural evidence, we believe the Laue peak shape evolution as shown in Fig. 4.25b, c indicates plasticity-assisted voiding. Or in other words, a crystal with many excess GND's could only rid itself of those excess dislocations by creating voids and then storing the excess dislocations in the voided volumes. This observation is thus another indication of how voiding could be both assisted by vacancy diffusion as well as by plasticity, especially at lower and intermediate temperatures. Indeed, this could be the explanation of the shift to the lower temperature as we observed in Fig. 4.23 in this present study. This is also the reason that Fig. 4.25c also shows in general the strain map of the whole line becomes much darker indicating much less stress. This could be the combination of voiding due to diffusion and plasticity. Perhaps just in general after 500 h of baking, significant vacancy diffusion has occurred, and more and more voids are created. Due to the availability of these voided volumes, it becomes easier for crystals with many excess dislocations to rid themselves of the excess dislocations and thus creates maximum strain release.

Numerical Modeling

Three-dimensional finite element modeling was carried out to examine the thermal stress relaxation behavior of Cu during the constant-temperature hold time at different temperatures. The model includes two levels of Cu lines connected by a via, with the metal structure embedded within the SiOx dielectric. All nitride barrier and etch stop layers are also included. The Cu material is taken to be elastic–plastic with strain hardening [141]; its time-dependent response is provided by a creep plasticity constitutive model based on the power-law creep [142]. All other materials are assumed to be elastic. The simulations include a first cooling step from the initial stress-free temperature (taken to be 350 °C), and a heating step to the targeted "test temperature" (or staying at 20 °C). One hundred hours of constant-temperature dwell then ensues. The primary objective is to examine the evolution of stress relaxation and deformation at the different temperatures, to offer insight into the X-ray microdiffraction experimental result.

The time-dependent FEM model shows significant stress relaxation through plastic deformation (creep) during thermal loading at the intermediate temperature of 200 °C. Stress relaxation also occurred during thermal loading at 350 °C but the initial stress values were low to begin with, therefore no significant creep was observed. The case of 20 °C loading did feature high magnitudes of thermal stress in Cu, but the temperature is too low for creep plasticity to take effect. Figure 4.26 shows the equivalent creep strains in Cu at the end of 100 h at (a) 20, (b) 200 and

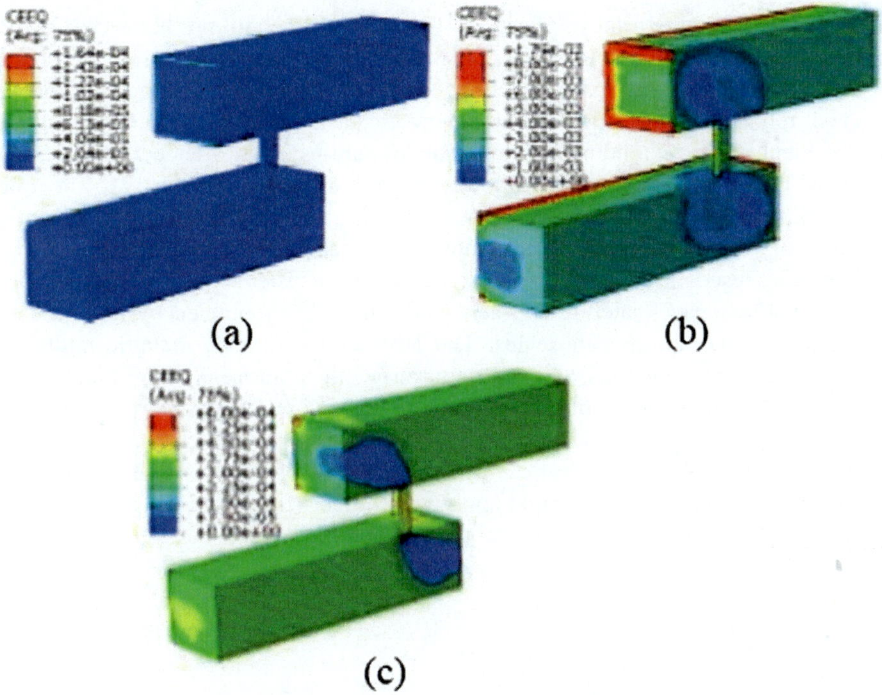

Fig. 4.26 Simulated equivalent creep strain in the Cu lines and via after 100 h at **a** 20 °C, **b** 200 °C and **c** 350 °C (Reproduced from Ref. [136] with permission from Elsevier)

(c) 350 °C. By comparing the strain magnitudes it is evident that the case of 200 °C shows the most significant creep.

Some preliminary evidences of a plasticity-assisted strain relaxation model were obtained from our experimental and modeling work. This can explain why SM worst case temperature is lower for smaller interconnect lines. A clear shift in the SM peak shift toward lower temperatures is observed as we go toward smaller linewidths (representing the more advanced technology nodes of 32 nm and beyond). This could have both technological and industrial importance as we move to even smaller technology nodes. The SM worse temperatures could then approach normal operating temperatures of devices and thus the reliability of the devices would become an important issue.

4.5 Summary

Recent cases in the important areas of microelectronics, micro/nano engineering systems and nanotechnology are highlighted here, in which the principles of extreme mechanics (XM) and/or the probing of the mechanics using cutting-edge materials characterization techniques have enabled emerging design for reliability (DfR).

In particular, the important implications for the high-technology industries in the modern world (semiconductor chips, MEMS, heterogeneous integration, etc.) have been emphasized—in terms of enabling extreme design/engineering, enhancing product reliability assessments where new materials and processed have been unprecedented and extending to novel functionalities of the devices/engineering systems. The first study case allows the advanced semiconductor industry to integrate the third dimension into their design—enhancing performance of microprocessors and memory devices used in our computers and smartphones. Advanced packaging technologies used in MEMS/heterogeneous devices/systems in extreme environments (such as space, satellite and aerospace) have been enabled by low-melting temperature materials in nanoscales. The advanced nanoscale metallic multilayered materials have enabled online structural health monitoring—important capabilities for the oil and gas industries, especially in remote areas or extreme environments (desert, arctic, deep oceans, etc.). Finally, the last study case contributes to better product reliability assessments which the industries (advanced semiconductor chips, IoT/Internet of Things) could apply with higher confidence—e.g. for product warranty, when often a lot of money could be at stake.

References

1. Banerjee, K., Souri, S.J., Kapur, P., Saraswat, K.C.: 3-D ICs: a novel chip design for improving deep-submicrometer interconnect performance and systems-on-chip integration. Proc. IEEE **89**(5), 602–633 (2001)
2. Knickerbocker, J.U., Andry, P.S., Buchwalter, L.P., Deutsch, A., Horton, R.R., Jenkins, K.A., Kwark, Y.H., McVicker, G., Patel, C.S., Polastre, R.J., Schuster, C.D.: Development of next-generation system-on-package (SOP) technology based on silicon carriers with fine-pitch chip interconnection. IBM J. Res. Dev. **49**(4.5), 725–753 (2005)
3. Savastiouk, S.: Semetech 3D ICs Workshop. San Diego, CA (2008)
4. Thompson, S.E., Sun, G., Choi, Y.S., Nishida, T.: Uniaxial-process-induced strained-Si: extending the CMOS roadmap. IEEE Trans. Electron Devices **53**(5), 1010–1020 (2006)
5. Budiman, A.S., Shin, H.A.S., Kim, B.J., Hwang, S.H., Son, H.Y., Suh, M.S., Chung, Q.H., Byun, K.Y., Tamura, N., Kunz, M., Joo, Y.C.: Measurement of stresses in Cu and Si around through-silicon via by synchrotron X-ray microdiffraction for 3-dimensional integrated circuits. Microelectron. Reliab. **52**(3), 530–533 (2012)
6. Okoro, C., Yang, Y., Vandevelde, B., Swinnen, B., Vandepitte, D., Verlinden, B., De Wolf, I.: Extraction of the appropriate material property for realistic modeling of through-silicon-vias using μ-Raman spectroscopy. In: 2008 International Interconnect Technology Conference, pp. 16–18. IEEE, June 2008
7. Liu, X., Chen, Q., Dixit, P., Chatterjee, R., Tummala, R.R., Sitaraman, S.K.: Failure mechanisms and optimum design for electroplated copper through-silicon vias (TSV). In: 2009 59th Electronic Components and Technology Conference, pp. 624–629. IEEE, May 2009
8. Budiman, A.S., Nix, W.D., Tamura, N., Valek, B.C., Gadre, K., Maiz, J., Spolenak, R., Patel, J.R.: Crystal plasticity in Cu damascene interconnect lines undergoing electromigration as revealed by synchrotron x-ray microdiffraction. Appl. Phys. Lett. **88**(23), 233515 (2006)
9. Budiman, A.S., Han, S.M., Greer, J.R., Tamura, N., Patel, J.R., Nix, W.D.: A search for evidence of strain gradient hardening in Au submicron pillars under uniaxial compression using synchrotron X-ray microdiffraction. Acta Mater. **56**(3), 602–608 (2008)

10. Lee, G., Kim, J.Y., Budiman, A.S., Tamura, N., Kunz, M., Chen, K., Burek, M.J., Greer, J.R., Tsui, T.Y.: Fabrication, structure and mechanical properties of indium nanopillars. Acta Mater. **58**(4), 1361–1368 (2010)
11. Budiman, A.S., Besser, P.R., Hau-Riege, C.S., Marathe, A., Joo, Y.C., Tamura, N., Patel, J.R., Nix, W.D.: Electromigration-induced plasticity: texture correlation and implications for reliability assessment. J. Electron. Mater. **38**(3), 379–391 (2009)
12. Budiman, A.S., Hau-Riege, C.S., Baek, W.C., Lor, C., Huang, A., Kim, H.S., Neubauer, G., Pak, J., Besser, P.R., Nix, W.D.: Electromigration-induced plastic deformation in Cu interconnects: effects on current density exponent, n, and implications for EM reliability assessment. J. Electron. Mater. **39**(11), 2483–2488 (2010)
13. Paik, J.M., Park, I.M., Joo, Y.C., Park, K.C.: Linewidth dependence of grain structure and stress in damascene Cu lines. J. Appl. Phys. **99**(2), 024509 (2006)
14. Okoro, C., Vanstreels, K., Labie, R., Lühn, O., Vandevelde, B., Verlinden, B., Vandepitte, D.: Influence of annealing conditions on the mechanical and microstructural behavior of electroplated Cu-TSV. J. Micromech. Microeng. **20**(4), 045032 (2010)
15. Lu, K.H., Zhang, X., Ryu, S.K., Huang, R., Ho, P.S.: Thermal stresses analysis of 3-D interconnect. In: AIP Conference Proceedings, vol. 1143, no. 1, pp. 224–230. American Institute of Physics, June 2009
16. Timoshenko, S.P., Goodier, J.N.: 1961. Theory of Elasticity, 2nd edn. McGraw-Hill Book Co. Inc.
17. Ganesan, S., Pecht, M.G. (eds.): Lead-Free Electronics. Wiley (2006)
18. Humpston, G., Jacobson, D.M.: Indium solders. Adv. Mater. Process. **163**, 45–47 (2005)
19. Yang, M., Li, M., Wang, C.: Interfacial reactions of eutectic Sn3.5Ag and pure tin solders with Cu substrates during liquid-state soldering. Intermetallics **25**, 86–94 (2012)
20. Borgesen, P., Bieler, T., Lehman, L.P., Cotts, E.J.: Pb-free solder: new materials considerations for microelectronics processing. MRS Bull. **32**(4), 360–365 (2007)
21. Shimizu, K., Nakanishi, T., Karasawa, K., Hashimoto, K., Niwa, K.: Solder joint reliability of indium-alloy interconnection. J. Electron. Mater. **24**(1), 39–45 (1995)
22. Kirschman, R.K.: Cold electronics: an overview. Cryogenics **25**(3), 115–122 (1985)
23. Hashimoto, K., Ochiai, M., Karasawa, K., Nakanishi, T.: Flip-chip interconnection technology for packaging of VLSI operated in liquid nitrogen. IEICE Trans. Electron. **74**(8), 2362–2368 (1991)
24. John, J., Zimmermann, L., De Moor, P., Van Hoof, C.: High-density hybrid interconnect methodologies. Nucl. Instrum. Methods Phys. Res. Sect. A: Accel. Spectrom. Detect. Assoc. Equip. **531**(1–2), 202–208 (2004)
25. Kwan, S., Andresen, J., Appel, J.A., Cardoso, G., Christian, D.C., Cihangir, S., Kendziora, C., Marinelli, M., Ruschman, M., Turqueti, M., Wong, M.L.: Study of indium and solder bumps for the BTeV pixel detector. In: 2003 IEEE Nuclear Science Symposium. Conference Record (IEEE Cat. No. 03CH37515), vol. 1, pp. 59–62. IEEE, Oct 2003
26. Kirschman, R.K., Sokolowski, W.M., Kolawa, E.A.: Die attachment for −120 °C to +20 °C thermal cycling of microelectronics for future mars rovers—an overview. J. Electron. Packag. **123**(2), 105–111 (2001)
27. Deppisch, C., Fitzgerald, T., Raman, A., Hua, F., Zhang, C., Liu, P., Miller, M.: The material optimization and reliability characterization of an indium-solder thermal interface material for CPU packaging. JOM **58**(6), 67–74 (2006)
28. Irving, S., Liu, Y.: Free drop test simulation for portable IC package by implicit transient dynamics FEM. In: 2004 Proceedings. 54th Electronic Components and Technology Conference (IEEE Cat. No. 04CH37546), vol. 1, pp. 1062–1066. IEEE, June 2004
29. Lall, P., Panchagade, D.R., Liu, Y., Johnson, R.W., Suhling, J.C.: Models for reliability prediction of fine-pitch BGAs and CSPs in shock and drop-impact. IEEE Trans. Compon. Packag. Technol. **29**(3), 464–474 (2006)
30. Lai, Y.S., Yang, P.F., Yeh, C.L.: Experimental studies of board-level reliability of chip-scale packages subjected to JEDEC drop test condition. Microelectron. Reliab. **46**(2–4), 645–650 (2006)

31. Tee, T.Y., Ng, H.S., Lim, C.T., Pek, E., Zhong, Z.: Impact life prediction modeling of TFBGA packages under board level drop test. Microelectron. Reliab. **44**(7), 1131–1142 (2004)
32. Darveaux, R., Tutlik, I.: Shear deformation of indium solder joints. In: InterSociety Conference, pp. 40–52 (1990)
33. Frenkel, R.E., Sherby, O.D., Dorn, J.E.: Activation energies for creep of cadmium, indium, and tin. Acta Metall. **3**(5), 470–472 (1955)
34. Lucas, B.N., Oliver, W.C.: Indentation power-law creep of high-purity indium. Metall. Mater. Trans. A **30**(3), 601–610 (1999)
35. Weertman, J.: Creep of indium, lead, and some of their alloys with various metals. Trans. Met. Soc. AIME **218** (1960)
36. Lee, G., Kim, J.Y., Burek, M.J., Greer, J.R., Tsui, T.Y.: Plastic deformation of indium nanostructures. Mater. Sci. Eng.: A **528**(19–20), 6112–6120 (2011)
37. Ali, H.P.A., Tamura, N., Budiman, A.S.: Probing plasticity and strain-rate effects of indium submicron pillars using synchrotron Laue X-ray microdiffraction. IEEE Trans. Device Mater. Reliab. **18**(4), 490–497 (2018)
38. Budiman, A.S., Lee, G., Burek, M.J., Jang, D., Han, S.M.J., Tamura, N., Kunz, M., Greer, J.R., Tsui, T.Y.: Plasticity of indium nanostructures as revealed by synchrotron X-ray microdiffraction. Mater. Sci. Eng.: A **538**, 89–97 (2012)
39. Tamura, N.: XMAS: a versatile tool for analyzing synchrotron X-ray microdiffraction data. In: Strain and dislocation gradients from diffraction: Spatially-Resolved Local Structure and Defects, pp. 125–155 (2014)
40. Van der Biest, O.O.: Dislocations and twinning in indium. Scr. Metall. **7**(12), 1279–1286 (1973)
41. Ungár, T., RRévész, Á., Borbély, A.: Dislocations and grain size in electrodeposited nanocrystalline Ni determined by the modified Williamson-Hall and Warren-Averbach procedures. J. Appl. Crystallogr. **31**(4), 554–558 (1998)
42. Ungár, T.: Dislocation model of strain anisotropy. Powder Diffr. **23**(2), 125–132 (2008)
43. Budiman, A.S., Narayanan, K.R., Li, N., Wang, J., Tamura, N., Kunz, M., Misra, A.: Plasticity evolution in nanoscale Cu/Nb single-crystal multilayers as revealed by synchrotron X-ray microdiffraction. Mater. Sci. Eng.: A **635**, 6–12 (2015)
44. Beerling, T.E., Wong, M.G., Ng, W.S., Arifin, J., Sun, J., Suriadi, A.B., Kawamura, N., HP Inc.: Electrical interconnect for an inkjet die. U.S. Patent 6,454,955 (2002)
45. Budiman, A.S., Illya, G., Handara, V., Caldwell, W.A., Bonelli, C., Kunz, M., Tamura, N., Verstraeten, D.: Enabling thin silicon technologies for next generation c-Si solar PV renewable energy systems using synchrotron X-ray microdiffraction as stress and crack mechanism probe. Sol. Energy Mater. Sol. Cells **130**, 303–308 (2014)
46. Tippabhotla, S.K., Radchenko, I., Rengarajan, K.N., Illya, G., Handara, V., Kunz, M., Tamura, N., Budiman, A.S.: Synchrotron X-ray micro-diffraction—probing stress state in encapsulated thin silicon solar cells. Procedia Eng. **139**, 123–133 (2016)
47. Tippabhotla, S.K., Radchenko, I., Song, W.J.R., Illya, G., Handara, V., Kunz, M., Tamura, N., Tay, A.A., Budiman, A.S.: From cells to laminate: probing and modeling residual stress evolution in thin silicon photovoltaic modules using synchrotron X-ray micro-diffraction experiments and finite element simulations. Prog. Photovolt. Res. Appl. **25**(9), 791–809 (2017)
48. Ali, I., Tippabhotla, S.K., Radchenko, I., Al-Obeidi, A., Stan, C.V., Tamura, N., Budiman, A.S.: Probing stress states in silicon nanowires during electrochemical lithiation using in situ synchrotron X-ray microdiffraction. Front. Energy Res. **6**, 19 (2018)
49. Yamada, S., Sakai, T.: In situ transmission electron microscopy observation of dislocation motion in 9Cr steel at elevated temperatures: influence of shear stress on dislocation behavior. J. Electron Microsc. **63**(6), 449–461 (2014)
50. Barra, F., Caru, A., Cerda, M.T., Espinoza, R., Jara, A., Lund, F., Mujica, N.: Measuring dislocation density in aluminum with resonant ultrasound spectroscopy. Int. J. Bifurc. Chaos **19**(10), 3561–3565 (2009)
51. Barra, F., Espinoza-González, R., Fernández, H., Lund, F., Maurel, A., Pagneux, V.: The use of ultrasound to measure dislocation density. JOM **67**(8), 1856–1863 (2015)

52. Burek, M.J., Budiman, A.S., Jahed, Z., Tamura, N., Kunz, M., Jin, S., Han, S.M.J., Lee, G., Zamecnik, C., Tsui, T.Y.: Fabrication, microstructure, and mechanical properties of tin nanostructures. Mater. Sci. Eng.: A **528**(18), 5822–5832 (2011)
53. Sivakami, R., Dhanuskodi, S., Karvembu, R.: Estimation of lattice strain in nanocrystalline RuO_2 by Williamson-Hall and size–strain plot methods. Spectrochim. Acta Part A: Mol. Biomol. Spectrosc. **152**, 43–50 (2016)
54. Khorrami, G.H., Zak, A.K., Kompany, A.: Optical and structural properties of X-doped (X = Mn, Mg, and Zn) PZT nanoparticles by Kramers-Kronig and size strain plot methods. Ceram. Int. **38**(7), 5683–5690 (2012)
55. Zhang, J.M., Zhang, Y., Xu, K.W., Ji, V.: General compliance transformation relation and applications for anisotropic hexagonal metals. Solid State Commun. **139**(3), 87–91 (2006)
56. Warren, B.E., Averbach, B.L.: The separation of cold-work distortion and particle size broadening in X-ray patterns. J. Appl. Phys. **23**(4), 497–497 (1952)
57. Groma, I., Ungár, T., Wilkens, M.: Asymmetric X-ray line broadening of plastically deformed crystals. I. Theory. J. Appl. Crystallogr. **21**(1), 47–54 (1988)
58. Ungár, T., Groma, I., Wilkens, M.: Asymmetric X-ray line broadening of plastically deformed crystals. II. Evaluation procedure and application to [001]-Cu crystals. J. Appl. Crystallogr. **22**(1), 26–34 (1989)
59. Ungár, T., Borbély, A.: The effect of dislocation contrast on x-ray line broadening: a new approach to line profile analysis. Appl. Phys. Lett. **69**(21), 3173–3175 (1996)
60. Williamson, G.K., Hall, W.H.: X-ray line broadening from field alumunium and wolfram. Acta Metall. **1**(1), 22–31 (1953)
61. Scherrer, P.: Göttinger Nachrichten. Math. Phys. **2**, 98–100 (1918)
62. Warren, B.E., Averbach, B.L.: The effect of cold-work distortion on X-ray patterns. J. Appl. Phys. **21**(6), 595–599 (1950)
63. Ungár, T., Ott, S., Sanders, P.G., Borbély, A., Weertman, J.R.: Dislocations, grain size and planar faults in nanostructured copper determined by high resolution X-ray diffraction and a new procedure of peak profile analysis. Acta Mater. **46**(10), 3693–3699 (1998)
64. Burek, M.J., Greer, J.R.: Fabrication and microstructure control of nanoscale mechanical testing specimens via electron beam lithography and electroplating. Nano Lett. **10**(1), 69–76 (2009)
65. Cahn, R.W.: Recrystallization of single crystals after plastic bending. J. Inst. Met. **76**(2), 121 (1949)
66. Nye, J.F.: Some geometrical relations in dislocated crystals. Acta Metall. **1**(2), 153–162 (1953)
67. Feng, G., Ngan, A.H.W.: Creep and strain burst in indium and aluminium during nanoindentation. Scr. Mater. **45**(8), 971–976 (2001)
68. Kunz, M., Tamura, N., Chen, K., MacDowell, A.A., Celestre, R.S., Church, M.M., Fakra, S., Domning, E.E., Glossinger, J.M., Kirschman, J.L., Morrison, G.Y.: A dedicated superbend X-ray microdiffraction beamline for materials, geo-, and environmental sciences at the advanced light source. Rev. Sci. Instrum. **80**(3), 035108 (2009)
69. Tamura, N., Celestre, R.S., MacDowell, A.A., Padmore, H.A., Spolenak, R., Valek, B.C., Meier Chang, N., Manceau, A., Patel, J.R.: Submicron x-ray diffraction and its applications to problems in materials and environmental science. Rev. Sci. Instrum. **73**(3), 1369–1372 (2002)
70. Olson, G.B.: Designing a new material world. Science **288**(5468), 993–998 (2000)
71. Magee, C.L., Wood, K.L., Frey, D.D., Moreno, D.: Advancing design research: a "Big-D" design perspective. In: ICoRD'13, pp. 41–60. Springer, India (2013)
72. Cross, J.O., Opila, R.L., Boyd, I.W., Kaufmann, E.N.: Materials characterization and the evolution of materials. MRS Bull. **40**(12), 1019–1034 (2015)
73. Holdren, J.P., Thomas, K., Cyrus, W., Laurie, L.: Materials genome initiative strategic plan. National Science and Technology Council (2014)
74. Holdren, J.P.: Materials Genome Initiative for Global Competitiveness. National Science and Technology Council, OSTP, Washington, DC (2011). There is no corresponding record for this reference [Google Scholar]

75. Drosback, M.: The materials genome initiative and materials innovation infrastructure. Presentation, Office of Science and Technology Policy, Washington, DC, 5 Sept 2013
76. Robinson, L.: Looking back while moving forward: Cyrus Wadia discusses the evolution of MGI. JOM **67**(5), 883 (2015)
77. European Commission: DAMADEI: design and advanced materials as a driver of European innovation report. http://www.damadei.eu/report/ (2016)
78. Ong, S.P., Cholia, S., Jain, A., Brafman, M., Gunter, D., Ceder, G., Persson, K.A.: The materials Application Programming Interface (API): a simple, flexible and efficient API for materials data based on REpresentational State Transfer (REST) principles. Comput. Mater. Sci. **97**, 209–215 (2015)
79. Isayev, O., Fourches, D., Muratov, E.N., Oses, C., Rasch, K., Tropsha, A., Curtarolo, S.: Materials cartography: representing and mining materials space using structural and electronic fingerprints. Chem. Mater. **27**(3), 735–743 (2015)
80. Trimarchi, G., Peng, H., Im, J., Freeman, A.J., Cloet, V., Raw, A., Poeppelmeier, K.R., Biswas, K., Lany, S., Zunger, A.: Using design principles to systematically plan the synthesis of hole-conducting transparent oxides: Cu_3VO_4 and Ag_3 VO_4 as a case study. Phys. Rev. B **84**(16), 165116 (2011)
81. Beyerlein, I.J., Demkowicz, M.J., Misra, A., Uberuaga, B.P.: Defect-interface interactions. Prog. Mater Sci. **74**, 125–210 (2015)
82. Greer, J.R., De Hosson, J.T.M.: Plasticity in small-sized metallic systems: intrinsic versus extrinsic size effect. Prog. Mater. Sci. **56**(6), 654–724 (2011)
83. Misra, A.: Mechanical behavior of metallic nanolaminates. In: Nanostructure Control of Materials, pp. 146–176. Woodhead Publishing (2006)
84. Wang, J., Zhou, Q., Shao, S., Misra, A.: Strength and plasticity of nanolaminated materials. Mater. Res. Lett. **5**(1), 1–19 (2017)
85. Anwar Ali, H.P., Radchenko, I., Zhou, J., Qing, L., Budiman, A.: Designing novel multilayered nanocomposites for high-performance coating materials with online strain monitoring capability. Proc. Inst. Mech. Eng. Part L: J. Mater.: Des. Appl. **233**(4), 664–675 (2019)
86. Ali, H.P.A.: Ph.D. Dissertation, Singapore University of Technology & Design (SUTD) (2018)
87. Kim, Y., Lee, J., Yeom, M.S., Shin, J.W., Kim, H., Cui, Y., Kysar, J.W., Hone, J., Jung, Y., Jeon, S., Han, S.M.: Strengthening effect of single-atomic-layer graphene in metal–graphene nanolayered composites. Nat. Commun. **4**(1), 2114 (2013)
88. Wang, J., Misra, A.: An overview of interface-dominated deformation mechanisms in metallic multilayers. Curr. Opin. Solid State Mater. Sci. **15**(1), 20–28 (2011)
89. Misra, A., Hirth, J.P., Hoagland, R.G.: Length-scale-dependent deformation mechanisms in incoherent metallic multilayered composites. Acta Mater. **53**(18), 4817–4824 (2005)
90. Hoagland, R.G., Hirth, J.P., Misra, A.: On the role of weak interfaces in blocking slip in nanoscale layered composites. Philos. Mag. **86**(23), 3537–3558 (2006)
91. Hoagland, R.G., Mitchell, T.E., Hirth, J.P., Kung, H.: On the strengthening effects of interfaces in multilayer fee metallic composites. Philos. Mag. A **82**(4), 643–664 (2002)
92. Anderson, P.M., Foecke, T., Hazzledine, P.M.: Dislocation-based deformation mechanisms in metallic nanolaminates. MRS Bull. **24**(2), 27–33 (1999)
93. Anderson, P.M., Li, C.: Hall-Petch relations for multilayered materials. Nanostruct. Mater. **5**(3), 349–362 (1995)
94. Misra, A., Hirth, J.P., Kung, H.: Single-dislocation-based strengthening mechanisms in nanoscale metallic multilayers. Philos. Mag. A **82**(16), 2935–2951 (2002)
95. Nix, W.D.: Mechanical properties of thin films. Metall. Trans. A **20**(11), 2217–2245 (1989)
96. Zhang, J.Y., Lei, S., Niu, J., Liu, Y., Liu, G., Zhang, X., Sun, J.: Intrinsic and extrinsic size effects on deformation in nanolayered Cu/Zr micropillars: from bulk-like to small-volume materials behavior. Acta Mater. **60**(10), 4054–4064 (2012)
97. Kim, Y., Budiman, A.S., Baldwin, J.K., Mara, N.A., Misra, A., Han, S.M.: Microcompression study of Al–Nb nanoscale multilayers. J. Mater. Res. **27**(3), 592–598 (2012)
98. Wang, J., Yang, C., Hodgson, P.D.: Strain gradients in Cu–Fe thin films and multilayers during micropillar compression. Mater. Sci. Eng.: A **651**, 146–154 (2016)

99. Carpenter, J.S., Misra, A., Uchic, M.D., Anderson, P.M.: Strain rate sensitivity and activation volume of Cu/Ni metallic multilayer thin films measured via micropillar compression. Appl. Phys. Lett. **101**(5), 051901 (2012)
100. Hattar, K., Misra, A., Dosanjh, M.R.F., Dickerson, P., Robertson, I.M., Hoagland, R.G.: Direct observation of crack propagation in copper–niobium multilayers. J. Eng. Mater. Technol. **134**(2) (2012)
101. Li, N., Wang, J., Misra, A., Huang, J.Y.: Direct observations of confined layer slip in Cu/Nb multilayers. Microsc. Microanal. **18**(5), 1155–1162 (2012)
102. Aydıner, C.C., Brown, D.W., Mara, N.A., Almer, J., Misra, A.: In situ x-ray investigation of freestanding nanoscale Cu–Nb multilayers under tensile load. Appl. Phys. Lett. **94**(3), 031906 (2009)
103. Mara, N.A., Bhattacharyya, D., Dickerson, P.O., Hoagland, R.G., Misra, A.: Ultrahigh strength and ductility of Cu–Nb nanolayered composites. In: Materials Science Forum, vol. 633, pp. 647–653. Trans Tech Publications Ltd. (2010)
104. Bhattacharyya, D., Mara, N.A., Dickerson, P., Hoagland, R.G., Misra, A.: Transmission electron microscopy study of the deformation behavior of Cu/Nb and Cu/Ni nanoscale multilayers during nanoindentation. J. Mater. Res. **24**(3), 1291–1302 (2009)
105. Mara, N.A., Misra, A., Hoagland, R.G., Sergueeva, A.V., Tamayo, T., Dickerson, P., Mukherjee, A.K.: High-temperature mechanical behavior/microstructure correlation of Cu/Nb nanoscale multilayers. Mater. Sci. Eng.: A **493**(1–2), 274–282 (2008)
106. Demkowicz, M.J., Hoagland, R.G., Hirth, J.P.: Interface structure and radiation damage resistance in Cu–Nb multilayer nanocomposites. Phys. Rev. Lett. **100**(13), 136102 (2008)
107. Brown, R.A.: Electrical resistivity of dislocations in metals. J. Phys. F: Met. Phys. **7**(7), 1283 (1977)
108. Basinski, Z.S., Dugdale, J.S.: Electrical resistivity due to dislocations in highly purified copper. Phys. Rev. B **32**(4), 2149 (1985)
109. Karolik, A.S., Luhvich, A.A.: Calculation of electrical resistivity produced by dislocations and grain boundaries in metals. J. Phys.: Condens. Matter **6**(4), 873 (1994)
110. Karasek, K.R., Bevk, J.: Dislocation resistivity in in situ formed Cu–Nb multifilamentary composites. Scr. Metall. **14**(4), 431–435 (1980)
111. Spitzig, W.A., Pelton, A.R., Laabs, F.C.: Characterization of the strength and microstructure of heavily cold worked Cu–Nb composites. Acta Metall. **35**(10), 2427–2442 (1987)
112. Grünberg, P.E.T.E.R., Schreiber, R., Pang, Y., Brodsky, M.B., Sowers, H.: Layered magnetic structures: evidence for antiferromagnetic coupling of Fe layers across Cr interlayers. Phys. Rev. Lett. **57**(19), 2442 (1986)
113. Parkin, S.S.P., Li, Z.G., Smith, D.J.: Giant magnetoresistance in antiferromagnetic Co/Cu multilayers. Appl. Phys. Lett. **58**(23), 2710–2712 (1991)
114. Sondheimer, E.H.: The mean free path of electrons in metals. Adv. Phys. **1**, 1–42 (1952)
115. Griffin, A.J., Embury, J.D., Hundley, M.F., Jervis, T.R., Kung, H.H., Scarborough, W.K., Walter, K.C., Wood, J., Nastasi, M.: Residual stress, mechanical behavior and electrical properties of Cu/Nb thin-film multilayers. MRS Online Proc. Libr. Arch. **382** (1995)
116. Wood, J.T., Griffin, A.J., Jr., Embury, J.D., Zhou, R., Nastasi, M., Veron, M.: The influence of microstructural scale on the combination of strength and electrical resistivity in copper based composites. J. Mech. Phys. Solids **44**(5), 737–750 (1996)
117. Budiman, A.S., Han, S.M., Li, N., Wei, Q.M., Dickerson, P., Tamura, N., Kunz, M., Misra, A.: Plasticity in the nanoscale Cu/Nb single-crystal multilayers as revealed by synchrotron Laue x-ray microdiffraction. J. Mater. Res. **27**(3), 599–611 (2012)
118. Budiman, A.S., Li, N., Wei, Q., Baldwin, J.K., Xiong, J., Luo, H., Trugman, D., Jia, Q.X., Tamura, N., Kunz, M., Chen, K.: Growth and structural characterization of epitaxial Cu/Nb multilayers. Thin Solid Films **519**(13), 4137–4143 (2011)
119. Uchic, M.D., Shade, P.A., Dimiduk, D.M.: Plasticity of micrometer-scale single crystals in compression. Annu. Rev. Mater. Res. **39**, 361–386 (2009)
120. Uchic, M.D., Dimiduk, D.M., Florando, J.N., Nix, W.D.: Sample dimensions influence strength and crystal plasticity. Science **305**(5686), 986–989 (2004)

121. Kim, Y.J., Qaiser, N., Han, S.M.: Time-dependent deformation of Sn micropillars. Mater. Des. **102**, 168–173 (2016)
122. Singh, G., Narayan, R.L., Asiri, A.M., Ramamurty, U.: Discrete drops in the electrical contact resistance during nanoindentation of a bulk metallic glass. Appl. Phys. Lett. **108**(18), 181903 (2016)
123. Miyajima, Y., Komatsu, S.Y., Mitsuhara, M., Hata, S., Nakashima, H., Tsuji, N.: Change in electrical resistivity of commercial purity aluminium severely plastic deformed. Philos. Mag. **90**(34), 4475–4488 (2010)
124. Pharr, G.M., Oliver, W.C., Cook, R.F., Kirchner, P.D., Kroll, M.C., Dinger, T.R., Clarke, D.R.: Electrical resistance of metallic contacts on silicon and germanium during indentation. J. Mater. Res. **7**(4), 961–972 (1992)
125. Ruffell, S., Bradby, J.E., Williams, J.S., Warren, O.L.: An in situ electrical measurement technique via a conducting diamond tip for nanoindentation in silicon. J. Mater. Res. **22**(3), 578–586 (2007)
126. Han, S.M., Bozorg-Grayeli, T., Groves, J.R., Nix, W.D.: Size effects on strength and plasticity of vanadium nanopillars. Scr. Mater. **63**(12), 1153–1156 (2010)
127. Han, S.M., Phillips, M.A., Nix, W.D.: Study of strain softening behavior of Al–Al_3Sc multilayers using microcompression testing. Acta Mater. **57**(15), 4473–4490 (2009)
128. Mara, N.A., Bhattacharyya, D., Hoagland, R.G., Misra, A.: Tensile behavior of 40 nm Cu/Nb nanoscale multilayers. Scr. Mater. **58**(10), 874–877 (2008)
129. Skomski, R., Enrech, M., Coey, J.M.D.: Resistance and structure of metallic multilayers. Nanostruct. Mater. **1**(4), 337–345 (1992)
130. Giancoli, D.C.: Physics Principles with Applications, 6th edn. (2005)
131. Matsuyama, H., Kouno, T., Suzuki, T., Shiozu, M., Ehara, H., Otsuka, S., Hosoda, T., Nakamura, T., Mizushima, Y., Shono, K., Miyajima, M.: Stress migration phenomena in narrow copper lines. In: 2006 IEEE International Integrated Reliability Workshop Final Report, pp. 28–30. IEEE (2006)
132. Ogawa, E.T., McPherson, J.W., Rosal, J.A., Dickerson, K.J., Chiu, T.C., Tsung, L.Y., Jain, M.K., Bonifield, T.D., Ondrusek, J.C., McKee, W.R.: Stress-induced voiding under vias connected to wide Cu metal leads. In: 2002 IEEE International Reliability Physics Symposium. Proceedings. 40th Annual (Cat. No. 02CH37320), pp. 312–321. IEEE, Apr 2002
133. Tamura, N., MacDowell, A.A., Spolenak, R., Valek, B.C., Bravman, J.C., Brown, W.L., Celestre, R.S., Padmore, H.A., Batterman, B.W., Patel, J.R.: Scanning X-ray microdiffraction with submicrometer white beam for strain/stress and orientation mapping in thin films. J. Synchrotron Radiat. **10**(2), 137–143 (2003)
134. Budiman, A.S., Shin, H., Kim, B.J., Hwang, S.H., Son, H.Y., Suh, M.S., Chung, Q.H., Byun, K.Y., Joo, Y.C., Caramto, R., Smith, L.: Comparison of mechanical stresses of Cu Through-Silicon Via (TSV) samples fabricated by Hynix vs. SEMATECH using synchrotron X-ray microdiffraction for 3-D integration and reliability. In: 2012 IEEE International Interconnect Technology Conference, pp. 1–3. IEEE, June 2012
135. Liu, X., Thadesar, P.A., Taylor, C.L., Oh, H., Kunz, M., Tamura, N., Bakir, M.S., Sitaraman, S.K.: In-situ microscale through-silicon via strain measurements by synchrotron x-ray microdiffraction exploring the physics behind data interpretation. Appl. Phys. Lett. **105**(11), 112109 (2014)
136. Morusupalli, R., Rao, R., Lee, T.K., Shen, Y.L., Kunz, M., Tamura, N., Budiman, A.S.: Critical temperature shift for stress induced voiding in advanced Cu interconnects for 32 nm and beyond. Procedia Eng. **139**, 32–40 (2016)
137. Burek, M.J., Jin, S., Leung, M.C., Jahed, Z., Wu, J., Budiman, A.S., Tamura, N., Kunz, M., Tsui, T.Y.: Grain boundary effects on the mechanical properties of bismuth nanostructures. Acta Mater. **59**(11), 4709–4718 (2011)
138. Feng, G., Budiman, A.S., Nix, W.D., Tamura, N., Patel, J.R.: Indentation size effects in single crystal copper as revealed by synchrotron x-ray microdiffraction. J. Appl. Phys. **104**(4), 043501 (2008)

139. McPherson, J.W., Dunn, C.F.: A model for stress-induced metal notching and voiding in very large-scale-integrated Al–Si (1%) metallization. J. Vac. Sci. Technol. B: Microelectron. Process. Phenom. **5**(5), 1321–1325 (1987)
150. Cadek, J.: Creep in metallic materials. Mater. Sci. Monogr. **48** (1988)
151. Shen, Y.L.: On the elastic assumption for copper lines in interconnect stress modeling. IEEE Trans. Device Mater. Reliab. **8**(3), 600–607 (2008)
142. Frost, H.J., Ashby, M.F.: Deformation Mechanism Maps: The Plasticity and Creep of Metals and Ceramics, pp. 600–607. Pergamon Press (1982)

Chapter 5
Latest Updates in Next-Generation Energy Technologies and Systems

The global demand for energy has grown rapidly over the past 40 years and it is widely accepted that 80–90% of the energy demand is currently met by burning fossil fuels. However, fossil fuels are a finite resource and also produce greenhouse gas (GHG) emissions which contribute to irreversible climate change. This has led to international governments forming the COP-21 Paris Agreement which aims to limit the increase of global temperatures to 2 °C above pre-industrial levels. Therefore, it is imperative to explore alternative energy resources such as advanced energy storage technology (such as the Li-ion battery) and solar photovoltaics (PV) in a bid to achieve energy security and environmental sustainability. Just like in the advanced microelectronics and related nanotechnologies (as discussed in the previous Chap. IV), the latest version of energy (generation, storage, etc.) technologies and systems involve advanced materials that are often in the submicron and nanometer scales, for performance purposes (as we will see in the first example of cutting-edge Li-ion battery anode design) as well as for cost reduction to improve its market competitiveness in the global energy market (as we will see in the second and third examples of solar PV technologies and systems). The creation of such small components again requires a thorough understanding of the mechanical properties of materials at these small length scales. Here, we propose to directly examine some of the effects that arise when small-scale materials are strained beyond its traditional limits during the manufacturing sequence of the devices and/or during operations of the overall systems in the field. In order to enable Design for Reliability (DfR) as well as design for cutting-edge performance and the cut-throat cost reduction in the global energy market, the principles of extreme mechanics (XM) in elasticity and especially in the limits of fracture were used in the following recent cases discussed in this chapter.

Silicon anode nanostructures have been the latest development in next-generation Li-Ion Battery (LIB). Nano-structuring of silicon has recently been widely accepted to produce LIB with much higher performance (such as energy density and capacity

A. S. Budiman, *Principles of Extreme Mechanics (XM) in Design for Reliability (DfR)*, Engineering Materials,
https://doi.org/10.1007/978-981-15-6720-9_5

retention), which originated partially from the higher surface/volume ratio in nanostructures, in addition to silicon elemental properties of very high charge capacity (~4200 mAh/g), compared to existing graphite-based anode whose charge capacity is roughly ten times smaller. However, silicon expands significantly during operation of the LIB device—leading to volumetric expansion of 300–400%, which causes extreme stresses, typically followed by fracture events, which would eventually result in the final catastrophic failure of the device. The unique and powerful capability of synchrotron X-ray microdiffraction was used here again to reveal the unique stress states in these vertical nanostructures as a result of large volume diffusion of lithium ions (sometime called "cloud of ions") involved in the operations of the battery (charging or recharging). Utilizing submicron-focused polychromatic synchrotron beam developed in the Beamline 12.3.2 at the Advanced Light Source (ALS), Berkeley Lab, the technique proved advantageous as a local probe of mechanical behavior (stress, elastic deformation) and plastic deformation in small-scale devices. Furthermore, with this facility, in-situ lithiation experiment was enabled, which is a rare opportunity that is not always practical with other characterization techniques. This capability enables us to investigate the evolution of the microstructure of the silicon nanowires as they deform due to the enormous clouds of ions of lithium coming in and out of the silicon nanowires (called lithiation and de-lithiation) associated with the charging and use of the battery, respectively. This is an important and crucial piece of information to enable silicon anode technology to be integrated in a commercially successful, next-generation LIB with much higher energy density (which would allow the much higher driving range typically required for electric vehicles, for instance). This will be discussed in the first section—Sect. 5.1: **Novel Silicon Nanowire Anode Technology for Next-Generation Lithium Ion Battery (LIB)**.

Two more similar recent cases in which materials at extremely small scales are utilized to enable novel and innovative energy technologies will be highlighted further following this section. The two cases are related, and both demonstrate the unique capability of the recent advances in the materials and mechanics characterization techniques as discussed in Chapter 3 to enable design and development of silicon-based solar photovoltaics (PV) technologies with extreme fracture strength and thus higher structural integrity. Section 5.2: **Enabling Thin Silicon Technologies for Next-Generation, Novel and Innovative Crystalline Silicon Photovoltaics Systems Design** will discuss how thinner and thinner silicon solar cells were successfully integrated into PV modules despite silicon being a very fragile material, let alone thin silicon as discussed in this section. Further, a novel type of encapsulant was shown to lead to lower mechanical stress concentrations in the thin, fragile silicon, and thus higher fracture resistance (so thinner, but stronger!). These fundamental insights of extreme fracture mechanics could potentially lead to Design for Extreme Reliability (DfXR) for silicon PV technologies, and are discussed in Sect. 5.3: **Low-Stress Encapsulants?—Enabling PV with Extreme Reliability**.

5.1 Novel Silicon Nanowire Anode Technology for Next-Generation Lithium Ion Battery (LIB)

Silicon is a promising material for lithium ion batteries. However, it expands by 300–400% upon lithiation, leading to fracture. Nano-structuring of silicon is expected to be a promising method to improve the mechanical strength of the silicon electrodes. In the present work, a unique battery test cell was designed and fabricated to study the in-situ stress evolution in the silicon nanowire electrode during electrochemical lithiation using synchrotron X-ray microdiffraction. The stress in the silicon nanowires at pristine state and during lithiation was evaluated using energy scans. The average stress in the pristine nanowires was found to be ~40 MPa tensile, which changed to ~325 MPa compressive upon lithiation. Further, the deviatoric stress state in the silicon nanowires during lithiation was evaluated using Laue diffraction, and the lithiated nanowires were found to be in triaxial stress state with high shear stresses. The technique and the findings provide new and more in-depth understanding of the stress evolution in the silicon nanowires during electrochemical lithiation.

5.1.1 Background and Motivation—Novel Silicon Nanowire Anode

Energy storage became an essential requirement in modern electric vehicle systems, portable electronic gadgetry, solar and wind farms, etc. There are several energy storage systems (batteries) available at different sizes and capacities, however, lithium ion battery (LIB) has been proven to be the most promising candidate because of its several qualities such as high efficiency, high energy density, limited self-discharge, longer life cycle even with high charging and discharging rates [1–3], which are important modern battery design parameters. Silicon has a very high charge capacity (~4200 mAh/g) theoretically, and it can be a promising replacement for the commercially used graphite electrodes, whose charge capacity (~372 mAh/g) is roughly ten times smaller than silicon. It further has a low discharge potential and attractive operating voltage [4, 5]. Along with these benefits, silicon also has a disadvantage, high volumetric expansion (300%) upon lithiation [6], leading to high stress, cracking, palletization and delamination from the current collector, leading to electrode failure [5, 7]. This is the main reason why silicon is still in the laboratory and not in commercial battery electrodes. Nano-structuring of silicon was proven to be effective in relaxing the high strain due to expansion and increase the surface area available for the chemical reaction [8, 9]. Silicon nanostructures were proved to be better than their bulk counterparts in capacity retention [10–12]. Different nanostructures such as nanoparticles, nanospheres, nanowires, nanotubes and nano-cones were experimentally proved to exhibit better cycling life and performance as LIB electrodes [4, 9, 11–21].

In the present study, silicon nanowires (SiNW) of uniform crystallographic orientation <100> , made of metal-assisted chemical etc. hing (MACE) process were used [22, 23]. SiNWs (and silicon nanospheres) have been widely studied due to simple geometry, leading to 1D or 2D kinetics and mechanics modeling. Crystalline SiNWs and nanospheres, upon progressive lithiation form a core–shell structure [24, 25], where the outer shell transforms from crystalline silicon (c-Si) to amorphous lithiated silicon (Li_xSi), upon reaction with Li metal ions and the inner crystalline core remains same. The boundary between the crystalline core and the lithiated silicon shell is characterized by a sharp lithiation reaction front [25]. These phase transformations and associated volumetric expansion of the lithiated phase, gives rise to mechanical stress in the nanowires. Based on these observations, several chemo-mechanical models, commonly known as "core–shell" lithiation models were developed to address the stress and fracture of the silicon nanoparticles/spheres [24–27] and SiNWs [28]. Studies were also performed to model the stress in the SiNWs, when constrained [29]. All these models are based on SEM/TEM observations of the lithiation process and lithiated silicon particles/wires.

In the current manuscript, direct in-situ stress measurement was performed using synchrotron X-ray microdiffraction as explained in the following sections. Compared to our recent earlier reports [30, 31], the present study reported an enhanced stress study detailing the complete stress states (the deviatoric tensor components) and their evolution during the in-situ electrochemical lithiation experiments. Synchrotron X-ray microdiffraction (μSXRD) has been proven to be effective for revealing insights of mechanical stress states and other mechanics considerations (such as plasticity, mechanisms preceding fracture events or final catastrophic failure) in small-scale crystalline structures in many important technological applications, such as microelectronics [32–37], nanotechnology [38–41] and photovoltaics [37, 42–44].

5.1.2 Experimental Methodology—Probing Stress in Extreme Expansion of Silicon Nanowires Anode

Experimental procedure developed to quantitively evaluate the stress states during electrochemical lithiation of SiNWs in a specially designed electrochemical cell, which facilitates an in-situ μSXRD probe was explained in this section.

Silicon nanowire samples and Electrochemical Cell

The SiNW samples used in this experiment were fabricated using a p-type silicon wafer of thickness 500 μm. The nanowires are oriented in <100> direction and 6–8 μm tall, standing on ~490 μm thick bulk silicon base as shown by the schematic in Fig. 5.1a and the SEM images in Fig. 5.1b. The samples were fabricated using metal-assisted chemical etching (MACE) process, at the Laboratory of Materials for Micro and Nano Systems, Massachusetts Institute of Technology (MIT). The MACE process is a simple, highly repeatable and scalable process for making silicon

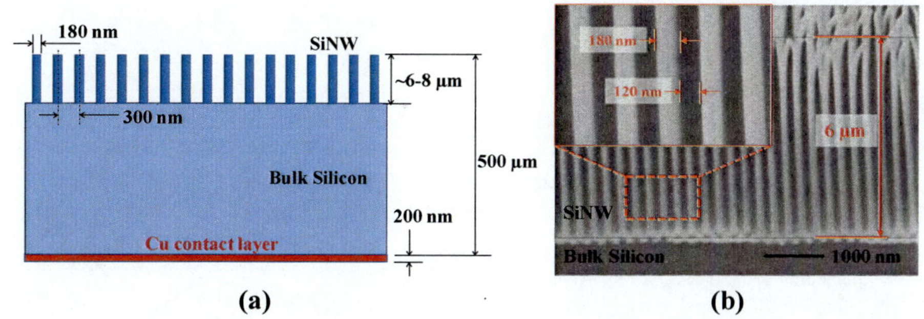

Fig. 5.1 **a** Schematic of the SiNW sample (dimensions not to scale) and the Cu back contact layer, **b** SEM image of pristine SiNWs, which was taken at tilt angle of 45°. The average height of SiNWs is ~6–8 μm (after tilt correction). Inset shows diameter and space between SiNWs (Reproduced from Reference [30] with permission from Elsevier)

nanostructures as explained elsewhere [22, 23]. The back side of the sample was deposited with a 200 nm thick copper layer for electric contact as shown in Fig. 5.1a.

A unique battery test cell, specially designed and fabricated, is the key enabler to perform the in-situ μSXRD experiment as explained in our previous reports [30, 31]. Such a special battery cell is required to enable the X-ray beam to reach the interior SiNW sample and come out (the diffracted X-rays from the SiNW sample) to reach the detector as shown in the μSXRD schematic diagram in Fig. 5.2a. And thus, enable the detector to record the in-situ μSXRD signal from the SiNWs during electrochemical lithiation in the cell. To achieve this task, the body of the cell (except the top metal cover) was made of an X-ray transparent material. In addition, a circular front window with transparent Kapton tape was fabricated to make the inner electrodes and other parts visible from outside as shown in Fig. 5.2b and c. This front window is essential to enable focusing adjustments of the sample stage for good diffraction and to closely monitor the interaction of the X-ray beam and the electrolyte (which may be a safety concern) from outside. The interior of the cell was tightly sealed using O-ring seals (Fig. 5.2b) and the cell was assembled in a Nitrogen-/Argon-filled glove box. The electrochemical cycling of the test cell from open circuit voltage (OCV) of ~3 V to 0.01 V was done on a single channel portable Potentio/Galvanostat (Bio-Logic Science Instruments).

The test cell thus described [30] is much larger in volume compared to the typical coin cells and hence requires large quantity of electrolyte to fill the empty space. Hence, a PTFE spacer was used to fill most of the cell space which is empty (Fig. 5.2b). However, the cell still requires much larger quantity of electrolyte (~10–20 ml) compared to a few drops in a typical coin cell. Filling the cell volume is necessary for the μSXRD experiment, explained in the following section. It requires the cell to be assembled at 45° angle with the base, the electrolyte displaces to the empty volume (if any) from the region between the electrodes, which may have affected the lithiation process. A more detailed description of the battery test cell to

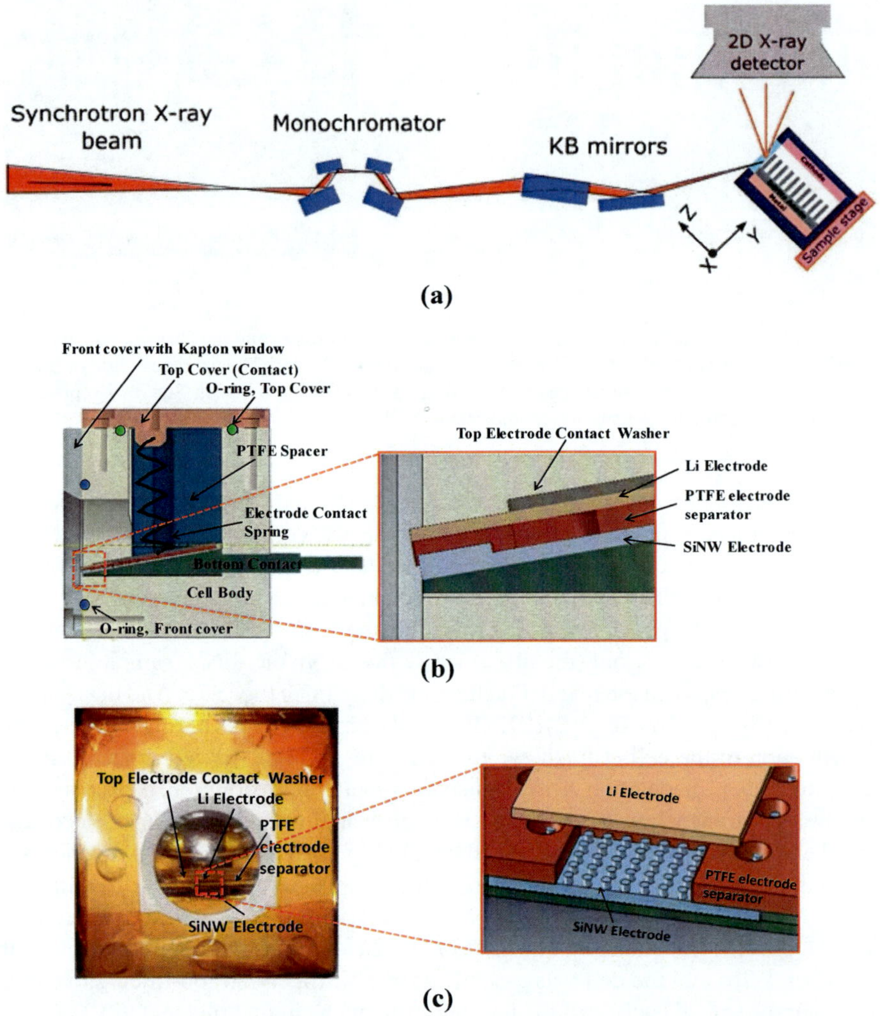

Fig. 5.2 **a** Schematic of the overall μSXRD experimental setup. **b** Schematic of battery test cell assembly with detailed view of electrodes in the inset. **c** Picture of the assembled battery test cell with schematic of the close view of SiNWs, Li-electrode and PTFE separator in the inset (Reproduced from Reference [30] with permission from Elsevier)

allow for this in-situ experiments could be obtained from our recent earlier article [30].

Synchrotron X-ray microdiffraction experiment and stress states in SiNWs

The μSXRD technique employs a focused polychromatic/monochromatic X-ray beam of submicron resolution, which is crucial to enable the study of stress states in SiNWs during electrochemical lithiation. The current experimental work was conducted using the μSXRD setup at beamline 12.3.2 of the Advanced Light Source, Lawrence Berkeley National Laboratory, Berkeley, California, USA [45]. This beamline provides ~0.8 μm spatial resolution, and high crystal orientation resolution of 0.01° [45–47]. The μSXRD setup at this beamline has been widely used to study microstructural changes, mechanical and residual stresses in many technologically important devices and structures. Through-Silicon Via (TSV) in the microelectronics, nanostructures such as Cu/Nb nanopillars and nanolayers for energy applications and crystalline silicon solar cells in photovoltaics, are well-known examples [32–44].

Figure 2a shows the schematic diagram of the μSXRD setup. The beamline end station (BL 12.3.2) produces a high energy (5–24 keV) polychromatic X-ray beam of ~0.8 μm diameter. A four-crystal monochromator in the setup is optionally engaged to get a monochromatic beam, whose cross-section is in elliptic shape with major axis ~10 μm/minor axis ~2 μm. The battery test cell was assembled on the sample stage as shown in the schematic in Fig. 5.2a. The sample stage can move in XYZ directions where Y is along SiNWs and Z is across the SiNWs as shown and X is perpendicular to the viewing plane. The sample stage was approximately adjusted to the focus of the X-ray beam on to the SiNW sample in the cell using a laser triangulation system available in the beamline setup [45].

The SiNW sample emits multiple diffraction signals from different crystalline planes (a Laue diffraction pattern) upon interaction with the polychromatic X-ray beam. A 2D X-ray hybrid pixel array (DECTRIS Pilatus 1 M) detector, mounted over the sample stage (Fig. 5.2a), captures the diffracted signal. Initial μSXRD scans were performed along the SiNWs (Y-direction) covering the entire thickness of the SiNWs electrode, i.e. a line scan from bottom (bulk Si) to top (tip of SiNWs). The resulting intensity variation of the diffracted signal along Y-direction was studied to approximately locate the SiNWs. This is because the bulk Si generates high intensity diffraction signal and the SiNWs at the same time generate weak diffraction signal, at least ten times lower in intensity than that of bulk Si.

Once the location of SiNWs was identified, they were scanned by both monochromatic and polychromatic X-rays before and after assembly of the battery cell. Further, an in-situ μSXRD scanning of the SiNWs was conducted during the electrochemical cycling of the test cell from open circuit voltage (OCV) to 0.01 V. During lithiation (cycling), the crystalline core of the partially lithiated SiNWs produces diffraction signal and the amorphous lithiated outer shell does not participate in the diffraction process. The polychromatic diffraction data was analyzed using the XMAS software [48]. Based on analysis of the initial polychromatic μSXRD scan from the SiNWs, the diffraction peak from the crystal plane $(\bar{3}\bar{1}5)$, energy ~9.716 keV in the case of Pristine SiNWs (before the battery cell assembly) was identified for further analysis using energy scan. Similarly, the diffraction peaks from the crystal planes $(\bar{5}\bar{1}7)$, energy ~14.08 keV and $(\bar{4}\bar{2}6)$, energy ~14.62 keV were identified in the case of SiNWs during lithiation, for further analysis using energy scan. The monochromatic scan was performed around the aforementioned approximate energy values (corresponding to the identified diffraction planes) with a step of 0.002 keV. From this monochromatic scan data analysis, the lattice plane spacing (d-spacing) of the diffracting planes, $d_{\mu SXRD}$ was evaluated for pristine SiNWs (before battery assembly) and lithiated SiNWs (during electrochemical cycling). The strain normal to the diffracting crystal planes, ε_{hkl}, was then calculated using the unstrained d-spacing, d_0, using the Eq. (5.1) below [48]. Thus, the stress acting normal to the diffracting planes can be calculated by multiplying the strain with the Young's modulus of SiNW (160 GPa). Since the focused monochromatic X-ray beam is an ellipsoid of several (~10 × 2) microns size, a single exposure covers several nanowires as shown in Fig. 5.3a and b, and hence, the resultant diffraction signal is an average from several nanowires and so are the strain and the stress.

$$\varepsilon_{hkl} = \frac{d_{\mu SXRD} - d_0}{d_0} \tag{5.1}$$

Apart from the monochromatic energy scans, white beam Laue microdiffraction scans were also performed during the lithiation process. The resulting Laue diffraction patterns were indexed using XMAS [48] to evaluate the deviatoric stress state in the SiNWs before and during the lithiation process. It should be noted that the size (diameter) of the polychromatic X-ray beam (white beam) is ~1 μm, much smaller than the size of the energy beam spot (Fig. 5.3b). Thus, the white beam spot covers fewer SiNWs compared to the energy scan. Hence, the diffraction signal from the white beam scan is also an average, but, from fewer nanowires. The Laue diffraction results provide evolution of shear stresses (the deviatoric stress tensor) during the electrochemical lithiation experiments in addition to the simple normal stresses in the diffracting planes already presented in our earlier reports [30, 31]. The results were discussed in the following section.

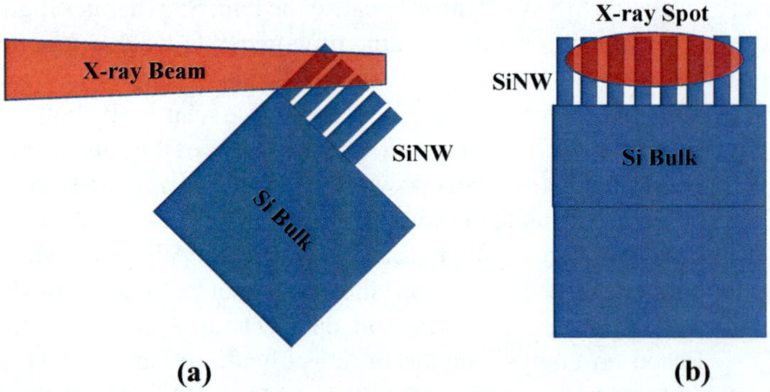

Fig. 5.3 Schematic closeup of the monochromatic X-ray beam focused on the SiNWs, **a** side view, **b** front view, the size of the X-ray spot is ~10 × 2 μm. The diffracting volume consists of several nanowires of different lengths, leading to averaging of measured parameters (Reproduced from Reference [30] with permission from Elsevier)

5.1.3 Enabling Novel Silicon Nanowire Anode for Next-Generation LIB

The location of the root (base) and tip of the SiNWs is identified by line scanning process. The energy scan intensities of the diffraction plane ($\bar{3}\bar{1}5$), at the root and tip of the nanowires before battery assembly is plotted in Fig. 5.4. The intensity of the SiNWs at the root is several orders high compared to the tip of the NWs. The respective diffraction peak images (at max. intensity) in Fig. 5.4 also show that the peak at the root is very bright and round due to high diffraction volume (influence from the bulk silicon below the root). The peak at the tip of SiNWs is dim and line-shaped due to low diffraction volume (due gaps between SiNWs). This observation is crucial to ensure a proper μSXRD scan of SiNWs during the in-situ battery cycling process. Besides, it also gives the stress state in the pristine SiNWs before the battery cell assembly.

The SEM image of the partially lithiated SiNW sample (after washing in fresh electrolyte solution and acetonitrile) is shown in Fig. 5.5a. The crystalline cores of the SiNWs are conical in shape, in line with the morphological study presented in our earlier report [31] and Yang et al. [28]. The conical shape of the crystalline cores is due to the non-uniform lithiation along the length of the nanowires, reducing from the tip to the root [28]. The zoomed-in images of Fig. 5.5a show crack-like facets (marked with red dotted lines) on the surface of the conical SiNW cores, especially near the tips. Figure 5.5b and c show bent and twisted SiNWs upon lithiation. These observations from the SEM images in the Fig. 5.5 could possibly indicate a complex three-dimensional stress state.

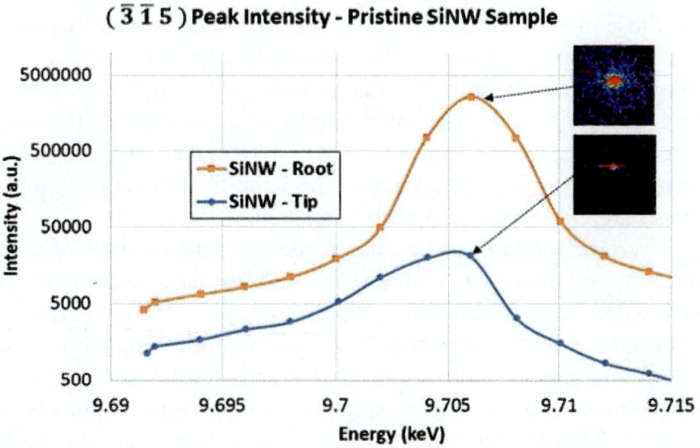

Fig. 5.4 Energy scan intensity plot of diffracting plane ($\bar{3}\bar{1}5$) from the root and tip of pristine SiNWs before battery cell assembly. Respective diffraction peaks at the max. Intensity were also shown. *Note* The intensity axis is in log scale (Reproduced from Reference [30] with permission from Elsevier)

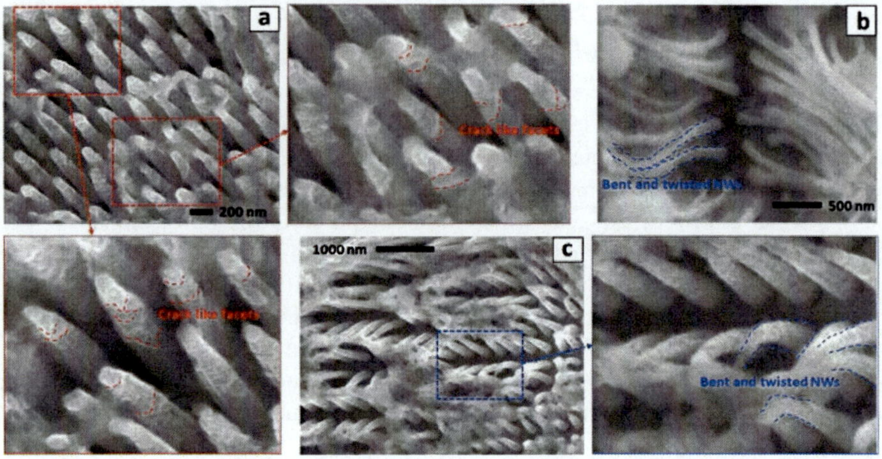

Fig. 5.5 SEM image of the lithiated SiNWs, **a** showing crystalline cores in conical shape. Zoomed-in images show crack-like facets on the conical surface of the NWs near tips. Almost every NW have them but only a few were marked for brevity. **b** and **c**—Bent and twisted SiNWs after lithiation (Reproduced from Reference [30] with permission from Elsevier)

Figure 5.6a shows stress evolution in the SiNWs during lithiation process, starting from the pristine state, calculated from the energy scan data. The stress in the pristine nanowires is ~40 MPa (tensile), and as the lithiation progressed, it turned compressive with magnitude between 250 and 350 MPa in different diffracting planes. This is in general consistent with the widely accepted core–shell lithiation models [24–27]. However, the magnitudes of the measured stress during lithiation is lower than that predicted by those models. Unlike those models, the magnitude of the stress almost remained constant during lithiation. These anomalies are because of averaging. As discussed earlier (Fig. 5.3), the stress state in the lithiated SiNW cores is longitudinally varying with high stress near the tip and low stress towards the root. In such a scenario, averaging of stress leads to a lower magnitude. Further, due to the loss of crystallinity, upon progressive lithiation, near the tip of the crystalline core, the average stress magnitude remained nearly constant.

The intensity variation of the diffraction peaks from the crystal planes $(\bar{5}\,\bar{1}\,7)$ and $(\bar{4}\bar{2}6)$ over the energy scan range is shown in Fig. 5.6b and c, respectively. While the low intensity, at the end of lithiation, indicates reduction in the diffraction volume (crystallinity) upon lithiation, the broadened and distorted intensity indicate high stress variation and crystal rotations in the diffracting volume. This is because, in a diffraction volume with no (or uniform) stress, the Bragg's condition is satisfied at a given energy (wavelength), giving a sharp intensity peak at a single energy (E) value, as given by the Eq. 5.2.

$$\lambda = \frac{hc}{E} = 2d\sin\theta \quad (5.2)$$

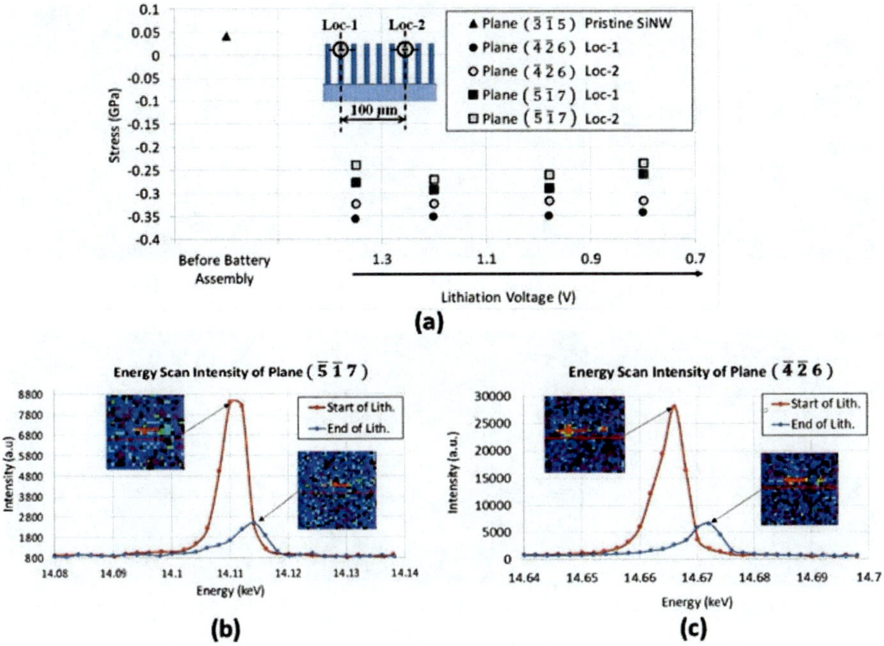

Fig. 5.6 **a** Stress evolution in the SiNWs during lithiation process from pristine state, evaluated from the energy scan. **b** Energy scan intensity variation plot of diffracting plane ($\bar{5}\bar{1}7$) at Loc-1 with the diffraction peak images at the max. intensity. **c** Energy scan intensity variation plot of diffracting plane ($\bar{4}\bar{2}6$) at Loc-1 with the diffraction peak images at the max. Intensity (Reproduced from Reference [30] with permission from Elsevier)

But, in the case of a diffraction volume that comprises of multiple stress states (different "d" spacing values) and crystal rotations (different "θ" values), the intensity peak is broadened, comprising a range of energy (E) values, where Bragg's condition is satisfied. This observation is in line with the finite element simulations performed by Yang et al. [28], considering non-uniform lithium ion diffusivities with respect to the crystal orientation, leading to a non-uniform lithiation around the circumference of the SiNWs. They further showed that such a circumferentially non-uniform lithiation causes multiple stress states, both tensile and compressive, within the crystalline core at any given cross-section along the length of the SiNWs. This is in addition to the longitudinal stress variation due to conical lithiation discussed earlier. Hence, the crystalline cores of the lithiated SiNWs contain multiple stresses at different points, a triaxial stress state, leading to high shear stresses. The crack-like facets and distortions near the tips of the crystalline cores seen in the SEM images in Fig. 5.5 indicate such non-uniform lithiation and thus, a triaxial stress state.

The polychromatic diffraction (Laue) patterns near the tip of the SiNWs (Loc-1 in Fig. 5.6a), before battery assembly and during lithiation, were analyzed using XMAS [48]. Figure 5.7a shows the Laue pattern image of the pristine SiNW sample before battery assembly. Similarly, Fig. 5.7b and c show the Laue patterns of lithiated SiNWs

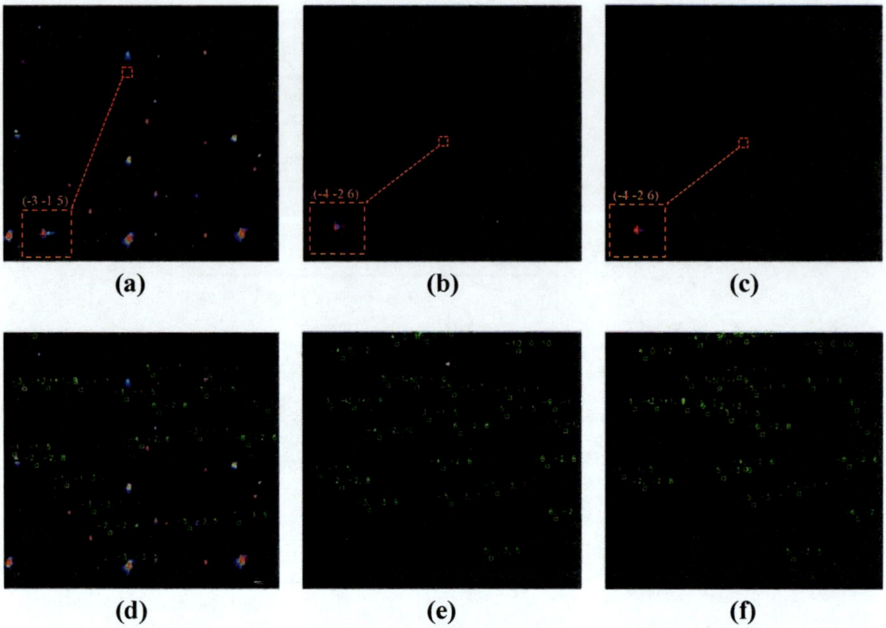

Fig. 5.7 Laue diffraction patterns (before indexing) of **a** Pristine SiNWs, before battery assembly, **b** Lithiated SiNWs @ 1.2 V, **c** Lithiated SiNWs @ 0.65 V (Zoomed-in images of the diffraction peaks of $(\bar{3}\bar{1}5)$ and $(\bar{4}\bar{2}6)$ in the insets). Indexed Laue diffraction patterns of **d** Pristine SiNWs, before battery assembly, **e** Lithiated SiNWs @ 1.2 V, **f** Lithiated SiNWs @ 0.65 V. *Note* the unindexed bright peaks in (d) are from the metals present in the sample stage (Reproduced from Reference [30] with permission from Elsevier)

at 1.2 and 0.65 V, respectively. Corresponding indexed Laue patterns are shown in Fig. 5.7d to f. In general, the Laue patterns generated by bulk monocrystalline silicon are characterized by bright and regular-shaped peaks as shown in [43]. In the case of SiNWs, the peaks are dim and odd-shaped, making indexing a tough job. Such circumstances made "special indexation" of the Laue pattern, an inevitable necessity, where we combined the automatic peak search routine with manual addition and removal of the peaks before indexation [48]. The unindexed bright peaks in the case of pristine SiNWs (Fig. 5.7d) are from the surrounding metals on the sample stage of the beamline.

The deviatoric stress tensor, which consists of the deviatoric components of the directional stresses and the shear stresses, was evaluated from the Laue pattern analysis. The resulting deviatoric stress components are shown in Fig. 5.8. The stresses in the lithiated nanowires are quite different from the pristine nanowires, clearly indicating a triaxial stress state, resulting from a strongly non-uniform lithiation in both longitudinal and circumferential directions of the nanowires. The magnitudes of the deviatoric components of the directional stresses are very low in the pristine nanowires and increased upon lithiation, which is in line with the findings of the

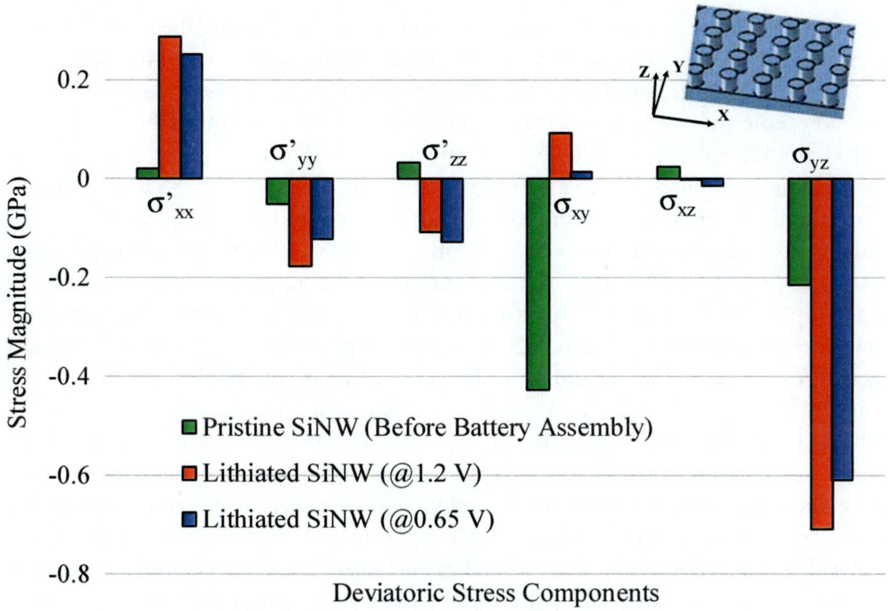

Fig. 5.8 Deviatoric stress components before and during lithiation (Reproduced from Reference [30] with permission from Elsevier)

energy scan analysis. Further, the directional components are in both compression and tension, indicating co-existent multiple stress states in the conical crystalline cores of the partially lithiated nanowires as shown in the simulations of Yang et al. [28]. The shear stresses across the nanowires in the pristine case are very high, may be due to the twisting, bending and clustering of the tips observed in the SEM image (Fig. 5.1b). After the initial lithiation (@ 1.2 V), these shear stresses across them were reduced, due to the lithiation of the tips, leading to loss of crystallinity and softening. However, the shear stresses in the planes parallel to the nanowires (YZ planes) increased significantly, may be, partly due to the non-uniform lithiation and conical crystalline core formation and partly due to the load from the expansion of the surrounding nanowires, as they are very close to each other (Fig. 5.1b). The crack-like facets, bent and twisted nanowires, visible in the SEM image of lithiated SiNWs (Fig. 5.5a, b and c) provide preliminary evidences for such possibilities. Further, the high shear stresses indicate more complicated deformation of the nanocrystals, constrained in the three dimensions by the amorphous parts of the nanowires. These constrained deformations of the nanowires lead to complicated stress states and their evolution during the electrochemical lithiation, especially the exact mechanisms preceding the fracture events, which are not explained earlier. Such a detailed understanding of the stress states of the silicon nanostructures could lead to novel insights to design the next-generation fracture-resistant nano-silicon electrodes. However, it should be noted that this polychromatic diffraction signal too is an average one over several nanowires, covering them at different lengths, and hence, the actual stresses in the single nanowire can be of much bigger magnitudes, leading to fractures.

The core–shell lithiation models in the literature consider spherical particles [24–27] and predict no shear stresses in the crystalline core upon lithiation. Yang et al. [28] simulated lithiation of a single nanowire, which does not include effect of surrounding nanowires in the electrode such as the sample (Fig. 5.1b) used in the current study. Even in our earlier report [31], the morphological evidence for conical lithiation pattern was shown, but, no information related to its effect on the stress states was presented. The present study shows that the nanowire cores are subjected to a complex stress state, with high shear stresses, along with increase of the directional stresses, during lithiation, which is both longitudinally and circumferentially uneven for a SiNW. The resolution of the stresses predicted could have been much better if the sample contain fewer and longer nanowires with larger pitch (center-to-center distance between consecutive NWs) compared to what was used in this work. However, we used the samples readily available at our disposal, which were actually made and optimized for some other study. Based on the current observations, the SiNW samples for the future studies can be optimized to suit the experiment requirements. The resolution also could have been much better if the X-ray beam is focused to ~100 nm size, but such a beamline would need a different setup, instrumentation and accessories altogether. Another factor, that could have affected the stress state in the lithiated SiNWs, is the high volume of the electrolyte in the test cell and longer distance between the lithium and silicon electrodes, compared those of a standard coil cell. These two factors may affect the rate of lithiation process and thus, the stress state at a given time during lithiation.

5.1.4 Summary

A unique lithium ion battery test cell design was presented in this work. Such a specialized test cell is essential for enabling the in-situ synchrotron scanning X-ray microdiffraction experiment of a silicon nanowire electrode during electrochemical lithiation to study the evolution of stress in the crystalline SiNWs. The test cell design was proven to be effective by successfully enabling the X-ray detector to record the microdiffraction signal from the SiNW sample.

The crystalline core of the SiNWs was found to be in compressive stress (up to ~325.5 MPa) during electrochemical lithiation using in-situ μSXRD energy scan analysis. However, the SEM image of the lithiated nanowires show conical shape with circumferential non-uniformity and crack-like facets, confirming the non-uniform lithiation pattern from the tip and also around the nanowires, indicating a triaxial stress state with high shear stresses in the SiNWs. The deviatoric stress evaluated from the Laue diffraction data analysis clearly show high shear stresses and significantly different stress state in the lithiated nanowire cores in comparison to pristine nanowires. Such stress states could lead to evolution of more complicated mechanisms leading to and especially preceding the final failure/fracture of the nanoscale electrodes of silicon. The current core–shell lithiation models do not account for such a complex stress state. Studies using this technique with tailor-made samples and advanced instrumentation would allow furthermore comprehensive experimental examination of the stress evolution in the SiNW electrodes during electrochemical lithiation. The findings of this research study help to further improve the understanding of the stress states in the SiNWs and could help to the design of fracture-free silicon nanostructures for next-generation LIBs.

5.2 Enabling Thin Silicon Technologies for Next-Generation, Novel and Innovative Crystalline Silicon Photovoltaics Systems Design

Recently, there has been a strong commercial push towards thinner silicon in the solar photovoltaic (PV) technologies due to the significant cost reduction associated with it. Tensile stress (normal, in-plane) and thus fracture of the silicon cells are increasingly observed and reported for products of crystalline solar cell technologies. In an effort to shed light on these topics, stress measurements and mapping of the solar cells in the vicinity of the most typically observed crack initiation locations using synchrotron X-ray microdiffraction technique were conducted and are reported in this section. The technique is unique as it has the capabilities to quantitatively determine stresses in silicon and to map these stresses with a micron resolution, all while the silicon cells are already encapsulated.

With this technique, we aim to gain fundamental understanding of the stress magnitudes as well as characteristics that could lead to crack initiation and propagation. We have thus far found evidences of both extrinsic (device related) as well as intrinsic (crystallographic) nature of silicon cracking, which further confirm that the control of mechanical stress is the key to enable thin silicon solar cell technologies in the coming years. This study represents an ongoing high impact technology research that addresses real and important fundamental materials issue facing the crystalline silicon solar PV industry and contributes directly to the industry drive to reduce cost of PV systems to grid parity.

5.2.1 Background and Motivation—Thin Silicon Solar Cell Technology

Silicon solar cell technology remains the most efficient and powerful way to commercially harness the sun's power for conversion to electricity. In order to keep improving the market adoption rate and thus providing solar energy that is cost-competitive with existing energy sources, the global silicon solar PV industry is aggressively reducing the cost of solar PV systems through innovations in materials and processes used in the manufacturing. One potential area of significant cost reduction is through the use of thinner silicon cells (<150 microns). However, coupled with other recent trends in the solar PV module technologies (thinner glass, lighter or no metal frames as well as increased use of some types of polymers for encapsulation of the cells), high stress and fracture of the silicon cells are increasingly observed and reported for products using thin silicon solar cell technologies. Cell-cracks immediately lower module efficiency, and can lead to premature aging of the entire package/module.

Studying the fracture mechanisms in thin silicon solar cells has thus increasingly become both an interesting topic scientifically as well as an important subject technologically during recent years, especially with the oncoming development of the new generations of solar PV renewable energy systems. More specifically, fractures in the thin silicon solar cells have been proposed to result from the high stress concentration area near the solder joints and busbars that provide interconnection between the cells [49–53]. Studying how stresses, especially in these areas, evolve during introduction of the solder joints in the solar cell assembly lines as well as later on during service and operational conditions/loading of the devices will be crucial in highlighting the propensity to fracture as well as its actual initiation and propagation mechanisms in the thin silicon solar cells. Crack propensity here is defined as the susceptibility of the material to initiate cracks and/or for the cracks to grow further. A material can be more susceptible to fracture when the stress level of the overall system is elevated due to the operational conditions or mechanical loading of the device/system. Therefore, the stress level in the material is approaching its fracture toughness which depends not only on the strength of the materials but also the ubiquitous presence of crack-like defects and their size. Quantitatively, this happens when the stress intensity, $\boldsymbol{K}$, has approached the materials' critical stress intensity, $\boldsymbol{K_c}$, or more generally, when the value of the strain energy release rate, $\boldsymbol{G}$, has approached the critical value of the strain energy release rate of the material, $\boldsymbol{G_c}$. $\boldsymbol{K_c}$ or $\boldsymbol{G_c}$ here is a measure of fracture toughness which depends on both the strength of the material and its inherent flaw sizes [49].

However, previous studies in the literature up to date on silicon cell fracture in c-Si solar PV technologies have been largely focused on the crack detection and imaging techniques [50, 51, 55–57]. Lacking the capacity to quantitatively characterize mechanical stress, which is the ultimate driving force of any crack initiation and propagation, these studies did not reveal much fundamental understanding of what affects the propensity to crack, the various failure modes, and, most importantly, the processes and loading conditions that lead to different states of mechanical stresses that ultimately cause different modes of the cracks. Synchrotron X-ray microdiffraction (μSXRD) has thus proved to be a suitable technique to elucidate stress and crack propensity in stringed silicon solar cell assemblies. This technique provides quantitative determination of stresses (normal, hydrostatic and shear) in silicon as well as in solder joint material with a micrometer resolution [42, 58]. It is non-destructive and the X-ray beam has reasonable penetration depth which enables one to do in-situ study of stresses in the silicon during cell operations or loading. This approach can thus provide us with the quantitative examination of stress and its evolution during the processes and operations of the devices in the realistic setting.

Stringed cell assemblies have always used solders and interconnects, and they have always been the high stress concentration areas in the assemblies (Fig. 5.9a). As the silicon cells get thinner (<150 microns), they acquire much higher sensitivities towards these high stresses and thus an increased propensity to cracks. Solders and metal interconnects induce high stresses in the silicon cells due to their typically very different coefficient of thermal expansion (CTE) values compared to that of silicon. Upon cooling from their typically high temperature introduction (solder reflow process), the CTE mismatch leads to high tensile in-plane stress in the silicon cell and this is what leads to crack initiation and propagation [51–53, 59]. Most cell cracking observed in PV products from various companies as well as those reported in the literature [50–53, 59] are indeed initiated from the solder joint areas and typically propagate following the metallization lines. It is thus our main focus in this study to measure stress in the silicon cell around solder joint area after lamination and cooling processes, as well as to investigate further the source of these stresses, i.e. the solder joint itself.

The rest of the section is divided into three sub-sections. Brief descriptions of the μSXRD technique used in the present work and sample preparation needed are in the Experimental sub-section. In the Results and Discussion sub-section, we present and discuss the stress measurement in the silicon solar cell around a solder joint as well as the microstructure evolution of the solder joint materials comparing between as-reflowed solder joint materials and after thermal cycles, and provide analysis how it could further aggravate the stress in the silicon near the solder joint area. The third sub-section summarizes the significance of the findings and concludes the present study.

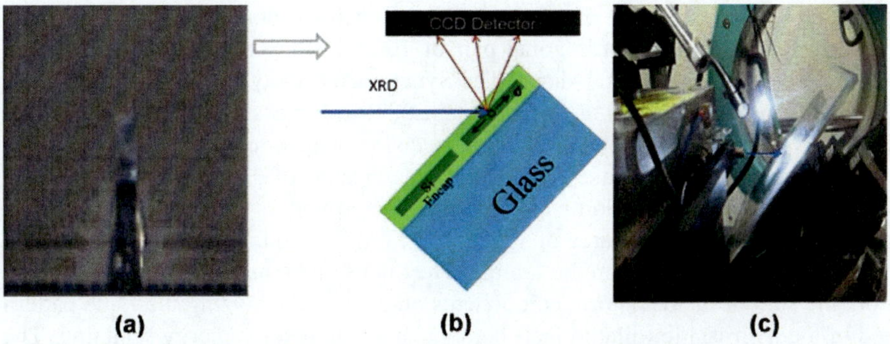

Fig. 5.9 A Optical image of silicon cell around solder joint, **b** The incoming synchrotron focused X-ray beam (the blue arrow) penetrates through the transparent backsheet layer (orange-colored) and the backside encapsulant layer (light green-colored) and is diffracted by the silicon crystal (red arrows) allowing its crystallographic information to be captured by the Charge-Coupled Device (CCD) detector (black-colored), which can then be translated into stress and, thus, crack propensity information. **c** The in-situ experimental using synchrotron X-ray micro-diffraction, where the encapsulated mini module sits on a XY piezoelectric sample stage tilted at 45° (Reproduced from Reference [42] with permission from Elsevier)

5.2.2 *Experimental Methodology—Probing Stress of Thin Silicon While Already Encapsulated*

A solar PV module typically is a sandwich of a glass, a front encapsulant layer, a stringed cell assembly, a back encapsulant layer and finally a backsheet. However, in this study, we built only a mini module, which satisfied our purpose to study the stress in the silicon after lamination process. Our mini module consists of one 125 × 125 mm monocrystalline solar cell covered by encapsulant layers on both sides, a transparent backsheet layer at the back side and a glass at the front side, laminated at high temperature using industry-typical processes and recipes (all materials are those typically used in the global PV industry) as shown in Fig. 5.9b (schematically, also showing how the in-situ X-ray microdiffraction experiment was set up). Figure 5.9c shows the actual mini module sitting on the sample stage in the X-ray microdiffraction experiment.

Stress measurements in the silicon around the solder joint were performed by synchrotron X-ray microdiffraction at beamline 12.3.2 at the Advanced Light Source (ALS) of the Lawrence Berkeley National Laboratory [58]. The stress measurement in the silicon solar cell here is unique and novel as it is done while the solar cell is already laminated (encapsulated by transparent polymers and glass) thus representing the actual solar PV system in operations in the field. This approach thus allows in-situ stress measurement of the silicon solar cell while the mini module is, for instance, being mechanically loaded or thermally cycled, to represent typical loading during real operations of the photovoltaic module system in the field. The polychromatic X-ray beam produced from a superconductor magnet source is refocused at the entrance of the experimental hutch by a 700 mm-long platinum-coated silicon toroidal mirror operating at a grazing angle of 4.5 mrad. Final focusing is done by Kirkpatrick-Baez mirrors consisting of an orthogonal pair of 100 mm-long tungsten-coated silicon substrate bent to an elliptical shape. The Synchrotron X-ray microdiffraction has 1 × 1 μm beam size, allowing one to obtain the local stress at any position on the sample. The sample sits on a XY high precision stage (range of ± 10 microns), which is mounted on a coarse XYZ Huber stage (range of ± 15 mm in XY and Z), see Fig. 5.9c. The diffraction patterns are collected with a DECTRIS Pilatus 1 M pixel area detector (active area of 179 × 169 mm), which is placed at a distance of approximately 140 mm from the sample. The total strain tensor is typically obtained from the sum of the deviatoric components obtained by analyzing the Laue pattern and of the hydrostatic (dilatational) components obtained by energy scanning. The current setup in the beamline 12.3.2 provides the capabilities to do both measurements (deviatoric as well as hydrostatic stresses).

Additionally, in the present work, this technique was also used to study the microstructure of the solder joint materials after thermal cycling (ex-situ). It is "ex situ" as we only compared here the X-ray diffraction patterns before versus after the thermal cycling and not ***during*** the thermal cycling itself. Investigation of the microstructural evolution in the solder joint materials could lead to fundamental understanding of controlling the stress in the silicon solar cells, as solder joint is the stress concentrator causing the crack initiation. For this part of the study, only soldered cells without encapsulation material were investigated. The solder joint material used here is the typical tin solder materials with a few (atomic) percentage of silver (typically used in the solar PV industry) with typical grain sizes in the range of 0.5–15 microns. The solder joint connects the silicon solar cells to the busbar (tin-coated copper ribbon Fig. 5.10c), which in turn, connects one solar cell to another solar cell (Fig. 5.10b). The thickness of the solder joint is a few tens of micrometers and its diameter is a few millimeters. The solder joint assembly was mounted in an epoxy, and the sample was cut and polished right through the center of the solder joint to allow cross-sectional examination using synchrotron X-ray microdiffraction.

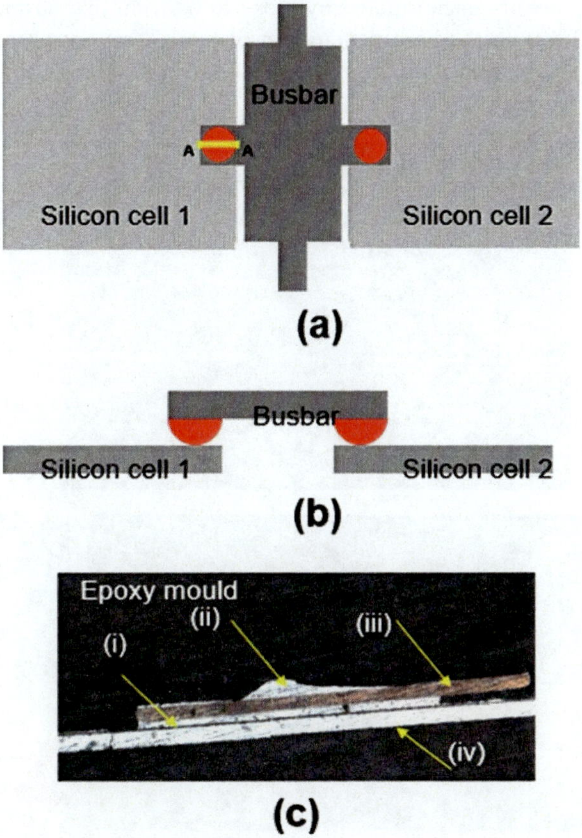

Fig. 5.10 **a**, **b** Two silicon cells with two solder joint interconnects. The red dots indicate the location of the solder joint material (the solder joint material is in between the silicon solar cell on the bottom and the busbar on the top). **c** A SEM image taken at the cross-section A–A as shown in (a) of the solder joint assembly which is embedded in the epoxy mold. The solder joint material is denoted by (i), where the one on top (ii) is solder material residue of the soldering process. The brown part is the busbar material (iii) and the flat part (iv) is the silicon solar cell (Reproduced from Reference [42] with permission from Elsevier)

The thermal cycling for this experiment was done separately for 2000 cycles, from –40 to 125 °C.

5.2.3 *Extreme Mechanics Characterization of Materials Enabling Thin Silicon Solar Cell Technology*

Synchrotron X-ray microdiffraction has been used to study and to map stress in single crystalline silicon wafer [42, 60], however, there are a few unique characteristics about the stress map in Fig. 5.11. First of all, as shown in Fig. 5.9, the silicon wafer here was already encapsulated by close to 0.5 mm of transparent polymers on each side. But with the high penetration of hard X-ray synchrotron radiation, we still could get the reflection from the silicon even though it was buried under

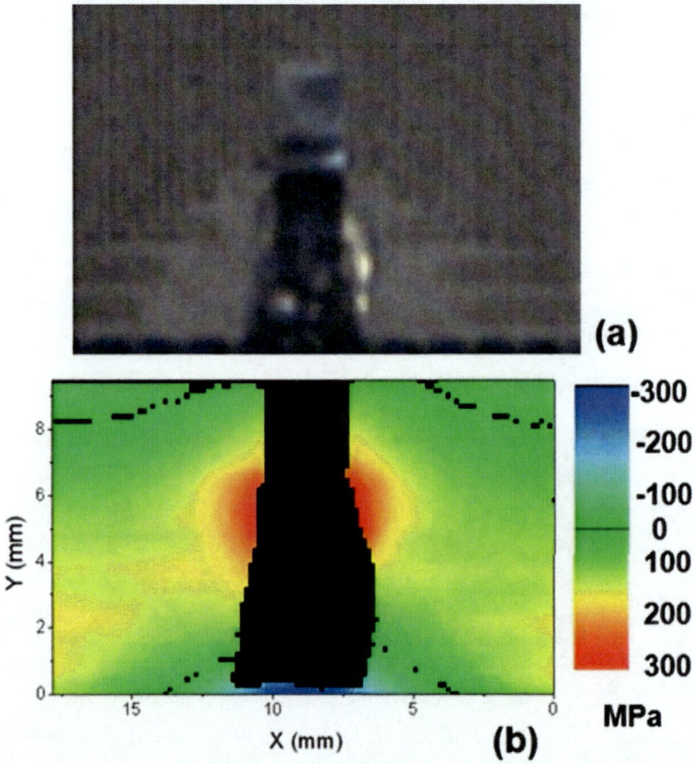

Fig. 5.11 **a** Optical image of a silicon solar cell around a solder joint of the mini module sample as illustrated in Fig. 1b and c. **b** Local curvature map of encapsulated silicon cell surrounding the solder joint (corresponding to the same area in (a)). Silicon under the solder joint (dark color in the map) is not mapped due to the X-ray's inability to penetrate through the thickness of solder joint (which is a few tens of microns) (Reproduced from Reference [42] with permission from Elsevier)

0.5 mm of encapsulant polymer material. This is a unique capability as it allows stress and microstructure evolution examination of silicon solar cells, for instance, during high temperature lamination process or even during the operation of the laminate, such as during thermal cycle or mechanical loading operation—a capability that has not hitherto been possible using charged particle microscopy techniques (Scanning Electron Microscopy and Transmission Electron Microscopy) or surface sensitive techniques (micro-Raman spectroscopy) and could lead to the unravelling of potentially important and critical insights about what is really happening in the solar cell during lamination or what the deformation mechanisms are in the silicon solar cells during thermal cycling or mechanical loading. This unique capability has been recently highlighted in the Berkeley Lab Science Highlight for the work that we did on solar cells in the Advanced Light Source (ALS) Berkeley Laboratory [44].

Secondly, the stresses here were actually determined not from the difference in lattice spacing before versus after the lamination process, but from the local curvature or, in other words, from the out-of-plane orientation of the Si (100) planes at any given pixel of the raster scanning. 3D crystal orientations of silicon were obtained from indexing the silicon Laue pattern. A methodology called "adaptive indexing" allows tracking the silicon reflections below the metallization layers thus enabling us to obtain the silicon 3D orientation information even after encapsulation. Curvature from all three directions (X, Y and Z) can thus be obtained. In that way, the stress analysis can be simplified, and there is no need to determine all the stress components, including the hydrostatic part, which often can be time consuming. As it has been already known, the stress in the single crystal silicon could be determined from the local curvature of crystal using Stoney's equation [61].

$$\sigma_{Si} = \frac{Y_m t_m^2}{6rt_{Si}(1-\nu_m)} = \frac{Y_m t_m^2 \theta}{6dt_{Si}(1-\nu_m)} \tag{5.3}$$

Stoney's Eq. (5.3) is used to calculate residual stress in bent material, where Y, t and ν are the Young's modulus, thickness and Poisson's ratio, respectively, the subscripts Si and m denote silicon cell and metallization/solder joint, respectively [61]. The local radius of curvature r is expressed as $r = d/\theta$, where d is the distance from center to the edge of the silicon chip.

We essentially use the Stoney's equation here but due to the localized geometry of our measurement, we adopted some of the modifications and approximations as detailed in [61]. In principle, however, as shown in Eq. (5.3), we could determine the stress in the material from how much it curves, or in other words, from the material's bending angle, θ, while other parameters which are geometrical parameters of the sample/experiment as well as material properties are known.

This Stoney's approach (with some modifications and approximations) could lead to potentially unique capability in studying stress in silicon solar cells in-situ during thermal cycling or mechanical loading or even during electrical/optical loading, which has not been hitherto possible or at least practical. By calculating how the curvature changes with the loading, the stress evolution could be determined almost in real time. This methodology could also help substantially in subtracting the diffraction pattern coming from the metallization and other layers on top of them, leaving behind only the diffraction pattern of the silicon substrate.

As we can see in Fig. 5.11a, there are metallization patterns and very thick solder joint along with busbar (total thickness ~0.5 mm). The X-ray beam could not penetrate through the very thick solder joint and busbar, and thus the dark region in the stress map indicating no information about silicon crystallography/stress was obtained from those pixels. But other than that, the X-ray beam was able to penetrate through the metallization layer and other materials, and thus we could obtain the silicon crystal stress everywhere else.

Our objective here is to measure the stress of the silicon in all areas including the one right under the solder joint. The stress of silicon due to the solder joint could be calculated using thermal mismatch or could even be simulated using FEM (Finite Element Method), but the actual stress on the actual solar cell (especially the one that has been encapsulated within the PV device/system package) has never been measured experimentally. This manuscript reports on the measurement that we managed to obtain the stress of the silicon solar cell around the solder joint, but not right under the solder joint because of the limitation we have mentioned above. Although the stress right under the solder joint still needs to be determined, we now know the stress (the magnitude, the components, whether or not it is hydrostatic or deviatoric, etc.) right around the solder joint, and thus allowing us to design the solder joint accordingly in such a way to minimize fracture occurrence in the silicon solar cells.

It is evident from the stress mapping in Fig. 5.11 that the stress state generally near the solder joint is in tensile (yellowish color) consistent with our expectation. The highest stresses seemed to be right around the solder joint (the red-colored regions) amounting to about 300 MPa. The effect of the metallization lines could still be seen on the left and right of the lower part of the scanning. There are subtle horizontal patterns following the metallization lines there as we can see in Fig. 5.11a. This is also expected as metallization lines are where the stresses are expected to be higher due to CTE mismatch between silicon and the metal (i.e. Cu in this case). But overall, we can see the effect from the solder joint dominates, and this is consistent with observations in the fields [51–53] as well as in the literature [59]. It was widely observed that majority of cracks originate from the solder joint areas indicating the highest stress area or stress concentration areas.

The X-ray scanning here was done in 10 microns step. The X-ray beam itself was approximately 1 micron in diameter. As a result, the data from a particular pixel is just from the area that is shined by X-ray. The 10 microns step size is sufficient as the metallization patterns in the solar cells do not have features that are typically smaller than 10 microns.

The micron resolution capability of the synchrotron X-ray microdiffraction allows the investigation of the local microstructure of the solder joint, which is a few tens of microns in thickness. Solder joint, as an interconnection between the solar cell and the environment, is usually designed to be compliant so that a large strain from the outside world, which can be caused by environments or unexpected mechanical

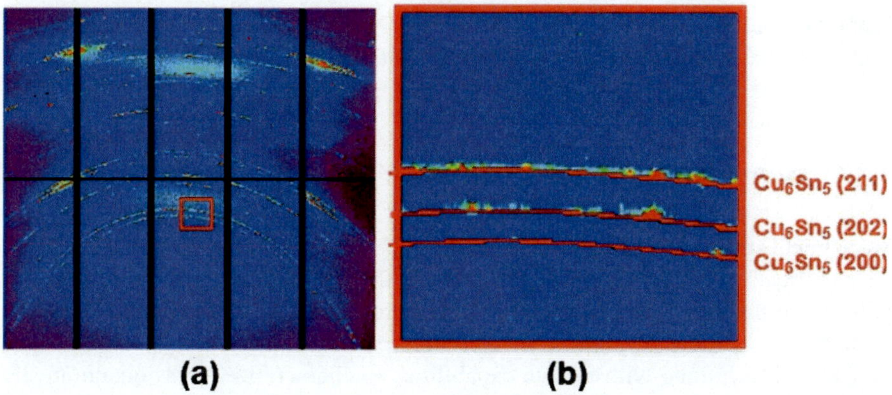

Fig. 5.12 **a** The powder diffraction pattern of the solder joint material obtained from X-ray microdiffraction resulting in the polycrystalline rings. **b** The enlargement of part (**a**) which is marked by red border lines; the indices, according to Cu6Sn5 crystal, which fit perfectly with the polycrystalline rings observed in our experiment, indicate the presence of Cu6Sn5 intermetallic compounds, among others, which were not present in the sample before the thermal cycles (Reproduced from Reference [42] with permission from Elsevier)

loading, can be damped/absorbed. Thus, the stress on the solar cells can be minimized. When a solder joint forms a brittle layer, its overall stiffness increases significantly as the brittle phase tends to be stiffer or more rigid [63, 64]. When a solder joint has effectively higher stiffness (thus lower compliance), it tends to transmit the stress more from the outside world to the solar cells increasing the stress and crack propensity inside silicon solar cell. Figure 5.12 shows the powder diffraction pattern of the solder joint material near the interface with Cu ribbon. The polycrystalline rings strongly indicate that due to the thermal cycling loading, a new intermetallic compound, Cu_6Sn_5, has formed as a result of activated interdiffusion between Cu and Sn (Fig. 5.12b). This compound has been known to form whenever Cu and Sn are in contact and at elevated temperatures [62]. Cu_6Sn_5 compound has also been known for its very brittle nature and thus could be particularly damaging and leading to higher stress and eventually producing fracture of the solder joint [62]. The presence of this layer/compound was not observed in the microstructure of the solder joint material in the bulk volume of the solder joint, far away from the interfaces with Cu ribbons (not shown in Fig. 5.12).

This is an observation that would not have been possible with a conventional X-ray diffraction due to its limited resolution. The lack of resolution would only provide the microstructure of solder joint as an aggregate from all parts of the solder joint material. And if the intermetallic compound had been, for example, small volumes as they tend to be, then its existence might not have even been detected at all. Obviously, the small volumes of this very brittle compound would be enough to lead to catastrophic failure of the overall PV system by initiating cracks which would be propagating through the solder joint material and disconnecting the whole assembly. This is again the unique capability of the synchrotron X-ray microdiffraction technique to study and characterize stress and fracture in solar PV system.

5.2.4 Summary

Mechanical stress and fracture are the origins of failures of the solar PV cells during their fabrication as well as their operations in the fields. Thus, it is very important to have the tools to characterize them in a more quantitative manner. Through the studies in this article, Synchrotron X-ray microdiffraction has thus proved to be a very useful technique with unique capabilities to characterize stress quantitatively and investigate fracture mechanisms in solar cells as well as modules. Especially once the solar cells are laminated, this tool has proved to be perhaps the only practical technique that could provide the capability to study the stress in the silicon solar cells and how it evolves with fabrication processes (during lamination, cooling, etc.) as well as with operations or loadings in the fields. This has been an even more important capability, as recently many researchers have started to observe the rather unexpected substantial effect of the mechanical properties of the encapsulant on the stress level, and thus ultimately on the cracking rate of the silicon solar cells in the fully integrated PV package. If the effect of the polymer material is important, then the silicon solar cell needs to be studied when it has already been encapsulated by it. Secondly, solder joint is the origin of majority of the silicon fracture. This technique provides the unique capability to study the local microstructure and how it evolves during processing or operations/loadings in the fields. This would enable, for instance, fundamental understanding of the origin and degradation mechanisms of the embrittlement of solder joint due to thermal cycling as shown in this manuscript. The exact understanding of the origin and the mechanisms of stresses in the thin silicon solar cells are the critical factors to enable the thin silicon (<150 microns) technology for lower cost next-generation PV systems.

5.3 Low-Stress Encapsulants?—Enabling PV with Extreme Reliability

In this section, we study the effect of two polymer encapsulations with different material properties, such as Young's modulus (E) and yield strength, on the residual stress of monocrystalline silicon solar cells. We observe through synchrotron X-ray microdiffraction that solar photovoltaic (PV) module laminated with encapsulants A (soft) and B (stiff) which have Young's modulus of 6.34 and 28.32 MPa, respectively, reveals distinct variations in residual stress of silicon. The stress of silicon measured near the solder (stress concentration region) showed a maximum quantitative value of 300 MPa with encapsulant A, whereas for the solar PV with encapsulant B, it showed a much higher value of 450 MPa. Further correlation of stress to fracture/crack initiation events of silicon was also understood using three-point bending tests. The result shows that with encapsulant A, crack initiation of silicon at a mean force of 1.2 KN is observed, whereas for the PV with encapsulant B, silicon cracked at much lower force of 0.3 KN. These studies confirm that encapsulant materials have a significant effect on the residual stress of silicon, which directly affects the working efficiency and reliability of the solar PV.

5.3.1 Background and Motivation—Thinner But Stronger Silicon Solar PV

Recently, there has been strong commercial push in silicon-based solar PV industries especially towards enabling thinner and thinner silicon solar cells (<150 μm) due to the significant cost reduction associated with it. Even though the module becomes thinner, the solar PV needs to have a lifetime operation of more than 20 years, to be considered as a cost-effective solution [52, 65]. In the field, the PV module is always under the influence of various external loading conditions, such as mechanical loads due to wind or snow, along with other environmental conditions, i.e. extreme temperatures, thermal cycles, etc. These external factors increase the built-in stress, in addition to the inherent residual stresses developed during the processing such as lamination, etc. Further, this eventually leads to fracture events, thereby affecting the efficiency and reliability of the solar PV module [66–68].

Currently, in the solar PV community, there is large interest and focus on understanding the role of lamination materials and processes, in addition to the soldering materials and processes [69]. The encapsulant material used in solar PV industry must satisfy several different requirements, which include high optical trans- mittance of incident light, good dielectric properties (electrically insulating), mechanical compliance to protect the solar cells from external mechanical loads and stresses induced by differences in thermal expansion coefficients, good adhesion to both glass and silicon solar cells [66]. Silicones as an alternative encapsulation material were experimented by Ketola et al. [70]. Their objective of using this material was based on high transparency offered in the UV-visible wavelengths, wide range of refractive indices and excellent tolerance to high optical flux. Hurter et al. [71] also assessed glass fiber reinforced plastics (GFRP) composite materials as encapsulation to study the electrical and thermal efficiency of the solar PV. Studies pertinent to solely effect of encapsulation material properties and its relation to building up of the PV residual stress are rather limited and have not been much reported in the literature. Recently, Dietrich et al. [69] have experimentally characterized the in-plane fracture strength of the solar PV module comparing two different encapsulation materials, namely, EVA having high and low stiffness properties. Their results clearly show that PV module encapsulated with high stiffness EVA cracked at lower load, whereas module resisted high load before initiation of crack, when encapsulated with the low stiffness EVA.

Similarly, Mickiewicz et al. [72] also analyzed the encapsulation material stiffness at cold environmental conditions and subsequently, its effect on the fracture of the solar PV module.

This preliminary evidence on the effect of laminating materials and processes motivated us to understand the basic mechanisms, thus opening the research gap for fundamental and quantitative investigation of the stress which is the driving force of fracture, therefore the subject of this paper. Studying how the effect of the encapsulant on the stress level after the lamination process will be crucial in highlighting the propensity to fracture in the thin silicon solar cells. Since it is the effect of the lamination material, we study the stress gradients of silicon cell while it is already encapsulated/laminated and thus the synchrotron X-ray microdiffraction is the unique technique, which allows us to map this evolution of stress [73]. The silicon wafer here was already encapsulated by 0.5 mm of transparent polymers on each side. But with the high penetration of hard X-ray synchrotron radiation, we still could get the reflection from the silicon even though it was buried under 0.5 mm of encapsulant polymer material. This is a unique capability as it allows stress and microstructure evolution examination of silicon solar cells, for instance, during high temperature lamination process, or even during the operation (in operando) of the laminate, such as during thermal cycle or mechanical loading operation. Also, this kind of study is not hitherto possible using charged particle microscopy technique such as SEM and TEM or surface sensitive techniques such as micro-Raman spectroscopy. This technique could further lead to unravelling of potentially important and critical insights about what is really happening in the solar cell during lamination or what the deformation mechanisms are in the silicon solar cells during thermal cycling or mechanical loading. Our recent study on solar cells has garnered attention at the Advanced Light Source (ALS), Berkeley Lab Science Highlight [74].

In this study, we use two different encapsulation materials, namely, an industry standard EVA (encapsulant A) and a commercially available thermoplastic (encapsulant B). Both the materials are then used as a lamination for the solar PV module. Then, solar PV module laminated with materials A and B are characterized for residual stress using synchrotron X-ray microdiffraction at Advanced Light Source (ALS) beamline 12.3.2. We further conducted three-point bending tests of the PV module to understand the fracture/crack initiation mechanisms.

5.3.2 Experimental Methodology—Probing Strength of Thin Silicon PV Laminates

A series of PV mini modules/laminates (one-cell coupons with commercially available monocrystalline silicon solar cells inside) were prepared using the two different encapsulants: an industry standard EVA (encapsulant A) and a commercially available thermoplastic (encapsulant B). Every laminated mini module consists of a glass, a front encapsulant layer, a 125 × 125 mm monocrystalline silicon solar cell, a back encapsulant layer and finally a backsheet. Stress measurements in the silicon around the solder joint were performed by synchrotron X-ray microdiffraction (beamline 12.3.2) at the Advanced Light Source (ALS) of the Lawrence Berkeley National Laboratory following the methodology that has been described earlier in this Chapter (Section 5.2.2 Experimental Methodology—Probing Stress of Thin Silicon While Already Encapsulate). Three-point bending tests of the PV module were also conducted to investigate the fracture. Materials are characterized using advanced mechanical testing methodologies as described below.

Tensile Tests

The tensile mechanical properties of encapsulants A and B at room temperature (25 °C) were measured using the Instron testing machine. The initial samples were 25 mm wide (W) with a total length of 150 (L) mm, which was kept constant for all the tests. An initial thickness of 0.42 (sample A) and 0.46 mm (sample B), respectively, were noted. For testing, aluminum tabs were bonded to the soft samples using an adhesive, which helped in preventing the erosion during gripping. An extensometer with a gauge length (lo) of 50 mm was used to measure the engineering strain (e = l/lo), as shown in Fig. 5.13a. Different displacement rates of 10, 100 and 200 mm/min were applied, and the corresponding load-displacement were recorded for 10 samples at each of this testing condition. The samples were strained to a value of 0.3 (mm/mm), which was close to the maximum limit of the extensometer displacement (25 mm) and thereby, the tests were stopped. The samples after the end of the tests are shown in Fig. 5.13b.

Dynamic Mechanical Analysis

Dynamic mechanical analysis (DMA) was also used to test the high temperature viscoelastic properties of the encapsulation materials A and B. An equal sample

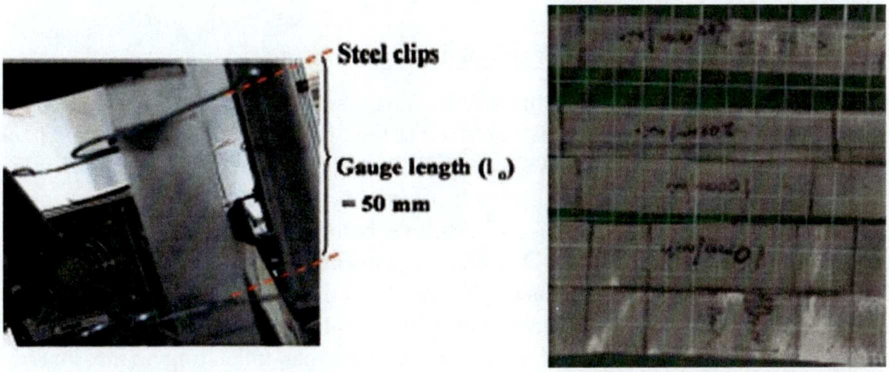

(a) Extensometer (steel clips) mounted at the gauge length.

(b) Samples at the end of the test.

Fig. 5.13 Tensile testing of encapsulation materials (Reproduced from Reference [73] with permission from Elsevier) length of 20 mm was used for this testing. Since the samples being soft and compliant, the tests were run using the tension mode set up. All the samples (five from each material) were subjected to a temperature range of 20–100 °C with a heating rate of 4 °C /min. The test condition was kept constant for all the samples.

In-Laminate Three-Point Bending Tests

An in-laminate three-point bending setup as shown in Fig. 5.14 was used to conduct the tests. In this test, a PV coupon is placed on two parallel supporting pins with a distance of 10 cm. A uniform quasi static loading is monotonically applied in the center of the glass by means of a loading pin (the maximum force of the load cell 50 ± 5% KN). The supporting and loading pins are mounted in a way, allowing the samples to freely rotate about axis parallel to the pin axis. As the load is increased monotonically, visual inspection was conducted via a camera, which transmits the magnified images to a computer to allow visual/manual detection of macro scale cracks in the silicon solar cells. At the first sight of macro scale cracks in the silicon solar cells, the loading force was noted and later indicated as "Characteristic Strength" of the encapsulated solar cells in Sect. 3.3.

5.3.3 Low-Stress Encapsulant Enabling Extreme Fracture-Resistant Silicon PV Modules

The results of quantitative residual stress of the silicon measured in the laminated PV module using the synchrotron X-ray source is presented followed by the mechanical properties of the two encapsulation materials A and B. Further, to experimentally simulate the environmental loading, such as snow, wind, etc., investigations were carried out using three-point bending test to characterize the fracture initiation of the solar PV module.

Residual Stress Map of the Silicon Solar Cells

The stresses on the silicon near the solder joint (stress concentration region) were determined from the out-of-plane orientation of the Si (100) planes at any given pixel of the raster scanning using XMAS code [58]. 3D crystal orientations of silicon were obtained from indexing the silicon Laue pattern. A methodology called "adaptive indexing" allows tracking the silicon reflections below the metallization layers thus enabling us to obtain the silicon 3D orientation information even after encapsulation. Curvature from all three directions (X, Y and Z) can thus be obtained. As we can see in Fig. 5.15a and b, the X-ray beam could not penetrate through the very thick solder joint and busbar, and thus the dark region in the stress map indicating no information about silicon crystallography/stress was obtained from those pixels. Apart from that, the X-ray beam was able to probe through the metallization layer and other materials, and thus we could obtain the silicon crystal stress everywhere else. The effect of the metallization lines could still be seen on the left and right of the lower part of the scanning. This is also expected as metallization lines are the regions where the stresses are expected to be higher due to mismatch of coefficient of thermal expansion (CTE) between silicon and the metal (i.e. Cu in this case).

Our objective here is to measure the stress of the silicon as the effect of each encapsulant material. The stress map of the silicon with encapsulant A is shown in Fig. 5.15a. It can be seen that the in-plane stress state (along X) of the silicon solar cell is in tensile (Red to green color, amounting about minimum 200 MPa). The nature of stress (tensile) in the silicon cell (in-plane, along X) agrees well with the stress-strain distribution map obtained through modeling as reported by Dietrich et al. [69]. The highest stresses seemed to be right around the solder joint (the red-colored regions) amounting to about 300 MPa. Figure 5.15b shows the stress mapping of the silicon with encapsulant B. Observations of stress state near the solder joint is similar with higher tensile than the one with encapsulant A. The highest stress was also found to be right around the solder joint (the red-colored regions) amounting to about 450 MPa. From these results, it shows that encapsulation material property has substantial effect on the residual stress of silicon. We further analyze the encapsulation material properties and its correlation to fracture, which will be discussed in the following sections.

Encapsulation Polymer Materials Properties

The representative engineering and true stress verus strain of samples A and B tested at a displacement rate of 100 mm/min is shown in Fig. 5.16a. The engineering strain directly measured from extensometer is converted to true strain $\varepsilon T = ln\ (1 + e)$. Similarly, the engineering stress σe calculated based on the original cross-sectional area (Ao) of the samples is used to obtain the true stress, $\sigma T = \sigma e\ (1 + e)$. The Young's modulus E obtained from the slope of the linear true stress region is shown in Fig. 5.16b with respect to various displacement rates. From the results of Young's modulus, i.e. stiffness versus displacement rate, it can be seen that sample A tested at displacement rate of 10 mm/min has lower mean modulus of 6.34 MPa compared to sample B which shows much higher value of 28.32 MPa, tested at the same condition. Moreover, the modulus values measured at different strain rates such as 100 and 200 mm/min also show similar behavior. The Young's modulus of sample A is 10.79 and 11.81 MPa, respectively, whereas sample B shows a value of 32.41 and 36.51 MPa for the same displacement rates of 100 and 200 mm/min. Further comparing the Young's modulus of samples A and B, it can be clearly observed that sample B is approximately three to four times stiffer as compared to sample A for all the displacement rate conditions.

From the DMA results, the storage modulus obtained from the high temperature tensile testing of both the encapsulation materials A and B are also shown in Fig. 5.16c. The results show that sample B has almost two orders (log scale) higher storage modulus compared to sample A at all temperatures from 20–100 °C. This also confirms our previous observations that the sample B is stiffer than A and does not show signs of significant softening even at high temperature (100 °C).

Fracture Observations

The characteristic fracture load of the solar PV obtained from the three-point bending tests of the samples from both group (20 from each) is shown in Fig. 5.17. The samples were subjected to multiple load steps of 30 N each and monitored for crack initiation through visual inspection. Visual inspection is not the ideal way here to detect cracks in the solar cells—an electro-luminescence (EL) should really be used. However, due to lack of facilities in our new university (SUTD), this method of crack detection/inspection was used to obtain a rough estimate of the fracture load. A load drop during the test was also a key parameter to check the crack initiation. When obvious crack propagation was found, the tests were stopped at those loads and the readings were noted. The results show a mean fracture load of 1.2 KN for PV laminated with encapsulation material A, whereas for specimens encapsulated with material B, the mean fracture load dropped considerably to 0.3 KN. These observations show that PV with encapsulant A can resist higher external loads, i.e. both mechanical and environmental before failure.

It is worth mentioning here that the silicon solar cells are back-contacted and the solder joints are on the bottom side in Fig. 5.14, although due to the thickness of the glass in this assembly (i.e. 3 mm), the whole thickness of the silicon solar cells here is under tensile stress. Cracks do tend to initiate in the bottom side of the silicon solar cells (due to slightly higher tensile stress), but they also tend to transmit through the thickness of the silicon due to single crystalline nature of the samples used here, thus appearing on the upper side almost the same time they were initiated on the bottom side, given the increasing bending load. The fact that the upper side of the solar cells are black-colored also provides better contrast for the crack detection. Furthermore, these results do suggest significant difference in the signals from the two groups (beyond statistical requirements). This gave us more confidence in the data. In addition, similar studies by other groups [66, 72] have also concluded that different encapsulation materials could lead to distinct fracture propensity of the PV mini modules with otherwise the same, nominal silicon solar cells inside.

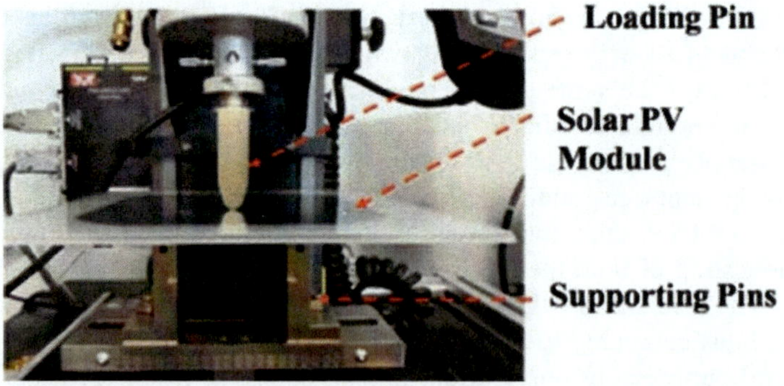

Fig. 5.14 Three-point bending test setup (Reproduced from Reference [73] with permission from Elsevier)

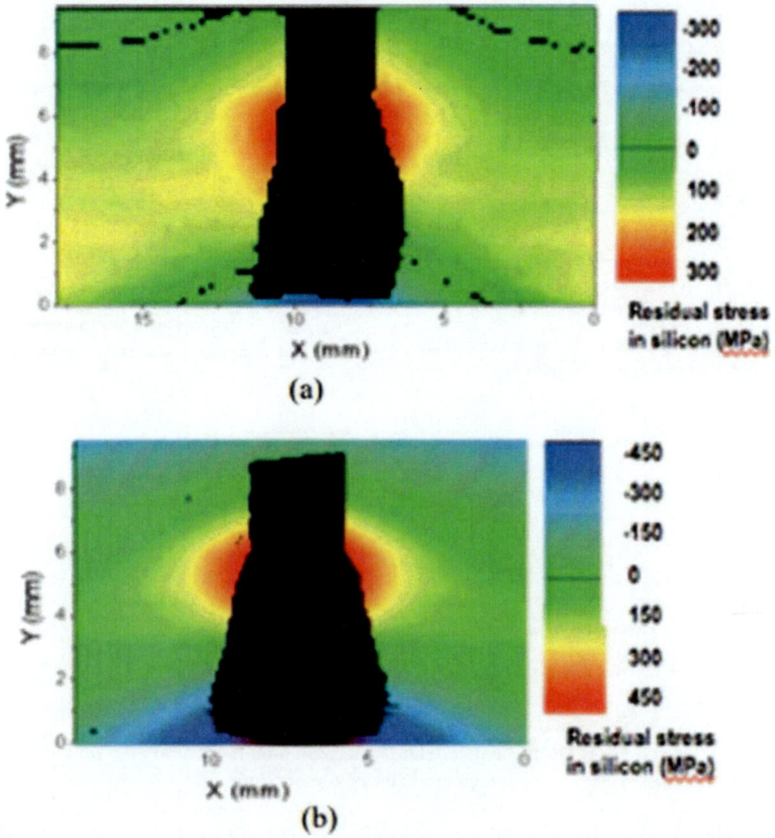

Fig. 5.15 Residual stress map of silicon cell near the solder joint of solar PV module, laminated with **a**. Material A, and **b**. Material B (Reproduced from Reference [73] with permission from Elsevier)

The laminated PV mini modules, encapsulated with these two materials A and B show distinct variations of residual stress in silicon near the solder region as revealed by Synchrotron X-ray microdiffraction (Fig. 5.15a and b) which is expected based on the different material stiffness property. From the mechanical properties of the encapsulation materials used in this study, an industry standard EVA (encapsulant A) and a commercially available thermoplastic (encapsulant B), it is distinctively clear from Fig. 5.16b that material B is approximately three to four times stiffer than A. Viscoelastic behavior (Fig. 5.16c) to understand these materials at near lamination processing temperature (150 °C), shows similar observations. The encapsulation material B is stiffer even at the maximum test temperature of 100 °C as compared to encapsulant A. During lamination (at high temperature of 150 °C), the lamination pressure might give rise to additional stress con- centration when the solar cells with very inhomogeneous topology (due to solder joints and metallization bars, etc.)

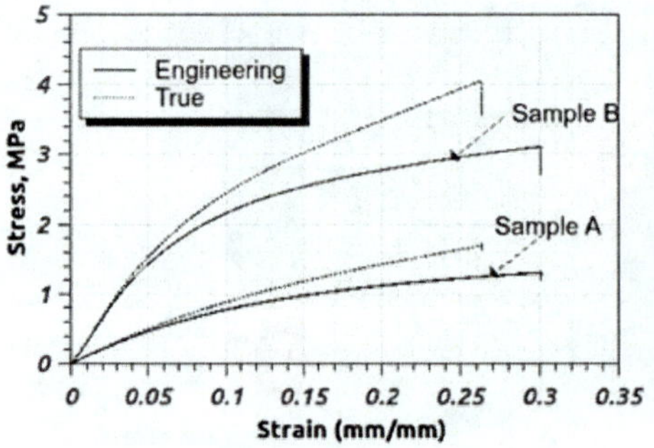

(a) Representative stress vs. strain curves at displacement displacement rate of 100 mm/min

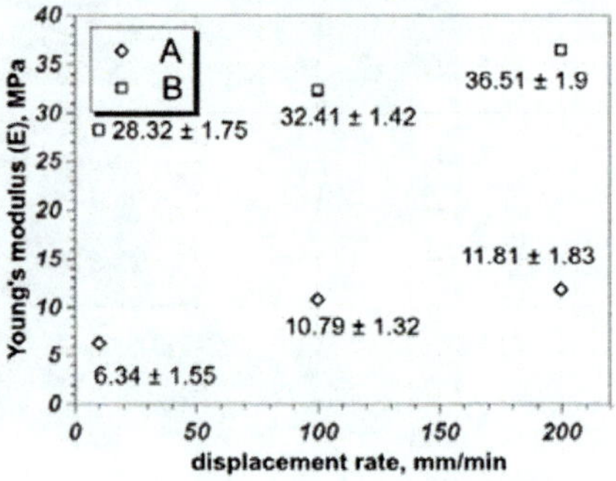

(b) Young's modulus at various rates.

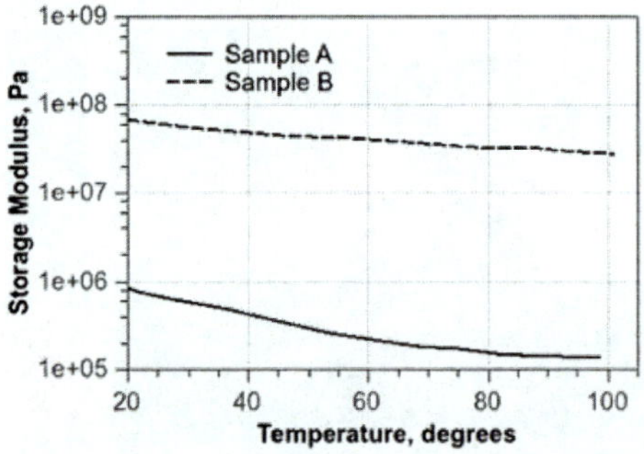

(c) Representative DMA (tension mode) curve of storage modulus vs. temperature

Fig. 5.16 Mechanical properties of the encapsulation materials A and B (Reproduced from Reference [73] with permission from Elsevier)

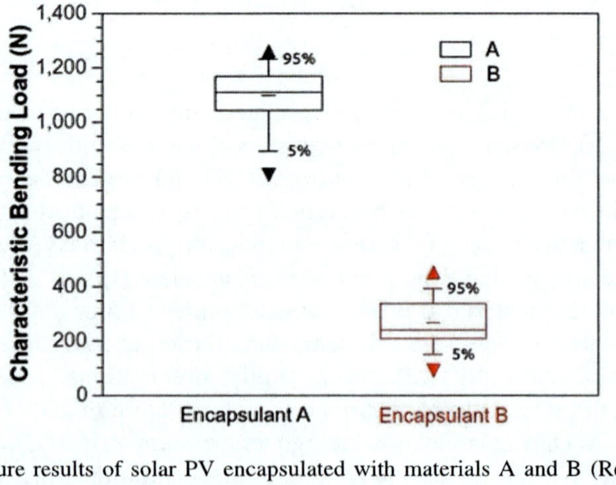

Fig. 5.17 Fracture results of solar PV encapsulated with materials A and B (Reproduced from Reference [73] with permission from Elsevier)

were pressed against a stiffer encapsulant (compared to a softer one). This could lead to much higher stress concentration (450 MPa) in the silicon solar cells of Samples B (with stiffer encapsulant), thus leading to much earlier crack initiation and propagation (at 0.3 KN of bending force). While for Samples A, due to its softer encapsulant, the pressure from the laminator might give rise to lower stress concentration (300 MPa) in the silicon solar cells and thus leading to much delayed crack initiation and propagation (at 1.2 KN of bending force). This could be one of the possible explanations, however, the exact mechanisms leading to the phenomenon observed in this study are a subject of our continued investigations.

The effect of the solder joint on the stress concentration on silicon solar cells is obviously noted, but as both groups of samples used nominally the same silicon solar cells (from the same production batch, etc.) with the same solder joints materials and processes (save for a minimum high-volume manufacturing variation), the effect of the solder joint on the stress concentration on silicon solar cells in both groups of Samples, A and B, would be nominally the same. Thus, the difference observed in terms of residual stress on silicon solar cells as well as the load at which fracture events start occurring could be solely attributed to the difference in the encapsulants. Further analyzing the fracture, residual stress indeed mimics like a pre-notch, which under a homogeneous loading, aids in the crack initiation and propagation. The in-plane stresses of the monocrystalline silicon were all tensile (Fig. 5.15a and b) and the highest compared to other components of the PV module [69]. Based on the bending theory, during the loading, the maximum in-plane tensile stresses accumulated during lamination will obviously increase from the solder (supporting pin side) and we suspect that crack would have initiated from there. This is supported based on results of Schroder et al. [68], who also showed that the silicon cracks from the bottom side during the bending tests.

5.3.4 Summary

Work reported in this section has presented the effect of lamination material property on the residual stress of monocrystalline silicon solar cell using advanced materials and stress characterization techniques. Solar PV laminated with encapsulation materials A (Soft) and B (Stiff) shows different residual stress of silicon cell (Fig. 5.15a and b). The residual stress of silicon with material A shows a maximum value of 300 MPa near the solder whereas with material B shows 450 MPa. The characteristic fracture load of the PV encapsulated with material A shows 1.2 KN compared to 0.3 KN for PV laminated with material B. The exact mechanisms leading to these findings are a subject of continued investigations, which could lead to a low stress encapsulation technology that would potentially have a great impact to the viability of the solar PV technology to replace the fossil-based energy sources.

References

1. Omar, N., Daowd, M., Bossche, P.V.D., Hegazy, O., Smekens, J., Coosemans, T., Mierlo, J.V.: Rechargeable energy storage systems for plug-in hybrid electric vehicles—assessment of electrical characteristics. Energies **5**(8), 2952–2988 (2012)
2. Lukic, S.M., Cao, J., Bansal, R.C., Rodriguez, F., Emadi, A.: Energy storage systems for automotive applications. IEEE Trans. Industr. Electron. **55**(6), 2258–2267 (2008)
3. Saw, L.H., Ye, Y., Tay, A.A.: Integration issues of lithium-ion battery into electric vehicles battery pack. J. Clean. Prod. **113**, 1032–1045 (2016)
4. Chan, C.K., Peng, H., Liu, G., McIlwrath, K., Zhang, X.F., Huggins, R.A., Cui, Y.: High-performance lithium battery anodes using silicon nanowires. Nat. Nanotechnol. **3**(1), 31–35 (2008)
5. Li, J.Y., Xu, Q., Li, G., Yin, Y.X., Wan, L.J., Guo, Y.G.: Research progress regarding Si-based anode materials towards practical application in high energy density Li-ion batteries. Mater. Chem. Front. **1**(9), 1691–1708 (2017)
6. Lee, S.J., Lee, J.K., Chung, S.H., Lee, H.Y., Lee, S.M., Baik, H.K.: Stress effect on cycle properties of the silicon thin-film anode. J. Power Sourc. **97**, 191–193 (2001)
7. Beaulieu, L.Y., Eberman, K.W., Turner, R.L., Krause, L.J., Dahn, J.R.: Colossal reversible volume changes in lithium alloys. Electrochem. Solid State Lett. **4**(9), A137–A140 (2001)
8. Bruce, P.G., Scrosati, B., Tarascon, J.M.: Nanomaterials for rechargeable lithium batteries. Angew. Chem. Int. Ed. **47**(16), 2930–2946 (2008)
9. Park, M.H., Kim, M.G., Joo, J., Kim, K., Kim, J., Ahn, S., Cui, Y., Cho, J.: Silicon nanotube battery anodes. Nano Lett. **9**(11), 3844–3847 (2009)
10. Lee, S.W., McDowell, M.T., Berla, L.A., Nix, W.D., Cui, Y.: Fracture of crystalline silicon nanopillars during electrochemical lithium insertion. Proc. Natl. Acad. Sci. **109**(11), 4080–4085 (2012)
11. Liu, X.H., Zhong, L., Huang, S., Mao, S.X., Zhu, T., Huang, J.Y.: Size-dependent fracture of silicon nanoparticles during lithiation. ACS Nano **6**(2), 1522–1531 (2012)
12. Ryu, I., Choi, J.W., Cui, Y., Nix, W.D.: Size-dependent fracture of Si nanowire battery anodes. J. Mech. Phys. Solids **59**(9), 1717–1730 (2011)
13. Shelke, M.V., Gullapalli, H., Kalaga, K., Rodrigues, M.T.F., Devarapalli, R.R., Vajtai, R. Ajayan, P.M.: Facile synthesis of 3D anode assembly with Si nanoparticles sealed in highly pure few layer graphene deposited on porous current collector for long life li-ion battery. Adv. Mater. Interf. **4**(10) (2017)
14. Zhou, X., Yin, Y.X., Wan, L.J., Guo, Y.G.: Facile synthesis of silicon nanoparticles inserted into graphene sheets as improved anode materials for lithium-ion batteries. Chem. Commun. **48**(16), 2198–2200 (2012)
15. Ma, T., Yu, X., Li, H., Zhang, W., Cheng, X., Zhu, W., Qiu, X.: High volumetric capacity of hollow structured SnO2@ Si nanospheres for lithium-ion batteries. Nano Lett. **17**(6), 3959–3964 (2017)
16. Ma, H., Cheng, F., Chen, J.Y., Zhao, J.Z., Li, C.S., Tao, Z.L., Liang, J.: Nest-like silicon nanospheres for high-capacity lithium storage. Adv. Mater. **19**(22), 4067–4070 (2007)
17. Kennedy, T., Brandon, M., Ryan, K.M.: Advances in the application of silicon and germanium nanowires for high-performance lithium-ion batteries. Adv. Mater. **28**(27), 5696–5704 (2016)
18. Shim, H.C., Woo, C.S., Hyun, S.: Silicon-carbon nanotube aerogel core-shell nanostructures for lithium-ion batteries with long-cycle life and high capacity. In *ECS Meeting Abstracts* (no. 3, p. 276). IOP Publishing (2016, September)
19. Zhang, S., Du, Z., Lin, R., Jiang, T., Liu, G., Wu, X., Weng, D.: Nickel nanocone-array supported silicon anode for high-performance lithium-ion batteries. Adv. Mater. **22**(47), 5378–5382 (2010)
20. Yang, H., Huang, S., Huang, X., Fan, F., Liang, W., Liu, X.H., Chen, L.Q., Huang, J.Y., Li, J., Zhu, T., Zhang, S.: Orientation-dependent interfacial mobility governs the anisotropic swelling in lithiated silicon nanowires. Nano Lett. **12**(4), 1953–1958 (2012)

21. Liu, X.H., Zheng, H., Zhong, L., Huang, S., Karki, K., Zhang, L.Q., Liu, Y., Kushima, A., Liang, W.T., Wang, J.W., Cho, J.H.: Anisotropic swelling and fracture of silicon nanowires during lithiation. Nano Lett. **11**(8), 3312–3318 (2011)
22. Chang, S.W., Chuang, V.P., Boles, S.T., Ross, C.A., Thompson, C.V.: Densely packed arrays of ultra-high-aspect-ratio silicon nanowires fabricated using block-copolymer lithography and metal-assisted etching. Adv. Func. Mater. **19**(15), 2495–2500 (2009)
23. Choi, W.K., Liew, T.H., Dawood, M.K., Smith, H.I., Thompson, C.V., Hong, M.H.: Synthesis of silicon nanowires and nanofin arrays using interference lithography and catalytic etching. Nano Lett. **8**(11), 3799–3802 (2008)
24. Zhao, K., Pharr, M., Wan, Q., Wang, W.L., Kaxiras, E., Vlassak, J.J., Suo, Z.: Concurrent reaction and plasticity during initial lithiation of crystalline silicon in lithium-ion batteries. J. Electrochem. Soc. **159**(3), A238–A243 (2011)
25. Huang, S., Fan, F., Li, J., Zhang, S., Zhu, T.: Stress generation during lithiation of high-capacity electrode particles in lithium ion batteries. Acta Mater. **61**(12), 4354–4364 (2013)
26. Chen, L., Fan, F., Hong, L., Chen, J., Ji, Y.Z., Zhang, S.L., Zhu, T., Chen, L.Q.: A phase-field model coupled with large elasto-plastic deformation: application to lithiated silicon electrodes. J. Electrochem. Soc. **161**(11), F3164–F3172 (2014)
27. Zuo, P., Zhao, Y.P.: Phase field modeling of lithium diffusion, finite deformation, stress evolution and crack propagation in lithium ion battery. Extreme Mechan. Lett. **9**, 467–479 (2016)
28. Yang, H., Fan, F., Liang, W., Guo, X., Zhu, T., Zhang, S.: A chemo-mechanical model of lithiation in silicon. J. Mech. Phys. Solids **70**, 349–361 (2014)
29. Lee, S.W., Lee, H.W., Ryu, I., Nix, W.D., Gao, H., Cui, Y.: Kinetics and fracture resistance of lithiated silicon nanostructure pairs controlled by their mechanical interaction. Nat. Commun. **6**(1), 1–7 (2015)
30. Tippabhotla, S.K., Radchenko, I., Stan, C.V., Tamura, N., Budiman, A.S.: Enabling the study of stress states using in situ μSXRD in the silicon nanowire anode during electrochemical lithiation in a specially designed Li-ion battery test cell. Proc. Eng. **215**, 263–275 (2017)
31. Ali, I., Tippabhotla, S.K., Radchenko, I., Al-Obeidi, A., Stan, C.V., Tamura, N., Budiman, A.S.: Probing stress states in silicon nanowires during electrochemical lithiation using in situ synchrotron X-ray microdiffraction. Front. Energy Res. **6**, 19 (2018)
32. Budiman, A.S., Nix, W.D., Tamura, N., Valek, B.C., Gadre, K., Maiz, J., Spolenak, R., Patel, J.R.: Crystal plasticity in Cu damascene interconnect lines undergoing electromigration as revealed by synchrotron x-ray microdiffraction. Appl. Phys. Lett. **88**(23), 233515 (2006)
33. Budiman, A.S., Hau-Riege, C.S., Baek, W.C., Lor, C., Huang, A., Kim, H.S., Neubauer, G., Pak, J., Besser, P.R., Nix, W.D.: Electromigration-induced plastic deformation in Cu interconnects: effects on current density exponent, n, and implications for EM reliability assessment. J. Electron. Mater. **39**(11), 2483–2488 (2010)
34. Budiman, A.S., Besser, P.R., Hau-Riege, C.S., Marathe, A., Joo, Y.C., Tamura, N., Patel, J.R., Nix, W.D.: Electromigration-induced plasticity: Texture correlation and implications for reliability assessment. J. Electron. Mater. **38**(3), 379–391 (2009)
35. Budiman, A.S., Shin, H.A.S., Kim, B.J., Hwang, S.H., Son, H.Y., Suh, M.S., Chung, Q.H., Byun, K.Y., Tamura, N., Kunz, M., Joo, Y.C.: Measurement of stresses in Cu and Si around through-silicon via by synchrotron X-ray microdiffraction for 3-dimensional integrated circuits. Microelectron. Reliab. **52**(3), 530–533 (2012)
36. Kim, B.J., Kim, J.H., Hwang, S.H., Budiman, A.S., Son, H.Y., Byun, K.Y., Tamura, N., Kunz, M., Kim, D.I., Joo, Y.C.: Microstructure evolution and defect formation in Cu through-silicon vias (TSVs) during thermal annealing. J. Electron. Mater. **41**(4), 712–719 (2012)
37. Radchenko, I., Tippabhotla, S.K., Tamura, N., Budiman, A.S.: Probing phase transformations and microstructural evolutions at the small scales: Synchrotron X-ray microdiffraction for advanced applications in 3D IC (integrated circuits) and solar PV (photovoltaic) devices. J. Electron. Mater. **45**(12), 6222–6232 (2016)
38. Budiman, A.S., Lee, G., Burek, M.J., Jang, D., Han, S.M.J., Tamura, N., Kunz, M., Greer, J.R., Tsui, T.Y.: Plasticity of indium nanostructures as revealed by synchrotron X-ray microdiffraction. Mater. Sci. Eng., A **538**, 89–97 (2012)

39. Budiman, A.S., Han, S.M., Li, N., Wei, Q.M., Dickerson, P., Tamura, N., Kunz, M., Misra, A.: Plasticity in the nanoscale Cu/Nb single-crystal multilayers as revealed by synchrotron Laue x-ray microdiffraction. J. Mater. Res. **27**(3), 599–611 (2012)
40. Kim, Y., Budiman, A.S., Baldwin, J.K., Mara, N.A., Misra, A., Han, S.M.: Microcompression study of Al-Nb nanoscale multilayers. J. Mater. Res. **27**(3), 592–598 (2012)
41. Budiman, A.S., Narayanan, K.R., Li, N., Wang, J., Tamura, N., Kunz, M., Misra, A.: Plasticity evolution in nanoscale Cu/Nb single-crystal multilayers as revealed by synchrotron X-ray microdiffraction. Mater. Sci. Eng., A **635**, 6–12 (2015)
42. Budiman, A.S., Illya, G., Handara, V., Caldwell, W.A., Bonelli, C., Kunz, M., Tamura, N., Verstraeten, D.: Enabling thin silicon technologies for next generation c-Si solar PV renewable energy systems using synchrotron X-ray microdiffraction as stress and crack mechanism probe. Sol. Energy Mater. Sol. Cells **130**, 303–308 (2014)
43. Tippabhotla, S.K., Radchenko, I., Song, W.J.R., Illya, G., Handara, V., Kunz, M., Tamura, N., Tay, A.A., Budiman, A.S.: From cells to laminate: Probing and modeling residual stress evolution in thin silicon photovoltaic modules using synchrotron X-ray micro-diffraction experiments and finite element simulations. Prog. Photovoltaics Res. Appl. **25**(9), 791–809 (2017)
44. Handara, V.A., Radchenko, I., Tippabhotla, S.K., Narayanan, K.R., Illya, G., Kunz, M., Tamura, N., Budiman, A.S.: Probing stress and fracture mechanism in encapsulated thin silicon solar cells by synchrotron X-ray microdiffraction. Sol. Energy Mater. Sol. Cells **162**, 30–40 (2017)
45. Kunz, M., Tamura, N., Chen, K., MacDowell, A.A., Celestre, R.S., Church, M.M., Fakra, S., Domning, E.E., Glossinger, J.M., Kirschman, J.L., Morrison, G.Y.: A dedicated superbend X-ray microdiffraction beamline for materials, geo-, and environmental sciences at the advanced light source. Rev. Sci. Inst. **80**(3), 035108 (2009)
46. MacDowell, A.A., Celestre, R.S., Tamura, N., Spolenak, R., Valek, B., Brown, W.L., Bravman, J.C., Padmore, H.A., Batterman, B.W., Patel, J.R.: Submicron X-ray diffraction. Nucl. Instrum. Methods Phys. Res., Sect. A **467**, 936–943 (2001)
47. Chen, X., Dejoie, C., Jiang, T., Ku, C.S., Tamura, N.: Quantitative microstructural imaging by scanning Laue x-ray micro-and nanodiffraction. MRS Bull. **41**(6), 445–453 (2016)
48. Tamura, N.: XMAS: A versatile tool for analyzing synchrotron X-ray microdiffraction data. In *Strain and dislocation gradients from diffraction: Spatially-Resolved Local Structure and Defects* (pp. 125–155) (2014)
49. Launey, M.E., Ritchie, R.O.: On the fracture toughness of advanced materials. Adv. Mater. **21**(20), 2103–2110 (2009)
50. Sander, M., Henke, B., Schwarz, H., Dietrich, S., Schweizer, S., Ebert, M., Bagdahn, J.: Characterization of PV modules by combining results of mechanical and electrical analysis methods. In *Reliability of Photovoltaic Cells, Modules, Components, and Systems III* (Vol. 7773, p. 777308). International Society for Optics and Photonics (2010, August)
51. Wendt, J., Träger, M., Mette, M., Pfennig, A., Jäckel, B.: The link between mechanical stress induced by soldering and micro damages in silicon solar cells. In *24th European Photovoltaic Solar Energy Conference* (pp. 3420–3423) (2009, September)
52. Gabor, A.M., Ralli, M., Montminy, S., Alegria, L., Bordonaro, C., Woods, J., Felton, L., Davis, M., Atchley, B,. Williams, T.: Soldering induced damage to thin Si solar cells and detection of cracked cells in modules. In *21st European Photovoltaic Solar Energy Conference* (pp. 4–8) (2006, September)
53. Nieland, S., Baehr, M., Boettger, A., Ostmann, A., Reichl, H.: Advantages of microelectronic packaging for low temperature lead free soldering of thin solar cells. In *22th European Photovoltaic Solar Energy Conference, Milan, Italy, September* (2007, September)
54. Lin, K.M., Lee, Y.H., Wang, L.K., Chen, L.W., Yang, S.Y., Chen, Y.C., Liu, D.C., Huang, M.Y., Wu, Z.C., Chen, C.P.: Electroluminescence observation of microcrack growth behavior of crystalline silicon solar modules fabricated by hot-air soldering technology. In *International Journal of Modern Physics: Conference Series* (Vol. 6, pp. 43–48). World Scientific Publishing Company (2012)
55. Demant, M., Glatthaar, M., Haunschild, J., Rein, S.: Analysis of luminescence images applying pattern recognition techniques. In *25th European Photovoltaic Solar Energy Conference and Exhibition/5th World Conference on Photovoltaic Energy Conversion* (pp. 1078–1082) (2010)

56. Haunschild, J., Glatthaar, M., Demant, M., Nievendick, J., Motzko, M., Rein, S., Weber, E.R.: Quality control of as-cut multicrystalline silicon wafers using photoluminescence imaging for solar cell production. Sol. Energy Mater. Sol. Cells **94**(12), 2007–2012 (2010)
57. McMillan, W., Trupke, T., Weber, J.W., Wagner, M., Mareck, U., Chou, Y.C., Wong, J.: In-line monitoring of electrical wafer quality using photoluminescence imaging. *25th EPVSEC/5th WCPEC, Valencia, Spain*, pp. 1346–1351 (2010)
58. Tamura, N., MacDowell, A.A., Spolenak, R., Valek, B.C., Bravman, J.C., Brown, W.L., Celestre, R.S., Padmore, H.A., Batterman, B.W., Patel, J.R.: Scanning X-ray microdiffraction with submicrometer white beam for strain/stress and orientation mapping in thin films. J. Synchrotron Radiat. **10**(2), 137–143 (2003)
59. Budiman, A.S., Illya, G., Handara, V., Caldwell, W.A., Bonelli, C., Kunz, M., Tamura, N. and Verstraeten, D.: Enabling thin silicon technologies for next generation c-Si solar PV renewable energy systems using synchrotron X-ray microdiffraction as stress and crack mechanism probe. Solar Energy Mater. Solar Cells **130**, 303–308 (2014)
60. Meier, R., Pander, M., Klengel, R., Dietrich, S., Klengel, S., Ebert, M., Bagdahn, J.: Reduction of soldering induced stresses in solar cells by microstructural optimization of copper-ribbons. In *Reliability of Photovoltaic Cells, Modules, Components, and Systems IV* (Vol. 8112, p. 811206). International Society for Optics and Photonics (2011, September)
61. Kim, B.J., Kim, J.H., Hwang, S.H., Budiman, A.S., Son, H.Y., Byun, K.Y., Tamura, N., Kunz, M., Kim, D.I., Joo, Y.C.: Microstructure evolution and defect formation in Cu through-silicon vias (TSVs) during thermal annealing. J. Elect. Mater. **41**(4), 712–719 (2012)
62. Handara, V.A., Radchenko, I., Tippabhotla, S.K., Narayanan, K.R., Illya, G., Kunz, M., Tamura, N., Budiman, A.S.: Probing stress and fracture mechanism in encapsulated thin silicon solar cells by synchrotron X-ray microdiffraction. Solar Energy Materials and Solar Cells**162**, 30–40 (2017)
63. Chen, K., Tamura, N., Tang, W., Kunz, M., Chou, Y.C., Tu, K.N., Lai, Y.S.: High precision thermal stress study on flip chips by synchrotron polychromatic x-ray microdiffraction. J. Appl. Phys. **107**(6), 063502 (2010)
64. Chiu, T.C., Zeng, K., Stierman, R., Edwards, D., Ano, K.: Effect of thermal aging on board level drop reliability for Pb-free BGA packages. In *2004 Proceedings. 54th Electronic Components and Technology Conference (IEEE Cat. No. 04CH37546)* (Vol. 2, pp. 1256–1262). IEEE (2004, June)
65. Mei, Z., Ahmad, M., Hu, M., Ramakrishna, G.: Kirkendall voids at Cu/solder interface and their effects on solder joint reliability. In *Proceedings Electronic Components and Technology, 2005. ECTC'05.* (pp. 415–420). IEEE (2005, May)
66. Kim, B.J., Lim, G.T., Kim, J., Lee, K., Park, Y.B., Lee, H.Y., Joo, Y.C.: Intermetallic compound growth and reliability of Cu pillar bumps under current stressing. J. Electron. Mater. **39**(10), 2281–2285 (2010)
67. Gabor, A.M., Ralli, M., Montminy, S., Alegria, L., Bordonaro, C., Woods, J., Felton, L., Davis, M., Atchley, B., Williams, T.: Soldering induced damage to thin Si solar cells and detection of cracked cells in modules. In *21st European Photovoltaic Solar Energy Conference* (pp. 4-8) (2006, September)
68. Sander, M., Dietrich, S., Pander, M., Ebert, M., Bagdahn, J.: Systematic investigation of cracks in encapsulated solar cells after mechanical loading. Sol. Energy Mater. Sol. Cells **111**, 82–89 (2013)
69. Sander, M., Dietrich, S., Pander, M., Schweizer, S., Ebert, M., Bagdahn, J.: Investigations on crack development and crack growth in embedded solar cells. In *Reliability of Photovoltaic Cells, Modules, Components, and Systems IV* (Vol. 8112, p. 811209). International Society for Optics and Photonics (2011, September)
70. Köntges, M., Kunze, I., Kajari-Schröder, S., Breitenmoser, X., Bjørneklett, B.: The risk of power loss in crystalline silicon based photovoltaic modules due to micro-cracks. Sol. Energy Mater. Sol. Cells **95**(4), 1131–1137 (2011)
71. Kajari-Schröder, S., Kunze, I., Eitner, U., Köntges, M.: Spatial and orientational distribution of cracks in crystalline photovoltaic modules generated by mechanical load tests. Sol. Energy Mater. Sol. Cells **95**(11), 3054–3059 (2011)

72. Dietrich, S., Pander, M., Sander, M., Schulze, S.H., Ebert, M.: Mechanical and thermo-mechanical assessment of encapsulated solar cells by finite-element-simulation. Proceedings of SPIE–The International Society for Optical Engineering (2008)
73. Ketola, B., McIntosh, K.R., Norris, A., Tomalia, M.K.: Silicones for photovoltaic encapsulation. In *23rd European Photovoltaic Solar Energy Conference* (pp. 2969–2973) (2008, September)
74. Hurter, W., Oosthuizen, G., Janse van Rensburg, N.: Investigating the effects of composite materials in solar cell encapsulation. International Conference on Competitive Manufacturing (2013)
75. Mickiewicz, R., Li, B., Doble, D., Christian, T., Lloyd, J., Stokes, A., Voelker, C., Winter, M., Ketola, B., Norris, A., Shephard, N.: Effect of encapsulation modulus on the response of PV modules to mechanical stress. In *Proceedings of the 26th European Photovoltaic Solar Energy Conference* (2011, September)
76. Narayanan, K.R. et al.: Proc. Eng. **139**, 76 (2016)
77. Budiman, A.S.: Enabling thin silicon technology. *Science Highlights, June* (2013)
78. MacDowell, A.A., Celestre, R.S., Tamura, N., Spolenak, R., Valek, B., Brown, W.L., Bravman, J.C., Padmore, H.A., Batterman, B.W., Patel, J.R.: Submicron X-ray diffraction. Nucl. Instr. Methods Phys. Res. Sect. A: Accelerat. Spectro. Detect. Assoc. Equip. **467**, 936–943 (2001)
79. Tamura, N., Celestre, R.S., MacDowell, A.A., Padmore, H.A., Spolenak, R., Valek, B.C., Meier Chang, N., Manceau, A., Patel, J.R.: Submicron x-ray diffraction and its applications to problems in materials and environmental science. Rev. Sci. Instrum. **73**(3), 1369–1372 (2002)
80. Tamura, N., MacDowell, A.A., Spolenak, R., Valek, B.C., Bravman, J.C., Brown, W.L., Celestre, R.S., Padmore, H.A., Batterman, B.W., Patel, J.R.: Scanning X-ray microdiffraction with submicrometer white beam for strain/stress and orientation mapping in thin films. J. Synchrot. Radiat. **10**(2), 137–143 (2003)
81. Tamura, N., Padmore, H.A., Patel, J.R.: High spatial resolution stress measurements using synchrotron based scanning X-ray microdiffraction with white or monochromatic beam. Mater. Sci. Eng., A **399**(1–2), 92–98 (2005)
82. Tamura, N., Kunz, M., Chen, K., Celestre, R.S., MacDowell, A.A., Warwick, T.: A superbend X-ray microdiffraction beamline at the advanced light source. Mater. Sci. Eng., A **524**(1–2), 28–32 (2009)
83. Budiman, A.S., Nix, W.D., Tamura, N., Valek, B.C., Gadre, K., Maiz, J., Spolenak, R., Patel, J.R.: Crystal plasticity in Cu damascene interconnect lines undergoing electromigration as revealed by synchrotron x-ray microdiffraction. Appl. Phys. Lett. **88**(23), 233515 (2006)
84. Budiman, A.S., Han, S.M., Greer, J.R., Tamura, N., Patel, J.R., Nix, W.D.: A search for evidence of strain gradient hardening in Au submicron pillars under uniaxial compression using synchrotron X-ray microdiffraction. Acta Mater. **56**(3), 602–608 (2008)
85. Feng, G., Budiman, A.S., Nix, W.D., Tamura, N., Patel, J.R.: Indentation size effects in single crystal copper as revealed by synchrotron x-ray microdiffraction. J. Appl. Phys. **104**(4), 043501 (2008)
86. Chen, K., Tamura, N., Valek, B.C., Tu, K.N.: Plastic deformation in Al (Cu) interconnects stressed by electromigration and studied by synchrotron polychromatic X-ray microdiffraction. J. Appl. Phys. **104**(1), 013513 (2008)
87. Budiman, A.S., Besser, P.R., Hau-Riege, C.S., Marathe, A., Joo, Y.C., Tamura, N., Patel, J.R., Nix, W.D.: Electromigration-induced plasticity: texture correlation and implications for reliability assessment. J. Elect. Mater. **38**(3), 379–391 (2009)
88. Budiman, A.S., Hau-Riege, C.S., Baek, W.C., Lor, C., Huang, A., Kim, H.S., Neubauer, G., Pak, J., Besser, P.R., Nix, W.D.: Electromigration-induced plastic deformation in Cu interconnects: effects on current density exponent, n, and implications for EM reliability assessment. J. Elect. Mater. **39**(11), 2483–2488 (2010)
89. Budiman, A.S., Li, N., Wei, Q., Baldwin, J.K., Xiong, J., Luo, H., Trugman, D., Jia, Q.X., Tamura, N., Kunz, M., Chen, K.: Growth and structural characterization of epitaxial Cu/Nb multilayers. Thin Solid Films **519**(13), 4137–4143 (2011)
90. Budiman, A.S., Han, S.M., Li, N., Wei, Q.M., Dickerson, P., Tamura, N., Kunz, M., Misra, A.: Plasticity in the nanoscale Cu/Nb single-crystal multilayers as revealed by synchrotron Laue x-ray microdiffraction. J. Mater. Res. **27**(3), 599–611 (2012)

91. Budiman, A.S., Shin, H.A.S., Kim, B.J., Hwang, S.H., Son, H.Y., Suh, M.S., Chung, Q.H., Byun, K.Y., Tamura, N., Kunz, M., Joo, Y.C.: Measurement of stresses in Cu and Si around through-silicon via by synchrotron X-ray microdiffraction for 3-dimensional integrated circuits. Microelect. Reliab. **52**(3), 530–533 (2012)
92. Kim, B.J., Kim, J.H., Hwang, S.H., Budiman, A.S., Son, H.Y., Byun, K.Y., Tamura, N., Kunz, M., Kim, D.I., Joo, Y.C.: Microstructure evolution and defect formation in Cu through-silicon vias (TSVs) during thermal annealing. J. Elect. Mater. **41**(4), 712–719 (2012)
93. Budiman, A.S., Lee, G., Burek, M.J., Jang, D., Han, S.M.J., Tamura, N., Kunz, M., Greer, J.R., Tsui, T.Y.: Plasticity of indium nanostructures as revealed by synchrotron X-ray microdiffraction. Mater. Sci. Eng. A **538**, 89–97 (2012)
94. Budiman, A.S., Narayanan, K.R., Li, N., Wang, J., Tamura, N., Kunz, M., Misra, A.: Plasticity evolution in nanoscale Cu/Nb single-crystal multilayers as revealed by synchrotron X-ray microdiffraction. Mater. Sci. Eng. A **635**, 6–12 (2015)

Chapter 6
Emerging Trends in Additively Manufactured Materials and Novel Flexible/Stretchable Conductor Technologies

Most recent advances in the Additive Manufacturing (AM) methodologies as well as other bottom-up fabrication approaches such as atomic or layer-by-layer deposition methodologies—combined with 3D multi-hierarchical microarchitecturing—have enabled development and design of advanced novel materials with extraordinary properties approaching extreme limits (compared to the usual operational/service range of the elemental materials), as well as unprecedented, new functionalities allowing emerging technologies, such as soft robotics, electronic skins and flexible devices, to become reality in the near future. In this Chapter, we have captured a portion of the broad spectrum of research, representing some of the most important recent advances in the fabrication of novel 3D architected polymer composites enabled by electrospinning-based AM technique and development of novel metallic stretchable materials made possible by atomic reconfiguration approaches.

Novel 3D architected fiber-based composite materials as have been discussed in previous Chapter II (Extreme Mechanics in Advanced, Emerging Materials) will be investigated further to even greater depths (involving extreme and tunable impact properties) in this chapter and explored further into exciting applications in key, emerging technologies. Inspired by natural structural materials, such as found in mantis dactyl clubs, the novel 3D architecture based on helicoidally aligned fiber-reinforced matrix geometries have good potential to play key roles in the emerging energy technologies, such as photovoltaics (PV) and Lithium Ion Battery (LIB). This will be the focus of discussions in the first section—Sect. 6.1: **Tunable Impact Resistance in Novel Composites with Helicoidal 3D Architecture Via Additive Manufacturing Methodologies**. The extreme and tunable mechanical properties, especially impact characteristics, of the novel materials are illustrated through their potential application for enabling lightweight silicon-based PV technologies. The preliminary evidence for basic feasibility of the application is presented and the novel impact tunability allowed by eXtreme Mechanics (XM) is proposed. In the following sub-section, we propose yet another emerging application to help enable low-cost 3D nanostructure anode with high performance (kinetics related, such as power density

A. S. Budiman, *Principles of Extreme Mechanics (XM) in Design for Reliability (DfR)*, Engineering Materials,
https://doi.org/10.1007/978-981-15-6720-9_6

and charging rate) and reliability (long life/cyclability). These novel materials indeed provide great insights and unique illustrations of how XM may enable further the Design for Reliability (DfR), or even Design for Extreme Reliability (DfXR) in the strategic technologies of the future.

The unravelling of the interfacial sliding mechanisms through atomic reconfigurations in the FCC/BCC nanolayered materials opens up uncharted territories in the design and development of novel structural materials and their unprecedented limits and new functionalities. The roles of the eXtreme Mechanics (XM) and the advanced experimentation techniques to enable the probing of the new mechanic aspects at the nano or mesoscales cannot be understated here. This is the focus of the intense discussion and debates presented in the subsequent Section—Sect. 6.2: **Enabling Stretchable Metallic Conductors through Atomic Reconfigurations in FCC/BCC Nanolayers**. Such extreme interfacial shearing insights could only be obtained and further investigated with cutting-edge *in-situ* small-scale mechanical testing inside an SEM (Scanning Electron Microscopy). In this section, we explore the potential application of the newly discovered phenomenon for enabling the extreme stretchable conductor technology which will be key to myriad of emerging future technologies, such as soft robotics, electronic skins and flexible energy devices/systems. The real exciting aspect here is the fact that XM is not only enabling extreme reliability or DfXR (Design for Extreme Reliability) now, but also finally stretching the functional capabilities of the materials beyond our imagination! The term "stretchable metals" here, after all, is almost an oxymoron!

6.1 Tunable Impact Resistance in Novel Composites with Helicoidal 3D Architecture Via Additive Manufacturing Methodologies

Strong and tough structural materials occurring in nature have long fuelled the search of a new class of strong biomimetic synthetic structural materials (SSMs). One such example in recent years is the rapid progress associated with the naturally occurring helicoidal composite structures inspiring the development of the tough helicoidally architectured synthetic structural composites (HA-SSCs). Most of the tough HA-SSCs have been fabricated using conventional composite resins and fibers, or relatively newer 3D printing materials. Only recently, we ventured further into expanding the possibilities of imitating helicoidal architecture in smaller scales (microns and tens of microns) while exploring the possibilities of utilization of various polymer-based composite formulas. In our previous research articles, we have established the near-field electrospinning as an additive manufacturing methodology to construct tough and resilient 3D helicoidal membranes with micron-sized polymer fibers. In this article, we advance to next level of developing full composite structures from electrospun helicoidally assembled micro ribbons. A key aspect explored in this article is the importance of surface treatment. The physical, thermal and mechanical experiments indicated superior adhesion between the components of composites leading to enhanced toughness and impact properties. We have further affirmed that reducing the helicoidal angular orientation of ribbons in the HA-SSCs can help further enhance the specific toughness and impact resistance of these composites. Furthermore, we found that the mechanical and physical properties of these composites can be tuned via several architectural features (helicoidal rotational angles, interface characteristics between layers and between ribbons) to suit the different strain-rate needs for various applications (combat army vest, sports gears such as helmets or novel polycarbonate-based solar photovoltaic (PV) modules). The experimental findings here suggest the importance of interfacial characteristics (between layers, and between ribbons), ribbon to matrix ratio and angular arrangement of the ribbons as design guidelines for achieving next-generation HA-SSC materials which are not only strong, but also extremely tough and impact-resistant.

6.1.1 Background and Motivation—Helicoidally Aligned Fiber Composite Materials

Naturally occurring strong, tough and resilient biomaterials have inspired material scientist to fabricate tough and strong synthetic materials by mimicking the architecture and design found in natural materials [1, 2]. It is fascinating how nature combines otherwise weak constituent materials in a hierarchical fashion to design stronger and tougher bio composites [3]. One such remarkable natural structural material is mantis shrimp's dactyl club, which consists of helically arranged chitin fiber in mineralized hydroxyapatite [4, 5]. It is no wonder that the superior mechanical properties and light weight of mantis shrimps's dactyl club have sparked a growth in the development of a range of biomimetic materials with similar mechanical properties [6]. Its architectural design enables mantis shrimp to smash its prey, enemies and predators with instantaneous force greater than 700 N (~2500 times the shrimp's own weight) with speed of 23 m/s or more within less than 3 ms, earning them the title for the fastest punchers ever known to humans [5]. Upon impact, the helical arrangement of the fibers helps to deflect the crack propagation continuously along distinct planes and directions, encountering periodic elastic modulus transitions between fiber and matrix material, thus avoiding catastrophic failure [7, 8]. Therefore, helicoidal architecture is a promising avenue for us to develop lightweight tough materials.

Helicoidally arranged synthetic structural materials (HA-SSMs) inspired by the mantis shrimp dactyl club have exhibited tremendous damage resistance [9–13]. However, most of these studies range in macroscopic scales (sub-millimeter and larger), using conventional fabrication practice with limited access to varieties of material composition [11, 14, 15]. In our previous publications [16, 17], we have demonstrated that helicoidally aligned fiber architectures (with fiber diameters in microns to tens of microns) can be synthesized using near-field electrospinning (both melt type and solution type) using a vast variety of polymeric materials, and that these fibers demonstrate desirable mechanical behaviors. In the context of the present manuscript, the key desirable properties are light weight, toughness and impact resistance of the material. Recently, Chen et al. have also performed electrospinning to develop nylon-epoxy HA-SSC which showed good mechanical properties [18]. Nevertheless, in the present manuscript, we have demonstrated a few key architectural parameters of HA-SSC that we can use to tune its mechanical properties (i.e. specific toughness, impact resistance, etc.) and further develop stronger and tougher composites for various applications. In the current section, we present the following studies:

1. We studied the interface characteristics between the ribbons as well as between layers of matrix/ribbon materials through investigating the surface contact angle as an indicator of penetration of matrix in the ribbon material.
2. We demonstrated the effect of orientation angles on the mechanical properties of the composite (specific strength, specific toughness, impact resistance, etc.).

For developing HA-SSCs using soft polymers, we employed near-field stage-controlled electrospinning as a 3D additive manufacturing technology to create helicoidally arranged Polyvinylidene fluoride-co-hexafluropropylene (PVDF-HFP) ribbons and further embedded them into polyvinyl alcohol (PVA) matrix via controlled solution dispensing procedure. Interaction of ribbons and matrix, and their compatibility play a very important role in fine tuning the mechanical properties of composites. We employ surface treatment to improve the adherence between ribbon-matrix interface and ribbon-ribbon interface. Within the tensile characteristics, we extensively elaborate the effects of surface treatment and angular offsets on HA-SSCs. We also evaluate the potential of using ribbon weight percentage as a tunable parameter for customizing toughness. Tuning of several factors enable us to adjust the mechanical properties of the HA-SSCs according to the demand of application. These can be scaled up by implementing real-world industrial settings enabled by electrospinning methodology. Extending the research on such HA-SSCs can open doors for usage in impact-resistant applications like body armors, sports or biomedical applications, along with novel lightweight design of solar PV (photovoltaics) modules/technologies based on polycarbonate (rather than glass-based).

6.1.2 Experimental Methodology—Electrospinning-Based Additive Manufacturing (Es-AM) and Advanced Materials Characterization Techniques

Materials

Polyvinylidene fluoride-co-hexafluropropylene (PVDF-HFP, MW = 400,000), polyvinyl alcohol (PVA, MW = 130,000), dimethylacetamide (DMAc) and acetone were purchased from Merck, Singapore and used without further purification or modification. The syringes and needles were purchased from Terumo, Japan. The syringe pump and the motorized liner XY stages were procured from Premier Solution, and Physik Instrumente, Singapore, respectively.

Electrospinning-based Additive Manufacturing (Es-AM)

A 35 wt% of PVDF-HFP solution was prepared by dissolving PVDF-HFP in 3:1 (wt/wt) ratio of DMAc and acetone. The solution was stirred overnight at 45 °C temperature using a hot-plate stirrer before using this solution for electrospinning. The rheological properties of the solution enabled formation of flatter ribbons instead of usual circular fibers. The obtained flatter morphology resulted in an increase in the interfacial adhesion due to increase in the contact area for between the ribbons and PVA matrix. A 5 wt% PVA solution was prepared by dissolving PVA in distilled water using a hot-plate stirrer. The solutions were sonicated using (Elmasonic) P60H for 15 min at 45 °C to remove the presence of any bubbles and to obtain a homogenous solution.

A home-built near-field solution-electrospinning setup was used for producing uniform and dense ribbons. PVDF-HFP solution was filled in a 5 ml syringe (Terumo) with 25G needle. A syringe pump (EQ-500SP-H Syringe Pump, Premier Solution Pte Ltd) dispensed the solution at a flow rate of 1 ml h^{-1}. A grounded aluminum collector plate was placed on top of a programmable motorized XY stages (PI High Precision XY stages model C-891). The distance between the collector plate and the needle was 7 mm. The collector to needle distance was increased by 0.5 mm after collecting each layer of ribbons. The ribbons were collected in a layer-by-layer manner at various angular offsets. Figure 6.1 illustrates the steps used to obtain the

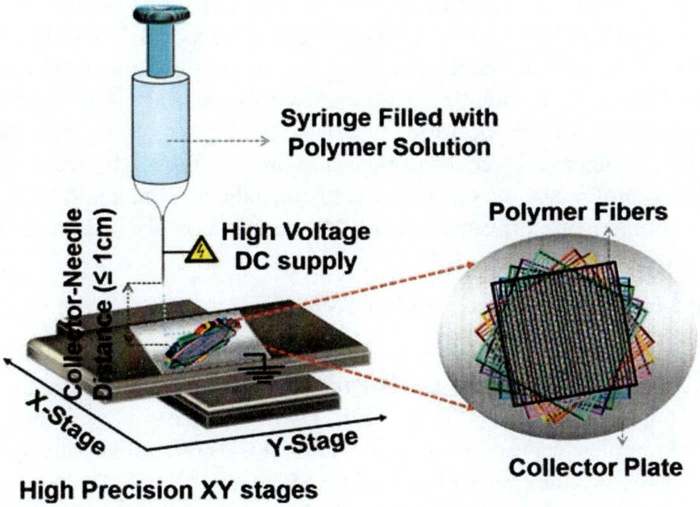

Fig. 6.1 Fabrication process: near-field electrospinning of helical PHR oriented at various angles and fabrication of the novel composites

three-dimensional structures. The electrospinning voltage of 2.4 kV was applied for the first four layers and later increased to 2.6 kV for the next four layers. The speed of motorized stage was set at 200 mm s^{-1}. After depositing the first layer of ribbons, the next layer was deposited such that their longitudinal axis was rotated by a fixed angle offset (45°; 15°; 30°; 40° and 90°). A three-dimensional structural network was then obtained by continuously stacking up to eight layers of ribbons one on top of each other. One set of electrospun ribbon membrane consisted of eight fused layers of aligned array of ribbons with angle of rotation starting from 0° and ending at the 8th multiple of the angle considered. For example, for 45° angular offset, the membrane consisted of the ribbon arrays patterned as 0°, 45°, 90°, 135°, 180°, 225°, 270° and 315°. The membrane was then allowed to dry in an oven at 50 °C for an hour to remove any residual solvent present in them. Three such eight-layered helicoidal membranes were manufactured and then stacked one on top of other to obtain a total of 24 layered three-dimensional helicoidal PVDF-HFP ribbons (PHR) membrane.

Fabrication of Composite

The composites were produced by stacking three sets of eight-layered PHR membranes and then embedding them in PVA matrix (PM). For this purpose, the PVA solution was filled in a syringe and the solution was sprayed on top of the three-dimensional helicoidal structure at a constant rate along the length of ribbon membrane (see Fig. 6.1). The samples were then left to dry at room temperature, and 75% humidity for 72 h. This produced PVDF-HFP ribbon (PHR) reinforced PVA matrix (PM) helicoidally architectured synthetic structural composites (HA-SSCs)—i.e. PHR-PMHC. Another set of samples were prepared by adding surfactant, Triton X-100 to the PVA solution. The presence of surfactant in PVA is expected to improve the interfacial adhesion between the ribbons and the matrix. This surface-treated PVDF-HFP ribbon (PHR) reinforced PVA matrix (PM) helicoidally architectured synthetic structural composites (HA-SSCs) were titled as sPHR-PMHC. The letter "s" in front of the sample name stands for the surface treatment associated with the samples. Table 6.1 details the notations used for naming of all samples in the present manuscript, with respect to their associated fabrication process parameters (angle layout, number of layers of ribbons and the ribbon weight percentage in the samples). The numbers at the end of the samples indicated the angle of PHR orientation. For comparison purposes, a pure PVA sample was also produced using the identical steps.

Advanced Materials Characterization Techniques

Hirox reflected light microscope was used to image the fracture region of the samples post mechanical tests. The microstructure of the samples was also examined using a scanning electron microscopy (SEM, JEOL, JSM-6700F) at an accelerating voltage of 5 kV. The samples were sputter-coated (JEOL JFC-1600) with a thin layer of gold (18 mA, 60 s) before they were examined by SEM.

A contact angle goniometer setup (KINO-optical contact angle and interface tension meter) was used to measure the contact angle (θ). The values reported were the averages of at least five measurements taken on different regions of the samples.

Table 6.1 Complete description of the notations used to describe samples, samples fabricated in this study, their stacking sequence and ribbon weight percentage

Notations used for naming samples	Description
PHR	PVDF-HFP Ribbons helicoidal membrane
PM	PVA Matrix
PHR-PMHC	Non-surface-treated helicoidal composite
sPHR-PMHC	Surface-treated helicoidal composite
sPHR-PM90	Surface-treated cross-ply composite

Sample Title	**Angle Orientation (°)**	**Number of Layers**	**Ribbon Weight Percentage (wt%)**
PM	–	0	0
PHR-PMHC45	45	24	88
sPHR-PMHC45	45	24	88
sPHR-PMHC15	15	24	60
sPHR-PMHC30	30	24	60
sPHR-PMHC40	40	24	60
sPHR-PM90	90	24	60

PVDF-HFP solution was casted on glass slides and dried in an oven for 1 h at 50 °C. PVA solution was fed into a 1 ml syringe with a 27.5G needle, and 1–2 μL droplets were placed on the surface of the PVDF-HFP films (PHF). For comparison purposes, contact angle made by PVA solution containing Triton X-100 on the PVDF-HFP was also determined.

Differential scanning calorimetry (DSC, DSC Q20 TA Instruments) was used to examine the melting temperature and crystallization temperature of the samples. The temperature was ramped at 3 °C min^{-1} from −60 to 250 °C, in nitrogen atmosphere with 40 ml min^{-1} flow rate.

Dynamic mechanical analysis (DMA, DMA Q800, TA Instruments) was used to understand the specific viscoelastic properties of the samples. The scanning temperature range was selected from −40 to 150 °C at strain ramp rate of 1% min^{-1} and 1 Hz frequency in multi strain mode.

The tensile deformation behavior of the samples was investigated using a tensile tester (MTS Criterion Model 43). The specimens for tensile test were trimmed using computer integrated laser cutting (GCC LaserPro Spirit GLS Hybrid) to avoid pulling or shear deformation of the samples. The samples were laser cut at 40% power control and 80 in s^{-1} motor speed. The ends of the samples were carefully glued to a cardboard sheet using superglue to ensure that the samples were gripped firmly during the tests [16]. The dimensions of the specimens were fixed at 25 mm × 5 mm (L × W). A 100 N load cell was used during the test and the samples were pulled at a rate of 1 mm min^{-1}. A minimum of six samples were tested for each type of composite.

A customized setup was used to determine the impact resistance of the samples. A standard steel ball bearing of 7.14 mm diameter and 1.4 g weight was selected and dropped on the sample. The height from which the ball was dropped was varied to determine the point of fracture of the cover slip that was placed under the samples [18]. The samples were adhered to a glass cover slip at the diagonal edge points, so as to minimize the effects on the measurements. A bare cover slip was chosen as reference (control group) to compare the impacts of the experimental samples. For each type of sample, a minimum of six impact tests were conducted.

6.1.3 *Additive Manufacturing Enabling Novel Materials with Helicoidal 3D Architecture*

Physical Characteristics

The optical microscope and SEM images shown in Fig. 6.2 reveal the microstructure of the helicoidally framed PHR samples fabricated using a near-field electrospinning technique. The SEM images demonstrate that the electrospinning led to the formation of broad ribbons. The broad ribbons offer better contact area than circular fibers which is expected to improve the interfacial adhesion between the ribbons and the matrix. The optical images shown in Fig. 6.2a1–e1 demonstrate the helicoidal arrangement

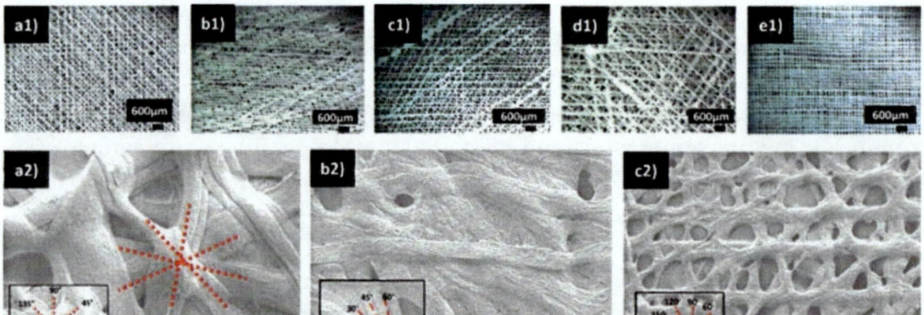

Fig. 6.2 Low magnification top-view optical microscope images (at 35X) of the 3D PHR membranes oriented at: (a1) 45°; (b1) 15°; (c1) 30°; (d1) 40°; and (e1) 90°. Higher magnification top-view SEM images of the 3D PHR membrane oriented at: (a2) 45° (at 270X); (b1) 15° (at 190X); (c1) 30° (at 75X); (d1) 40° (at 130X); and (e1) 90° (at 140X)

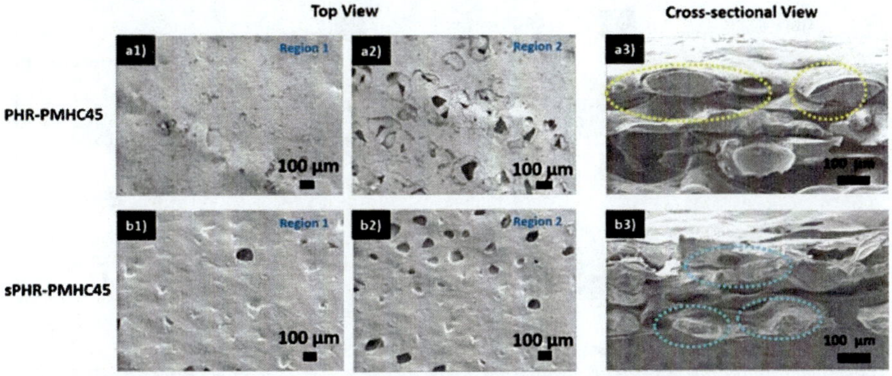

Fig. 6.3 SEM images of the composites PHR-PMHC45 and sPHR-PMHC45: (a1) top view of PHR-PMHC45 taken at region 1 (at 80X); (a2) top view of PHR-PMHC45 taken at region 2 (at 65X); (a3) cross-sectional view of PHR-PMHC45 (at 110X); (b1) top view of sPHR-PMHC45 taken at region 1 (at 90X); (b2) top view of sPHR-PMHC45 taken at region 2 (at 70X); (d) cross-sectional view of sPHR-PMHC45 (at 95X)

of ribbons within the PHR sample. Figure 6.2a2–e2 also displays the SEM images of the 24 layered PHR sample. It is clear from the SEM images presented in Fig. 6.2 that the ribbons are deposited at various angles within the PHR sample leading to the formation of helicoidal structure. The red dotted lines in Fig. 6.2a2–e2 depict the angular sequence between each layer of ribbon arrays. For example, Fig. 6.2a1, a2 show that the PHR sample is obtained by orienting the layers of ribbons by 45°. Similarly, Figs. 6.2b1 and 6.2b2, Figs. 6.2c1 and 6.2c2 and Figs. 6.2d1 and 6.2d2 show PHR samples obtained by orienting the layers of ribbons by 15, 30° and 40°, respectively. Figure 6.2e1, e2 display the cross-ply structure obtained for the PHR.

Figure 6.3 shows the SEM images of the PHR-PMHC45 (Figures a1–a3) and sPHR-PMHC45 (Figures b1–b3) samples. It is evident from Fig. 6.3a1, a2 that PVA matrix does not penetrate the porous regions (gaps/pockets) of the ribbon membrane in the PHR-PMHC45 sample. On the other hand, it is evident from the Fig. 6.3b1, b2 that the PVA solution with surfactant penetrated into the porous regions of the ribbon membrane and it uniformly coats the surface of the ribbons. This is further supported by the cross-sectional image of the PHR-PMHC45 as shown in Fig. 6.3a3. The cross-sectional image of PHR-PMHC45 shows that PVA is moderately adhered to the ribbons. The highlighted region in the Fig. 6.3a3 shows that PVA mainly deposits on the top surface of the ribbons leaving many porous regions between the ribbons unfilled. However, the cross-sectional SEM image of the sPHR-PMHC45 shown in Fig. 6.3b3 indicate improved interfacial interaction between the ribbons and the matrix. The highlighted region in Fig. 6.3b3 demonstrates that the PVA solution coated the ribbons uniformly and percolated within the ribbons. The porous regions appear to be better filled with PVA as compared to PHR-PMHC45. Contact angle measurements are performed to demonstrate that the presence of surfactant in PVA helps the PVA to uniformly wet the PVDF ribbons.

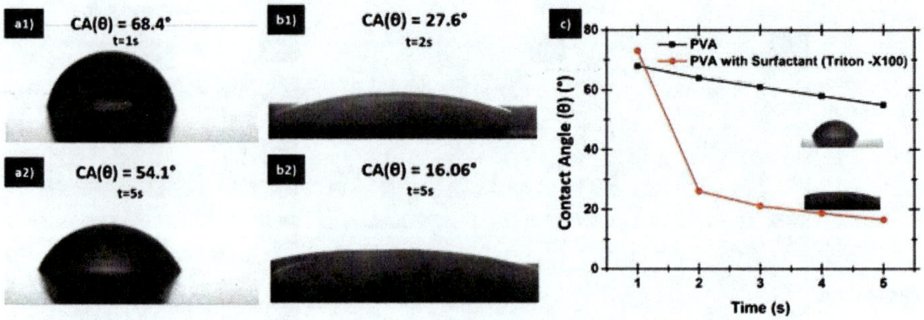

Fig. 6.4 Contact angle analysis: (a1) Contact angle made by PVA droplet on PHF at t = 1 s; (a2) Contact angle made by PVA droplet on PHF at t = 5 s; (b1) Contact angle made by PVA incorporated with surfactant Triton X-100, droplet on PHF at t = 2 s; (b2) Contact angle made by PVA incorporated with surfactant Triton X-100, droplet on PHF at t = 5 s; (c) Plot showing the contact angle vs time curves of PVA with and without surfactant when dropped on PHF

PVDF is hydrophobic due to the presence of fluorine, whereas PVA is hydrophilic and soluble in water [19, 20]. Blending PVA and PVDF improves the hydrophilicity of the PVDF/PVA membrane [21]. To demonstrate this, a droplet of PVA solution with and without surfactant is placed on the surface of the PHF sample and the contact angle made by the droplet is recorded. Figure 6.4a1, a2 display the contact angle made by the droplet of PVA solution on the surface of the PHF. The contact angle values are recorded as a function of time. The θ value at t = 1 s is determined to be ~68°. After 5 s, θ value recorded is 54°, which indicates that the PVA droplet wets the PHF sample (see black curve in Fig. 6.4c). On the other hand, when surfactant is added to the PVA solution, the contact angle values made by the droplet on the PHF is seen to be ~54° and ~16° after 1s and 5s, respectively (see Fig. 6.4b1, b2, and red curve in Fig. 6.4c). This indicates that the wettability can be improved by adding the surfactant to the PVA solution. Triton X-100 has a hydrophilic tail and a hydrophobic tail, with no charges. This reduces the surface tension at the interface of the membrane and the liquid, giving excellent wetting property. These results demonstrate that PVA with Triton X-100 can uniformly wet the ribbons and display improved adhesion between the matrix and the ribbon material.

Viscoelasticity

Following this, the thermal behavior of the samples is determined from the DSC scans recorded for the samples. The DSC scan of PM sample indicates an endothermic peak at 226 °C (black curve in Fig. 6.5a), corresponding to the melting temperature (T_m). The melting temperature recorded for the PM sample is in good agreement with the melting temperature values reported in literature [22, 23]. On the other hand, T_m of PHR is determined to be between 140 °C and 145 °C (see red curve in Fig. 6.5a).

The DSC scans of the composites are also recorded and presented in Fig. 6.5. It is evident from Fig. 6.5 that the composite samples, both PHR-PMHC45 (blue curve in Fig. 6.5a) and sPHR-PMHC45 (green curve in Fig. 6.5a), demonstrate two melting

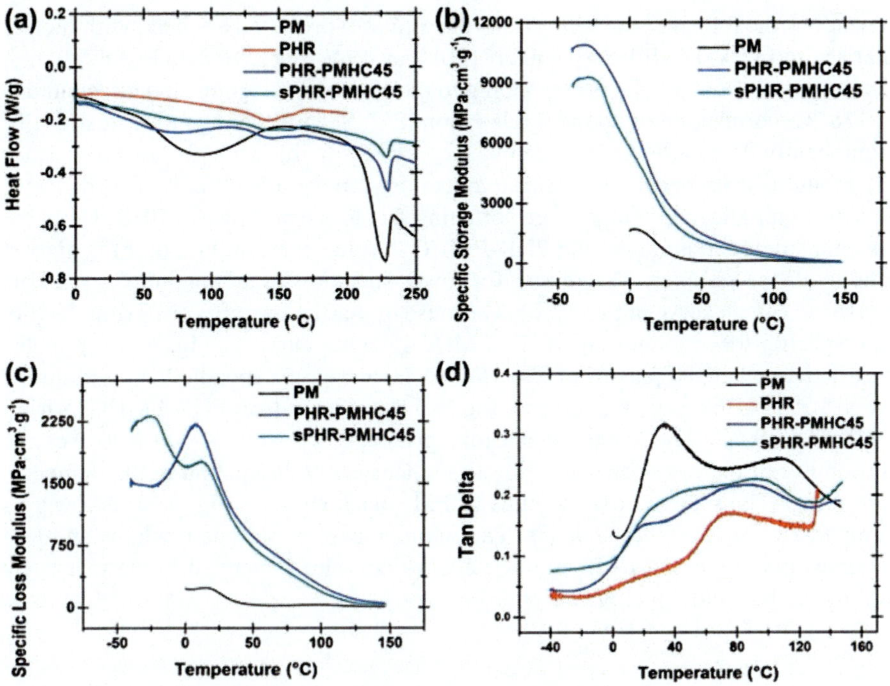

Fig. 6.5 Thermal and viscoelastic properties of PM, PHR, PHR-PMHC45 and sPHR-PMHC45: **a** Differential scanning calorimetry (DSC) thermal scans. Dynamic mechanical analysis (DMA) viscoelastic property: **b** storage modulus plot; **c** loss modulus plot; **d** tan delta plot

peaks. The first melting peak is seen to be at ~142.5 °C and the second peak is seen to be in the range of 228 to 230 °C. The first melting peak corresponds to the PVDF-HFP phase, while the second peak corresponds to PVA matrix. The crystallization peak of the surface-treated composite shift to lower temperature direction and broaden than compared to its counterparts in the DSC cooling curve, suggesting slight decrease in crystallinity. This indicates that the presence of Triton X-100 has a plasticizing effect on the PVA and PVDF-HFP composite [24, 25]. Although the crystallinity of the composites is lower than the bulk counterparts, the HA-SSCs display greater thermal and viscoelastic properties. In sPHR-PMHC45, the helicoidal architecture and the interfacial adhesion (due to presence of Triton X-100) between the components seem to play substantially more dominant role than crystallinity in reinforcing the thermal and viscoelastic properties of the composites.

The viscoelastic properties of the PM, PHR and the composites are shown in Fig. 6.5 (b, c and d, respectively). Over the entire temperature range, the specific storage modulus and the loss modulus values of composites are higher than those found for PHR and PM. The specific viscoelastic properties of the material are determined by dividing the modulus by the density of PVA. The two transition temperatures at around −40 °C and 71 °C (red curve in Fig. 6.5d), correspond to the glass transition (the β phase) and relaxation (the α phase) process associated with molecular chains in PVDF-HFP, respectively [26]. For PVA (black curve in Fig. 6.5d), first peak is observed at 38 °C corresponding to the α phase relaxation, and a second one at 118 °C, corresponding to the β relaxation [27]. Specific storage modulus for PM significantly drops after 25 °C (see black curve Fig. 6.5b), while for the composites the modulus drop occurs at higher temperature (above 50 °C) and then the curve flattens out, indicating temperature softening due to the presence of PHR. Here, the specific storage modulus of the PHR-PMHC45 (blue curve in Fig. 6.5b) is highest among all the samples. The reason for lower specific storage modulus for sPHR-PMHC45 (green curve in Fig. 6.5b) is because of plasticizing effect of Triton X-100. The specific loss modulus for PHR-PMHC45 (blue curve in Fig. 6.5c), remains higher than other samples from 0 to 150 °C. However, the specific loss modulus of the sPHR-PMHC45 (green curve in Fig. 6.5c) is higher than PHR-PMHC45 from −40 °C to 0 °C. This could be because presence of Triton X-100 increases the flowability of the composite resulting in higher energy dissipation in that temperature range. The specific loss modulus of PM (black curve in Fig. 6.5c) is lower as compared to all the other samples. The improvement of storage modulus and loss modulus may be associated with the network complex generated in the matrix by adding the helicoidally aligned ribbons.

6.1.4 Additive Manufacturing Enabling Materials with Extreme and Tunable Impact Properties for Novel Lightweight Photovoltaics (PV) Technology

These 3D architected synthetic materials enabled by electrospinning-based Additive Manufacturing (AM) methodology inspired by natural structural materials found in bones, nacre, teeth, shells and mantis shrimp's dactyl club have recently been reported to exhibit superior mechanical properties (tensile strength, toughness and fracture resistance) and especially also impact characteristics [2–5, 16–18]. These remarkable impact performances are attributed to the specific hierarchical arrangement of the individually mechanically weak constituents within the microstructure [4, 5]. The helicoidally aligned materials dissipate energy from the impact region through quasi-plastic compressive responses, thereby providing a fracture toughening barrier against catastrophic propagation of micro-cracks when subjected to repeated impacts [5]. The phenomenon can be further explained by assuming that the crack propagates straight into neighboring layers but encounters variation in elastic modulus due to the helicoidally laid fiber layers. This variation prevents crack propagation along the straight-line, thus improving the damage-tolerance and higher efficiency of damage/energy absorption of the helicoidal aligned structures [24–26].

Such efficient dissipation of damage/energy from a pointed impact loading and the ability to deflect the crack propagation sideways might enable a superior encapsulation polymer material in lightweight PV modules [2]. The layered geometry consisting of 3D helicoidally aligned fibers would absorb the impact energy efficiently and pass down very little energy to the fragile silicon solar cells, thus enabling novel lightweight PV module design (based on polymer front and backsheets) with enhanced impact resistance and structural integrity/reliability (especially against silicon cell cracks). Our aim in this sub-section is to present the evidence for the basic feasibility of the proposed concept—i.e. using 3D architected layered polymer structures consisting of helicoidally aligned fibers to provide protection or more effective encapsulation of the silicon solar cells (in lightweight PV module design) especially against crack initiation/propagation due to impact loading (from hailstorms, for instance). Building on our previous research investigations on the novel materials [16, 17, 25–27], we extend our methodologies to enable this feasibility study for the application of lightweight PV technologies. By enhancing the impact resistance and overall structural integrity of lightweight PV modules, we might accelerate the pace of adoption of lightweight PV technology needed for building application in urban areas as well as for independent low-cost energy supplies in remote, underdeveloped areas. In addition, lightweight PV module design would allow PV integration with curved surfaces or surfaces with contours (such as for automobiles or boats) thus enabling more esthetic design for PV to be incorporated in urban structures or building in cities [27].

Impact-Resistant 3D Helicoidally Architected Polymer Composites Enabling Next-Generation Lightweight Silicon Photovoltaics (PV) Module Design and Technology

Lightweight Photovoltaic (PV) modules are important for certain segments of the renewable energy markets—such as exhibition halls, factories, supermarkets, farms, etc. In addition, it is critical for certain geographies, like archipelagic Indonesia, which has many remote areas spread in thousands of separate islands. High transportation and installation costs of heavy conventional silicon PV modules for these applications would render these projects uneconomical and unattractive to business interests. However, lightweight silicon-based PV modules have their own set of technical challenges or concerns. Among them, one which is the subject of this sub-section is the lack of impact resistance, especially against hailstorms in the deep of winter in four-season countries. Even if the front sheet may be made sufficiently strong and impact-resistant, the silicon cells inside remain fragile and very prone to impact loading. It will lead to crack, which in due time will degrade performance (power output) substantially. A 3D helicoidally architected fiber-based polymer composite has recently been found to have excellent impact resistance, inspired by the multi-hierarchical inner structures of the mantis shrimp's dactyl clubs [25, 26]. In that previous work, our group has demonstrated that via electrospinning-based additive manufacturing methodologies, weak polymer material constituents could be made to exhibit much enhanced toughness and impact properties. In the following sub-section, we have extended our demonstration that the 3D architected fiber-based polymer composite has also been effective in absorbing the impact energy upon the silicon solar cells, in the manner similar to hailstorm impacting the PV modules. We have shown that silicon cells put under such 3D architected polymer layers break at substantially higher impact load/energy (compared to those put under the standard PV encapsulation polymer material). This could lead to the development of new type of PV encapsulant materials for next-generation lightweight PV module and technology with excellent impact resistance.

Impact Testing

Impact tests were carried out by the ball-dropping method, i.e. to find out at what height the PV solar cell lying under the sample would break. A similar methodology has been used by Chen et al. in their study of electrospun nylon-epoxy-based HA-SSCs [18], as well as in our own group's previous reports [25.26].

A customized setup was used to determine the impact resistance of the samples. A standard steel ball bearing of 7.14 mm diameter and 1.4 g weight was selected and dropped on the samples. The height from which the ball was dropped was increased with increments of 1 cm to determine the point of fracture of the silicon solar cell that was placed under the samples. A DC powered digital oscilloscope (Yokogawa Model DL1620-200MS/s 200 MHz) with 0.5 V/division resolution and a charge meter (Kistler type 5015) with the sensor sensitivity of –4.060 pC/N was used to measure the impact force. The samples were adhered to a silicon solar cell at the diagonal edge points, so as to minimize the effects on the measurements. A silicon solar cell without any sample adhered to it is used as a control sample to compare with silicon solar cell protected with samples. The silicon solar cells used is a typical commercially available monocrystalline type without any interconnect (new cell was used for each ball-dropping experiment/test).

Such an experimental methodology could be used to evaluate qualitative impact resistance of the composites—in the absence of a more quantitative impact fracture mechanics methodology, especially for such small-scale samples. Most standardized testing methodology for fracture mechanics such as charpy test, izod test and ballistic test are mostly suitable only for large-scale (bulk) samples [18, 25]. In addition to finding the fracture initiation height, the specific potential energy required to break the silicon solar cell placed beneath the samples is calculated following the methodology fully described in our previous reports [25, 26]. All the values of height and the calculated specific potential energy at which the silicon solar cell breaks are compared to gain deeper insights about the impact properties of bare silicon solar cell (without any protection), silicon solar cell under EVA (Ethylene Vinyl Acetate—typical encapsulant materials in PV technology), silicon solar cell under HA-SSCs (15° and 45°).

Enhanced Impact Resistance of Novel Lightweight PV Module Design Enabled by 3D Helicoidal Architecture

Silicon Photovoltaic (PV) cells are delicate and are easily broken upon impact, destroying them completely. A customized setup was used to determine the impact resistance of silicon solar cells (see Fig. 6.6) [8]. The performance of the samples was determined by the ability of the samples to protect a silicon solar cell placed underneath the samples. The samples were adhered to a silicon solar cell at the diagonal edge points, so as to minimize the effects on the measurements. A bare silicon

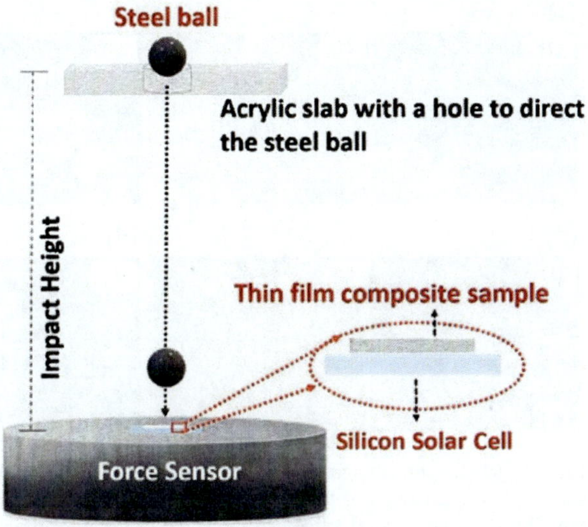

Fig. 6.6 Schematic illustration of the small-scale impact test setup [25, 26]

solar cell (without any protection layer) was chosen as a reference (control group) to compare the impact characteristics of the experimental samples. For each type of samples, a minimum of six impact tests was conducted. All the samples used for impact tests had a cross-sectional area of ~1.1×1.1 cm^2. A standard steel ball of 7.14 mm diameter and 1.4 g weight was selected and dropped on the sample under gravity with zero initial speed. The height from which the ball was dropped was initially increased with increments of 5 cm to determine the point of fracture of the PV solar cell that was placed under the samples. When the tentative fracture height for the samples was identified, then the tests were carried with 1 cm increments for careful estimation of the final fracture height. Table 6.1shows results of impact properties of bare silicon solar cell (without any protection), silicon solar cell under EVA (Ethylene Vinyl Acetate—typical encapsulant materials in PV technology), silicon solar cell under HA-SSCs (15° and 45°) when impacted with the steel ball. The error bars in Table 6.2 shows range from the mean impacted height accumulated by

Table 6.2 Impact resistance properties of the samples protecting silicon solar cells

Samples	Height at which the silicon cell breaks (cm)	Specific Gravitational Potential Energy (10^{-2} Jcm^3/g)
Silicon solar cell	25 ± 5	–
EVA on Si cell	50 ± 4	7.3 ± 0.5
HA-SSC15 on Si cell	69 ± 2	9.4 ± 0.2
HA-SSC45 on Si cell	82 ± 4	11.2 ± 0.2

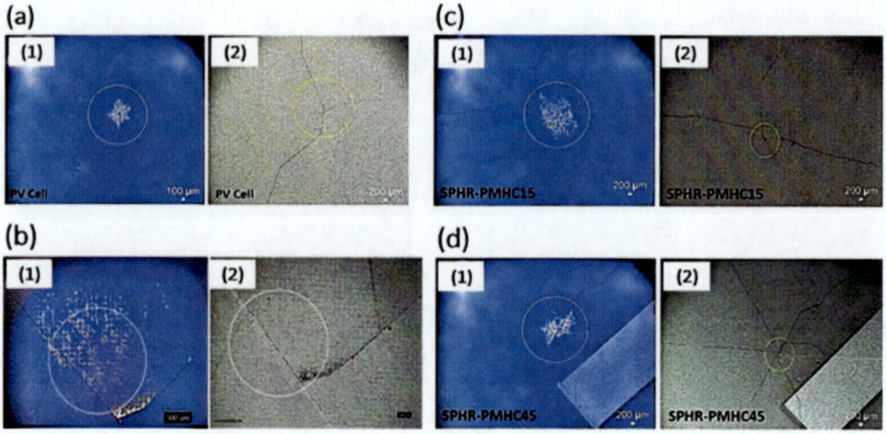

Fig. 6.7 Microscope images taken after impact test: (1) front and (2) back of silicon monocrystalline solar cells for the cases of **a** bare silicon cell, **b** silicon cell under EVA layer, **c** silicon cell under HA-SSC15 sample and **d** silicon cell under HA-SSC45 sample—after impacted with a steel ball. These images were taken from a representative silicon solar cell broken at the nominal height as shown in Table 6.2. The circles on the silicon solar cell surfaces indicate the approximate impact contact area with the steel ball during impact test

performing experiments on at least six samples for each test condition.

The optical microscope images after the impact testing of the representative silicon solar cells (front and back surfaces) are shown in Fig. 6.7. Each time the ball-dropping experiment was done, a new silicon solar cell was used. All silicon monocrystalline solar cells used in the present study came from the same batch of 200 cells. Within the expected manufacturing variability specifications, all the silicon solar cells used in the present impact test/experiment may reasonably be assumed to have uniform mechanical strength, structural integrity and fracture toughness. Thus, any differences we see after the impact experiments (as shown in Table 6.2, as well as in the following Fig. 6.7) are associated with the protection/encapsulation samples (either EVA or HA-SSCs or nothing) placed on top of the silicon solar cells.

Both the results shown in Table 6.2 (the fracture height/specific potential energy data and the ensuing statistical analysis) as well as the images shown in Fig. 6.7 clearly suggest that HA-SSC composite materials could provide better protection or encapsulation of the bare silicon solar cell against impact loading, compared to the PV standard EVA encapsulation layer. Both the HA-SSC composite materials (HA-SSC15 and HA-SSC45) allow substantially higher fracture heights and the associated specific potential energies (significantly beyond the experimental error bars/uncertainties) before the silicon solar cells under them start breaking or initiating/propagating catastrophic fracture events upon the ball-dropping impact testing experiments.

Receiving the impact energy from the ball-dropping test, the EVA would just comply due to its very low elastic modulus and thus the energy (and damage) was passed down to the silicon cells. The fragile silicon material is especially prone with such pointed impact loading, and thus crack occurs and extends through the thickness of the silicon solar cell almost instantaneously at relatively low height of 50 ± 4 cm. This is also evident from the images shown in Fig. 6.7b where the impact energy was absorbed mostly by creating just one main line (through the thickness of the cell) of very long crack (almost across the whole area of the cell—which is often considered to be the more serious concern in terms of its effect to electrical power output degradation).

In contrast, upon receiving the impact energy, both the HA-SSC membranes owing to its helicoidally aligned fiber-reinforced layered structures would dissipate the impact energy very efficiently and deflect the crack sideways following the helical path of the fiber directions instead of breaking instantaneously in a straight fracture line across the thickness of the membranes. In fact, the crack transitions between the ribbons and the matrix material at different angles in each layer—layer by layer. Every time the crack tip extends to the next layer, it keeps finding the fiber aligned at different angle of orientation and thus gets deflected sideways in every layer, and it gradually loses its primary driving force/energy to extend through the thickness of the membrane. This avoids the catastrophic failure of the HA-SSC membrane materials, but also absorbed more energy to deflect the crack along different directions and planes in every layer, and due to frequent encounter with laminate modulus changes, thus passing much less impact energy/damage down to the silicon solar cell underneath it. These results are consistent with the load dissipation mechanism we proposed in our previous reports [25, 26] of this novel 3D helicoidally fiber-aligned composite material, which is especially effective and significantly more efficient in dissipating energy such that only when the steel ball is dropped from the height of 69 ± 2 cm and 82 ± 4 cm for the HA-SSC15 and HA-SSC45, respectively, it would break the silicon solar cell underneath it.

The efficient load dissipation mechanism and effective absorption of damage/energy are also evident from the images shown in Fig. 6.7. Upon finally breaking the silicon solar cells underneath them at much higher heights, both the HA-SSC membrane materials allow several crack lines on the silicon cells indicating highly dissipated impact energy and transfer of load to the sideways (evident by the crack lines at different angles on the surfaces of the silicon cells) as received by the silicon cells. These cracks appear to follow the preferred crystallographic directions of (***110) associated with the weakest crystallographic planes of (111) in typical silicon monocrystalline wafers/solar cells, as has been reported widely in the literature in both PV and other silicon-based energy devices [28, 29].

These crack deflecting mechanisms as evident from the impact testing in the present study as well as complete mechanical and microstructural characterization studies conducted in our previous reports [25, 26] suggest also higher toughness and higher material resistance against crack propagation. Upon impacted on the surface, the ribbons of the HA-SSC composites assembled at different angles from one layer to the next would orient themselves along the direction of the applied force. The helicoidal ribbons during the impact deformation demonstrate efficient transferring of the load to the neighboring ribbons, gliding and pulling the ribbons continuously until the breaking point. This explains the higher energy absorbed by the HA-SSC samples. The combined effect of the structural design of the membrane and difference in the moduli lead to higher specific toughness to the HA-SSCs [25, 26].

The HA-SSC45 samples seem to be the most effective in deflecting crack or impact damage along different angular directions continuously. It is evident from the fracture height data that shows fracture of the silicon solar cells occurs at 82 ± 4 cm—the highest among the materials tested in the present study. The crack continuously encounters difference in modulus of ribbon and matrix material. In HA-SSC45, the crack continues along a ribbon, encounters a discrepancy in modulus due to the presence of the matrix material, continues to impregnate the matrix, then again encounters another ribbon laying along different plane and direction and deflects from its original path to follow another path. The crack proliferates along different directions and planes, twisting and turning its way inside and out of the ribbons and matrix phases, thus further stretching the ribbons without yet breaking them into the final catastrophic events of failure of the samples. It follows that this continuous deflection of cracks due to the helicoidal network of ribbons in the composite materials would require higher energy to break and thus pass minimum amount of impact energy/damage to the silicon solar cells underneath them.

However, our impact experiments suggest slightly different picture for the HA-SSC15. Fracture of the silicon solar cells underneath these samples occurs at lower heights (69 ± 2 cm) and thus lower specific potential energy (indicating lower impact energy/damage absorption rate). In these samples, the ribbons are closer to each other as compared to the other HA-SSC composite films, due to smaller angular rotation. When crack starts forming, it encounters much more frequent changes in directions and planes. The path of a straight crack is reduced when the crack encounters a change in the direction much more often as compared to the composites with a larger angle of ribbon arrangement. When the angles between layers are larger, the crack gets a straight path to propagate in the matrix until the next ribbon hinders it. Due to smaller angular rotation, the crack encounters much more frequent fluctuations in the modulus of phases, thereby it should be delaying the fracture events even more and absorbing even higher energy. These results were shown in our previous report [26] when we put a piece of glass under the materials. The results there [26] showed that the 15 deg helicoidally aligned samples could absorb much higher impact energy compared to the 45 deg samples. As we followed the same full procedure of the ball-dropping experiment with [25, 26], including the same size of the steel ball, we believe this deviation (in the impact damage absorption capacity between the 15 vs 45 HA-SSC samples) may be associated with the silicon monocrystalline solar cells we put under the HA-SSC samples. Compared to glass slip, the silicon monocrystalline samples have crystallographic dependence of mechanical properties including fracture preferred occurrences. However, the full correlation between the helicoidal alignment orientation and silicon monocrystalline solar cell fracture remains to be fully further investigated and beyond the scope of this sub-section.

It is thus evident from reports in the literature [25, 26] as well as from findings illustrated above that 3D helicoidal architecture proves to be effective in enhancing impact resistance of the novel materials. The potential application in the next-generation lightweight silicon-based PV technology has been unravelled from the findings and ensuing discussions in this sub-section. Other applications include military combat vest materials as well as advanced sports gears. All these applications rely on having different or even graded impact properties of the materials in service. Hence, the tunable impact properties enabled by the additive manufacturing as shown in this section will play important roles in key, emerging future technologies.

6.1.5 Additive Manufacturing Enabling Novel 3D Architected Anode Materials for Next-Generation Lithium Ion Battery (LIB) Technology

Yet another novel area of emerging application for this 3D architected materials might be in the next-generation energy storage technologies, such as Lithium Ion Battery (LIB). Anode design is a crucial factor for the further development of lithium ion battery (LIB) technology, especially in terms of achieving high charging rate performance, as typically needed in electric vehicles (EV) [30–32]. 3D nanostructures have recently been identified as the key to dramatically enhanced kinetics and performance of LIB [28]. However, existing fabrication methodologies for 3D nanostructures are inherently expensive—using advanced lithography techniques or other nanopatterning methodologies, as well as using standard additive manufacturing approaches. In the current sub-section, we propose a unique and low-cost fabrication methodology to enable the 3D nanostructured and even 3D nanoarchitected anode materials and design. Electrospinning (as an integrated additive manufacturing methodology) as has been shown in this chapter of the book may be used to produce low-cost 3D-printed fiber structures to allow for the novel anode design. This would pave the way to achieve the order of magnitude increase in battery kinetics and performance as needed to realize the vision of the sustainable transportation (with electric vehicle).

Energy storage has become an essential necessity in our modern life—from laptop and smart phone, to solar and wind farms and the latest trend is for the electrification of transportation, i.e. electric vehicles. Although many energy storage systems or technologies are available at different sizes and capacities, lithium ion battery (LIB) has been proven to be the leading choice of the market due to its several important qualities such as high efficiency, high energy density, limited self-discharge, longer life cycle even with high charging and discharging rates. For anode material, the most commercially used is graphite, whose charge capacity is 372 mAh/g, whereas silicon has a very high charge capacity (4200 mAh/g) theoretically and it can be a promising replacement for the existing anode technology which has roughly ten times smaller charge capacity than that of silicon. Silicon also has a low discharge potential and attractive operating voltage [30]. However, it has a disadvantage: its high volumetric expansion (300–400%) upon lithiation [31], leads to high stress/strain, cracking, palletization and delamination from the current collector, resulting in electrode failure [31, 32]. Nano-structuring of silicon was proven to be effective in relaxing the high strain due to expansion and increase the surface area available for the chemical reaction [32]. Silicon nanostructures have been shown to be superior to their bulk counterparts in capacity retention [30]. Different nanostructures such as nanoparticles, nanospheres, nanowires, nanotubes and nano-cones were experimentally proved to exhibit better cycling life and performance as LIB electrodes [28, 30].

Most recent reports [28, 30–32] suggest strongly that nano-structuring proved to be very effective in alleviating the main issues with next-generation anode technologies. For instance, 1D (one-dimensional) nanostructures, in the form of nanowires, have been demonstrated with enhanced charging rate [28, 30]. This is associated with the increase in the rate processes or kinetics of the Li intercalation and diffusion. In addition, nano-structuring was also proven to be effective in relaxing the high strain due to expansion (during charging, large flux of lithium ions absorbed in anode could lead to significant expansion of anode, and thus extremely high mechanical stress/strain which could potentially cause preliminary failure) and increasing the surface area available for the chemical reaction [31, 32]. If a nanowire anode (which is a 1D nanostructure) has been shown to be highly effective in enhancing performance (power density, charging rate, etc.) of the LIB device, due to its nanostructures (i.e. large surface areas for fast absorption and diffusion of lithium ions during lithiation or de-lithiation) to or out of anode materials, then certainly even more could be expected out of 3D nanostructures.

Key to this high charging rate performance is the anode design and technology (structures, geometries, etc.), in addition to the material itself [31, 32]. Today's mainstream lithium ion batteries have used graphite as anode material as it exhibits several optimal qualities, such as structural stability and low electrochemical reactivity (with respect to lithium ions). The basic idea here was to design carbon-based anode structures and geometries that maximizes the kinetics of the Li intercalation and diffusion. 3D nanostructures have thus been identified as the ultimate anode design for kinetics performance [28]. However, existing fabrication methodologies for 3D nanostructures are prohibitively expensive—using advanced lithography techniques or other nanopatterning methodologies.

Unlike traditional AM (Additive Manufacturing) techniques, electrospinning allows definition of individual fibers (dimensions, architectures, etc.) but produce them in batches (i.e. not individually, which makes regular AM very expensive, takes a long time to integrate into real engineering systems). But so far Es-AM (electrospinning-based additive manufacturing) produces largely random 3D fiber structure and not nanoscale [13–17]. Only very recently (2020), our group (joint collaboration between Singapore University of Technology and Design and La Trobe University, Australia) just reported successful fabrication of 3D fiber-based nanostructures—i.e. submicron fibers (several hundreds of nm) aligned densely and uniformly (not random) and stacked layer to layer and finally forming 3D nanolattices/grid structures [25, 26, 33]. This was enabled by a clever modification of the deposition platform of the Es-AM technique, such as reported in our most recent publication, essentially self-assembling and self-stretching the fibers into the smaller scales [25]. Thus, by further engineering of the methodology, we may reasonably expect to achieve real nanometer scaled fibers (~tens of nm). Such low-cost 3D nanostructures could pave the way to high charging rate battery performance, while allowing mass production and scalability.

However, despite the advantages of electrospinning in producing fiber-reinforced composites, the use of the conventional electrospinning setup has had limited success in producing controlled three-dimensional architectures. Recently, some studies have explored the melt-electrospinning setup to deposit fibers in a controllable manner [16, 17]. In the melt-electrospinning writing setup, the electrospinning voltage and the distance between the electrospinning nozzle and collector are lowered. As the polymer jet ejected from the electrospinning nozzle does not experience whipping instabilities, it ensures that the fibers can be deposited in a precise manner to obtain three-dimensional fiber patterns [17]. Thus, such an electrospinning-based methodology could be considered as an additive manufacturing (AM) approach, as it lays down one fiber material at a time in a precise and controllable manner [16–18, 25, 26]. With this approach, 3D fiber-based architectures were recently reported and demonstrated using the melt-electrospinning writing setup [16, 17, 33], such as shown in Fig. 6.8. The three-dimensional grid patterns were produced using melt-electrospinning based on polycaprolactone (PCL) material, which was chosen in this study due to its low melting point (~60°C) and thus the compatibility with the melt-electrospinning setup [16, 17, 33]. These three-dimensional fiber grid structures were then embedded in polyvinyl alcohol (PVA) matrix, which was chosen as the matrix material due to its water solubility and ability to form films. Since the fibers were melt extruded out of the nozzle and then deposited on a collector, the surface of the fibers tended to fuse with the fibers on top of which they were deposited. This also ensured that the fibers present at the interconnecting junctions within the grid bind strongly with one another. Embedding these fiber structures within the matrix demonstrated improvement of the structural properties of the architecture [16, 17, 33].

The above 3D-printed fibers were however in hundreds of micron scales as evident from Fig. 6.8 above, so the 3D grid structures are not nanostructures. Sahay et al. (2020) [25] have most recently demonstrated the successful fabrication of such 3D structures in the ~ 1 μm scales, as a further development of the electrospinning-based methodology. Specifically, Sahay et al. (2020) [25] demonstrated the unique use of parallel plates for deposition platform to allow significant improvement in the fiber diameter reduction (from ~hundreds of μm to ~1 μm range), while maintaining uniform and dense alignment of the fibers. With this setup, further engineering and optimization of the technique may be expected to lead to even further significant (1–2 order/s of magnitudes) reduction in fiber diameters into the tens of nanometer scales.

In addition to fiber dimension reduction, the current Es-AM setup is also capable of creating different structural design of the 3D architectures. 3D fiber grid patterns were produced in Agarwal et al. (2020) [26], as illustrated in Fig. 6.3D fiber helicoidal alignments were produced in Sahay et al. (2020) [25] and Agarwal et al. (2019) [16] and a few other references [34–36], as further illustrated in Fig. 6.9. These 3D structural architectures may affect substantially the kinetics in lithium ion intercalation and diffusion in LIB device, cycling lifetime/reliability of the anode (associated with stress/strain evolution) as well as structural integrity of overall anode design.

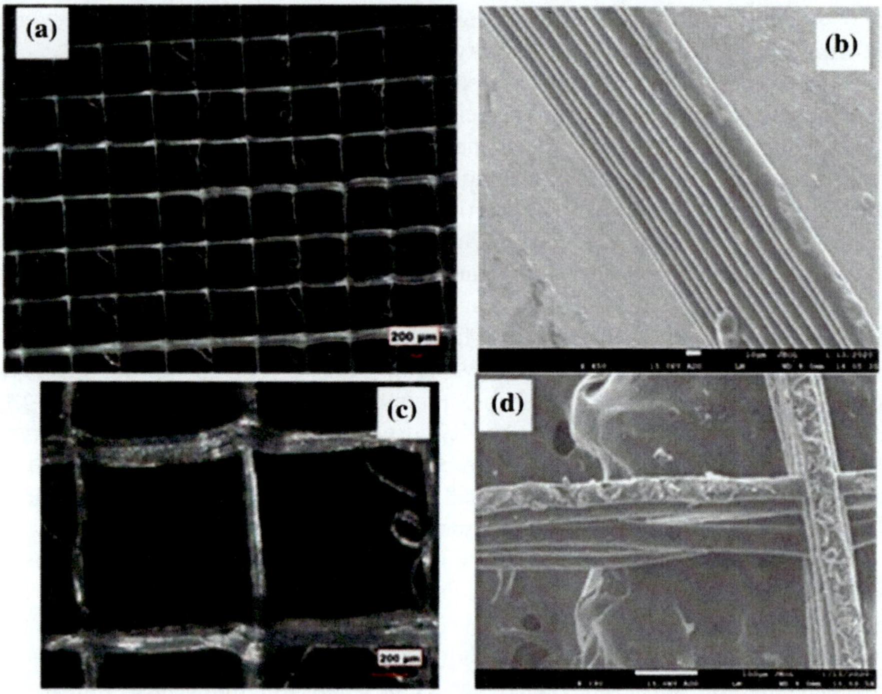

Fig. 6.8 **a** Optical microscope image of the polycaprolactone (PCL) fibers printed as grid structures using melt-electrospinning writing method; **b** scanning electron microscope (SEM) image of the cross-section of the grid structure (scale bar 10 μm). The fibers are deposited on top of each other in a controlled manner; **c** optical microscope image of the PCL fibers demonstrating the cross-over points of the grid structures; and **d** SEM image demonstrating the surface morphology of the fibers (scale bar 100 μm) [33]. Reproduced from Ref. [33] with permission from MDPI

Thus, the novel 3D architected fiber-based composite materials as have been discussed in further depths in this chapter have good potential to play key roles in the emerging 3D nanoarchitected anode design that are required to enable the next-generation energy storage technologies. In this sub-section, we propose a unique additive manufacturing methodology using an electrospinning-based process to help enable low-cost 3D nanostructure anode with high performance (kinetics related, such as power density and charging rate) and reliability (long life/cyclability). These novel 3D nano-architectures and their emerging applications in both PV and LIB technologies—both are key energy technologies to enable the next industrial revolution based on clean technology!—are enabled only by the knowledge of eXtreme Mechanics (XM) and the XM probing methodologies as presented in earlier chapters of this book. These novel materials indeed provide great insights and unique illustrations of how XM may enable further the Design for Reliability (DfR), or even Design for Extreme Reliability (DfXR) in the strategic technologies of the future.

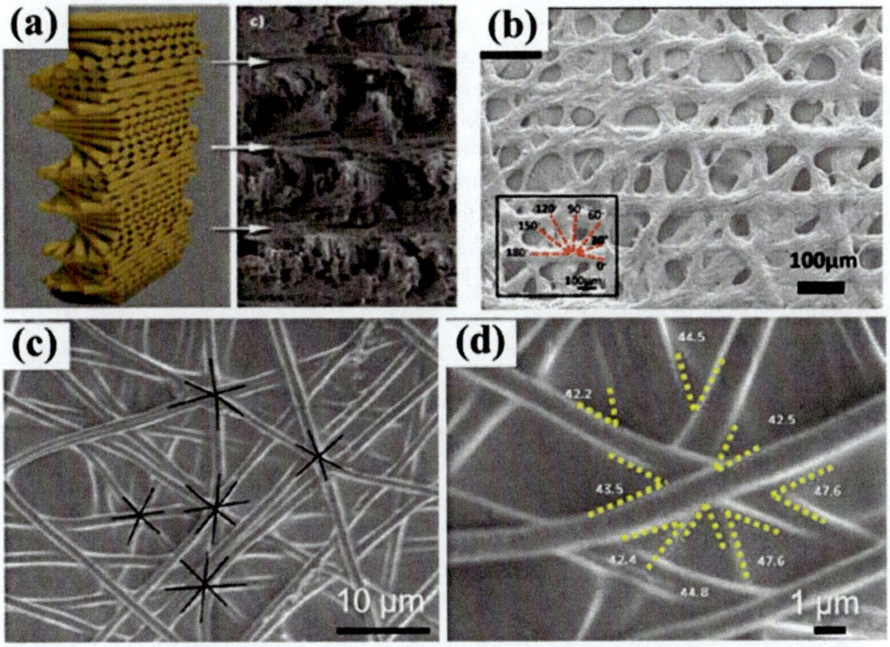

Fig. 6.9 **a** 3D model of helicoidal alignment as well as cross-sectional scanning electron microscope (SEM) image of *Odontodactylus scyllarus,* Mantis Shrimp's dactyl club, the bio-inspiration of the 3D helicoidal architecture [16]; **b** top-view SEM image showing helicoidal fiber alignment (fiber diameters in the ~ tens of μm range) [17]; **c** top-view SEM image showing helicoidal fiber alignment (fiber diameters in the single digits to ~1 μm range) [25]. Reproduced from Ref. [25] with permission from MDPI

6.2 Enabling Stretchable Metallic Conductors Through Atomic Reconfigurations in FCC/BCC Nanolayers

Metallic FCC-BCC nanolayers, such as Cu/Nb, have received wide attention due to their extraordinary mechanical properties as well as the unique self-healing capacities due to the interface characteristics. Most recently, the materials have also been shown to exhibit significant and tunable interfacial sliding mechanisms (based on defect structures in the interface). The significant interfacial sliding is all along while maintaining full contact between the layers, and thus one could expect negligible resistance increase upon straining, which would be attractive for stretchable metallic conductor technology. The interfacial sliding has been modeled with some combination of diffusional and displacive mechanisms—the extreme extents of which are afforded by the nanoscale layering in the materials. The exact mechanisms continue to be fully investigated with *in-situ* mechanical testing allowing the direct observation of the interfacial sliding events inside a SEM (Scanning Electron Microscopy), as well as many other characterization techniques. In this section, we aim to describe how we could harness the unique capacity for atomic reconfigurations in the FCC/BCC nanolayers to enable stretchable metallic conductors. This approach of using atomic reconfigurations (instead of the structural reconfigurations in existing metallic stretchable technologies) is novel and, due to the associated diffusional and displacive mechanisms strictly in the interface (thus maintaining contact all the while) could lead to potentially new, breakthrough metallic stretchable materials—stretchable (and recoverable) without compromising the electrical conductivity upon significant mechanical deformation, as well as upon long operational duration (durability). Stretchable conductors are important part of stretchable electronics, which could lead to many important technologies such as artificial skin, muscle, limb as well as soft robotics and human–machine interfaces.

6.2.1 Background and Motivation—Novel Metallic Stretchable Conductor Materials

Atomic reconfigurations in metallic materials have been of particular interests scientifically for many years [37–44]. However, in the past few years, the interests have been rejuvenated especially technologically with the recent advances in the fabrication and manipulation of materials at the nanoscale [45–48]. One particular material system of interest here is the metallic FCC-BCC nanolayer system, such as Cu/Nb nanolayers. They have received wide attention due to their extraordinary mechanical properties [48–51] as well as the unique self-healing capacities [52–54] stemming from their interface characteristics. The significant atomic reconfigurations allow, for instance, for the recently discovered extraordinary deformability with strains of up to 24% while maintaining homogenized deformation in micropillar compression experiments [54, 55]. Most recently, the materials have also been shown to exhibit significant and tunable interfacial sliding mechanisms (based on defect structures in the interface) [56, 57]. The significant interfacial sliding is all along while maintaining full contact between the layers, and thus one could expect negligible resistance increase upon straining, which would be attractive for stretchable metallic conductor technology.

In this section, we propose to harness the unique capacity for interfacial sliding [56, 57] in FCC/BCC metallic nanolayers, such as Cu/Nb [48–55, 58, 59] to enable novel stretchable metallic conductors. Metallic stretchable technologies to date involve mostly nanowires [34, 60–62], carbon nanotubes (CNTs) [35, 36, 63, 64] and graphene [65–67], however, the resistance increase upon deformation is still significant [34, 61, 68, 69]. For example, at 100% strain, the resistance of silver nanowires (which is among the highest conductivity material) embedded in elastomer increases by a factor of 20–35 [34, 61]. Similarly, at 120% strain, the resistance of CNTs film embedded in elastomer increases by a factor of 8 [68, 69].

Similar to the above-mentioned existing technologies, the FCC/BCC metallic nanolayers will allow significant deformability (in this case afforded by the interfacial sliding mechanisms). The key advantage here is that all along the interface, the nanolayers remain in full contact with one another and thus the structure maintains its conductivity by atomic reconfiguration (rather than structural reconfiguration, as in the case of nanowires, CNT and other metallic grid/nanostructures) during mechanical deformation. The proposed approach here is based on the hypothesis that interfacial sliding can be controlled in the FCC/BCC metallic nanolayers, such as Cu/Nb, and thus could be tuned to allow the deformability as well as the associated needed parameters, such as straining or stretching rates for specific applications (soft robotics vs. artificial limb, for instance).

These metallic stretchable materials with extraordinary potential to maintain conductivity upon significant deformation can be obtained and tuned by relatively simple fabrication processes. Cu/Nb nanolayers, for instance, have been produced using PVD (Physical Vapor Deposition) [50, 70] and Accumulative Rolling Bonding (ARB) [71, 72] methodologies. Such processes have allowed effective atomic engineering of the interfaces, especially in FCC/BCC metallic nanolayers, for instance, creating Cu/Nb interfaces with different orientation relationships (OR) and defect structures [50, 70]. These differences in turn lead to markedly different mechanical properties of the materials, including how much the nanolayers could slide against each other [56, 57], which we aim to use for the application of metallic stretchable conductors. As the interfacial sliding in the FCC/BCC nanolayer occurs through mostly atomic reconfiguration of the materials in the interfaces (i.e. through combination of diffusional and displacive mechanisms), the paths of conductivity between the nanolayers are fully maintained upon significant mechanical deformation. Furthermore, combined with smart design of the nanolayers using interdigitated bridges or grids, the basic structure could provide high stretchability while maintaining significant levels of its conductivity. Further, more specifically, this study will investigate how the structure through fundamental investigations and smart designs could achieve substantial recoverability (upon stretching) and elucidate the associated recovering mechanisms. These mechanisms, in addition to the interfacial sliding mechanisms, are likely to be novel and thus could lead to potentially new, breakthrough metallic stretchable materials technology—through atomic reconfiguration (rather than structural reconfiguration as in many existing technologies).

Having many interfaces in the FCC/BCC nanolayers with tunable interfacial characteristics in each of the interfaces, this unique structure could potentially lead to very efficient (high rate) as well as highly tunable energy dissipation mechanisms, and thus could allow significant stretchability while maintaining most of the conductivity. We will explore the effects of interfacial characteristics, such as orientation relationship and defect structure, as well as layer individual thickness and the possible size effects of the substructures on the interfacial sliding and stretchable properties of the resulting materials, as well as the associated deformation mechanisms. These are likely to be novel and ground-breaking. Our findings so far have indeed suggested that we can obtain control of interfacial sliding mechanisms leading to unprecedented stretchability and tunable stretching/straining rates. Furthermore, the nanometer size (few nanometers to few tens of nanometers) of the layers could lead to further superior mechanical properties, as known in the community as "smaller is stronger" phenomena. This will play an important role if the stretchable conductors are needed to also provide some structural functions, while the high surface area to volume ratio afforded by the nanolayers will further be useful to release strain and thus delay damage events. Even just for the stretchable functionality alone, the capacity to arrest crack propagation through interfacial deformation could lead to more robust and reliable/durable stretchable conductor materials technology.

The proposed study is highly original, scientifically as well as technologically. Scientifically, atomic reconfigurations associated with interface-based deformation have enabled recent scientific discoveries, involving the findings of extraordinary mechanical properties [48–51] and radiation damage-tolerance [53, 54] in Cu/Nb nanolayers making them withstand extreme environments (strain, strain rate/shock extremes [58, 59], self-healing[39,41] and radiation damage [52, 53]) such as in nuclear energy [48–53]. Beyond Cu/Nb nanolayered systems, the intriguing possibility of frictionless gliding of one solid surface on another has been predicted for certain incommensurate interfaces in crystals, based on Aubry's solution to the Frenkel-Kontorova model [37–40] of a harmonic chain in a periodic potential field. Incommensurate structures hold a special place in the science of crystalline materials and especially in the study of the exact roles of defects in influencing the overall response of the crystal upon mechanical deformation. Incommensuration arises when translational symmetries with different but incompatible repeat patterns interact, such as interfaces between two different crystals. Even single-phase materials can exhibit incommensurate structures, for example, grain boundaries with a misorientation that brings a periodic array of atoms in one grain into contact with an array of an incommensurate periodicity in the other grain [44]. Obviously, such incommensurate structures have important consequences for the behaviors predicted for such interfaces. Technologically, superglide, superlubricity or hypofriction—the absence of static friction—are particularly interesting properties predicted for incommensurate boundaries for a certain range of atomic interactions [39–44, 73] that could lead to many engineering applications. Although these properties have been observed for surfaces in contact in several studies on friction [74–77], experimental verification for grain boundaries of bulk materials is difficult to obtain and only one unequivocal experiment (using Au nanopillars) has been reported to date [45]. The possibility of superglide at an incommensurate grain boundary has been theoretically predicted but is not expected for those incommensurate grain boundaries that can be described by a network of localized dislocations [78].

The Cu/Nb nanolayered systems represent a platform in which gliding of one solid surface on another could be further studied. In fact, such gliding on the interfaces between the Cu and Nb nanolayers has been predicted using molecular dynamics [48], and many extraordinary properties reported for Cu/Nb nanolayers have since been attributed to such capacity for the interfaces to move against each other and for atomic reconfigurations [48–57]. Experimentally, our group (ASB) at Singapore University of Technology and Design (SUTD) has recently demonstrated the direct observation of such interfacial sliding events using an *in-situ* microbending experimental setup inside a SEM (Scanning Electron Microscopy) [56, 57] as shown in Fig. 6.10. Furthermore, it was found that the extent of the interfacial sliding could

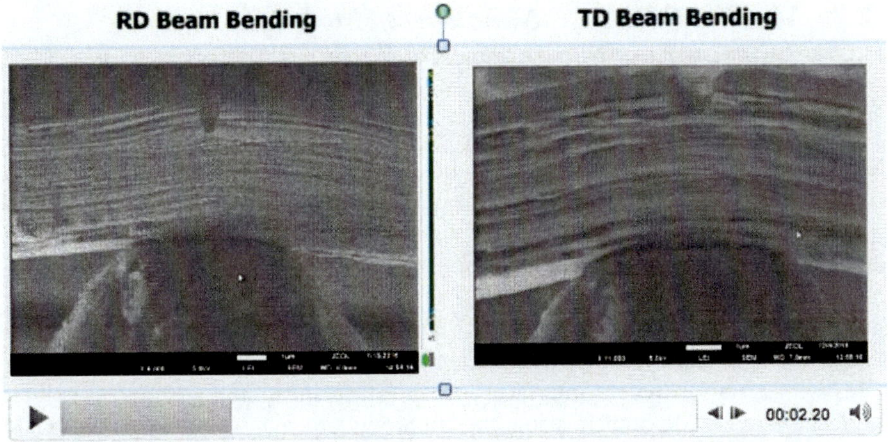

Fig. 6.10 Significantly different interfacial sliding extents and mechanisms in Cu/Nb nanolayers (at similar displacement; the experiments used displacement-control) following on differences in interface defect structures achieved through atomic engineering of interfaces (based on the rolling direction/RD and the transverse direction/TD of the Accumulative Rolling Bonding/ARB process) [56]

be controlled via the interface characteristics (orientation relationships) and defect structures (densities, interfacial shear capacity, etc.) [56].

Further, technologically, these interfacial sliding and the associated atomic reconfiguration mechanisms, such as found in Cu/Nb nanolayers, could lead to extraordinary and first-of-its-kind stretchable metallic conductor technology (maintaining high level of conductivity upon significant mechanical deformation, as well as upon long operational duration) with both stretchability and stretching rates that could be tuned according to the needs of the applications. The control of stretchability, straining rate and even the durability (performance over time) could be achieved through quantitative understanding of the deformation mechanisms involved in the interfacial sliding phenomena—be it through the microstructural-mediated mechanisms (diffusional or displacive), as well as via the geometrical design of the mechanical interdigitated grids/structures. These mechanisms are likely to be novel and potentially ground-breaking in terms of the new technologies they could enable. These novel mechanisms would be unraveled with cutting-edge materials characterization techniques (such as *in-situ* synchrotron X-ray nanodiffraction and *in-situ* straining technique inside a SEM) in combination with computational modeling capability (Finite Element Analysis combined with other more advanced molecular modeling).

6.2.2 Metallic FCC/BCC Nanolayered Materials System—Enabling Interfacial Sliding Through Atomic Reconfigurations

Combining materials into nanolayer composites (nanolaminates) is one of the more promising approaches to design new structural materials. A high density of interfaces in the nanolaminates may alter their deformation mechanisms [48–51, 79]. Nanolaminates can be classified into four groups based on these deformation mechanisms: metal–metal, metal-hard phase, metal-metallic glass and metal-polymer nanolaminates [80]. Out of these, plastic deformation of metal–metal nanolaminates is governed by dislocation-interface interactions [79]. As for metal–metal nanolaminates with a semi-coherent interface, low interface shear strength increases their interface barrier to slip transmission, thus improving the nanolaminate strength and ductility [52–55, 81–84]. Hence, a low interface shear strength is preferable, in order to enhance plasticity in metal–metal nanolaminates with a semi-coherent interface.

Nanoscale multilayered materials provide a unique mean to design interfaces for specific properties and functionalities [48–54]. Examples that have been much reported include nanoscale strengthening of materials [48–51], extending deformability through interface recovery [54, 55] and radiation damage-tolerance in which voids and other microstructural defects are effectively absorbed and annihilated in the many interfaces of nanolayers [52, 53]. Recently, nanolayers have been investigated through atomic engineering of interfaces to achieve superior fracture toughness and unraveling novel strategies on how to delay or arrest crack propagation from leading to final catastrophic events of failure of the materials [56, 57]. Very little work up to date has been reported focusing on the detailed role of interfaces, especially in how they could influence the fracture events and eventually enhance the fracture toughness of the nanoscale multilayered materials. Our group has recently performed *in-situ* microfracture bending tests inside a SEM, as shown in Fig. 6.11, on a notched clamped beam of Cu/Nb nanolayers and discovered a unique deformation mechanism of interfacial sliding [56, 57]—which allows significant delay of the crack propagation, and thus enhance fracture properties of the nanomaterials. However, in

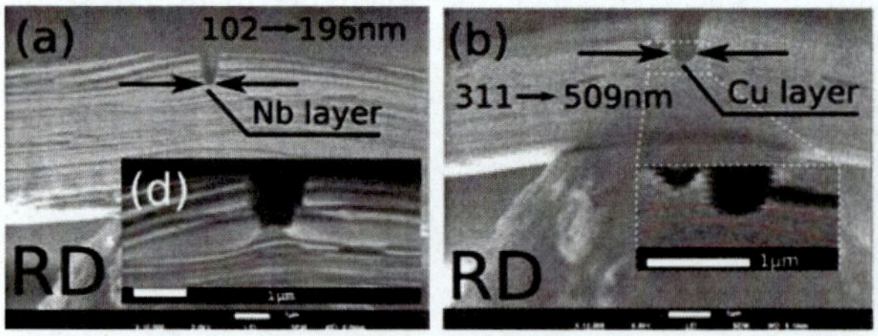

Fig. 6.11 Comparison of the notch widening between the rolling direction (RD) and the transverse direction (TD) beams—larger extent of interfacial sliding in TD beams is evident. Minimum (initial) and maximum notch widths are indicated in all figures. The materials of the layers at the notch tip was determined using low-angle back-scatter in the SEM [56]

this study, we propose to use the interfacial sliding phenomena—and the possibilities that we could control and thus tune them to a large extent—to allow atomic reconfigurations in metallic stretchable conductor technology (which has mostly hitherto been done through structural reconfiguration—nanowires, mechanical grids, etc.).

The mechanical properties of nanoscale metallic multilayered materials, in particular, Cu/Nb, have been studied extensively over the past decade for their extraordinary flow strength [48–50], thermal stability [51] and resistance to radiation and shock extremes [58, 59]. These enhanced properties have been attributed to the interfaces between two constituent layers as the individual layer thickness shrinks to the nanoscale. The established Kurdjumov–Sachs (K-S) oriented Cu/Nb interface is an atomically flat interface with a low energy ordered configuration despite the 11% lattice mismatch between Cu and Nb atoms [52, 70]. However, despite this large lattice mismatch, these Cu/Nb multilayers with nanoscale individual layer thicknesses outperformed their bulk constituents on various mechanical properties due to the unusually high density of interfaces and low density of defects accumulated in the layers during experimentations [83, 84]. Additionally, these interfaces were found to contain excellent sites for the recombination and annihilation of defects leading to a "self-healing" interface [53, 71].

The interfaces in multilayers can be varied with different crystallographic orientations through their fabrication process. Traditionally, multilayered films are prepared by physical vapor deposition (PVD) using either magnetron sputtering or electron beam evaporation up to a few microns on a silicon or sapphire substrate [83]. For Cu/Nb FCC/BCC multilayers made by PVD on silicon $< 100 >$, the $\{111\}$ Cu || $\{110\}$ Nb, K-S or Nishiyama-Wasserman (N-W) crystallographic orientation interfaces are created. Accumulative roll bonding (ARB) is a more recent technique that involves repeated operations of rolling, cutting and restacking and bond to form a single sheet of material [71, 72]. This process allows very large strains to be applied without changing the original dimensions of the sample, creating faceted, defect-free, ordered structures. The individual layer thickness can be easily refined and controlled with increasing strain. Cu/Nb ARB interfaces are found to be oriented $\{112\}<111>$ Cu || $\{112\}<110>$ Nb throughout the ARB sample [85]. At the nanometer scale, the ARB process results in severely elongated grains with widths of at least 10–50 times the thickness of the nanolayer [86] along the rolling direction (RD) of the ARB process. Figure 6.10 shows the very different extent of interfacial sliding phenomena that has been observed on the same Cu/Nb ARB samples—just bent on two different directions (the RD and the transverse direction/TD) [56].

FCC/BCC nanolayers experience three types of dislocation-mediated mechanisms upon deformation based on their individual layer thicknesses [69]: Hall–Petch strengthening through dislocation pile-ups (layer thickness > 100 nm), confined layer slip where dislocation loops glide parallel to the layers (~5 nm < layer thickness < 100 nm) and slip transmission across the interface (layer thickness < 5 nm). These plasticity mechanisms are well reported and discussed, especially for PVD-fabricated Cu/Nb samples [39–44, 58, 59]. ARB represents a way to alter the interface characteristics significantly [68, 69], as also shown in our group's most recent discovery

[56] (Fig. 6.10). Different defect structures along the two different directions (RD vs. TD) have been attributed to the markedly different extent of interfacial sliding.

There is a significant difference in the notch widening for most of the TD and RD beams, as shown in Fig. 6.11, which indicates extents of interfacial sliding. The maximum relative notch width increase is from 50 to 100% for RD beams; for TD beams it can exceed 400%. The notch widening for the RD beams is clearly associated with large plastic deformations within the layers at the notch tip, as can be seen from Fig. 6.11a, b. The deformations differ depending on the material of the layer at the notch tip. If the Nb layer is located at the notch tip, its plasticity is mainly governed by shear stresses at the tip corners, as shown in Fig. 6.11a, d, leading to a crack formation (Fig. 6.11d). In contrast, the more plastic Cu layers neck in the middle of the notch, as shown in Fig. 6.11b. Actual values of the plastic deformation around the tip cannot be precisely determined using *in-situ* SEM, but the average tensile plastic strain in the layer at the notch tip is comparable to the relative notch width increase. For example, the observed average plastic strain for the beam shown in Fig. 6.11b is more than 50%. Such a large plastic strain is similar to that in thin metallic films on flexible substrates [87]. If the dislocations do not cross the interface between the layers but can escape to the surface, then the plasticity of the layer may increase since layer necking is constrained [87]. Hence, the plastic strain limit is much larger in comparison with the standalone layers, and depends on the interface adhesion strength, the interface dislocation crossing barrier (at the applied strain) and the critical tensile strength of the layer.

Plasticity near the notch tip alone cannot explain why the notch widening in TD beams can be more than six times larger than in RD beams. We propose that the notch widening here is due to the layer sliding at the interface near the notch tip (interface shear). This hypothesis is consistent with previously reported molecular dynamics (MD) simulations that predicted the anisotropy of interface shear strength in Cu/Nb ARB [88]. These simulations predicted that in the most common {112}/{112} interface in Cu/Nb ARB [49], shear is not possible along the RD (interface shear strength is infinite), whereas the shear strength is 1.2 GPa along the TD. This large difference in the interface shear strength can explain why the interface shear is observed in the TD beams, but not as much in the RD beams.

However, the interface-based plasticity alone may not be sufficient to increase the notch width by 400%. Our investigation continues but from literature, we know upon large deformation and deformation rates, for instance, extreme shock loading, the Cu/Nb nanolayers have been reported to respond by diffusional mechanisms, such as nanovoid formation and their further growth and even coalescence around the interfaces [58, 89]. Upon high strain rate (0.01/s) and high temperature, Cu/Nb nanolayers, especially with small individual layer thicknesses (<10 nm), have been reported to respond through mostly dislocation climb [90] (due to constraints from the interfaces), which requires diffusional mechanisms. Through these known mechanisms (dislocation-mediated mechanisms—such as confined layer slip, dislocation pile-up and dislocation transmission through mostly gliding motion) and other possible mechanisms (such as diffusional deformation as well as diffusion-enabled dislocation movements), we propose that large deformation and deformation rates could be achieved and even controlled and tuned to a large extent. These mechanisms involve atomic reconfigurations around the interfaces. By quantitatively understanding the exact roles of each specific mechanism leading to the interfacial sliding, we could achieve high deformation (i.e. stretchability) and deformation rate (i.e. stretching rate) to suit the needs of the stretchable electronic applications.

6.2.3 Novel Stretchable Metallic Conductors—Structural Versus Atomic Reconfigurations

The recent excitements around deformable or stretchable electronics, flexible transparent displays/screens and their integration into the human body have allowed the development of multifunctional materials that are stretchable, transparent, wearable and/or biocompatible while meeting the energy or power requirements. Stretchable electronics in which circuits are built-in stretchable substrates can accommodate large mechanical deformation without loss of their electronic performance [91, 92]. Owing to their good deformability that can intimately conform to the unique characteristics of organisms and the human body, stretchable electronics are considered promising for biomimicry systems including artificial skin, muscle, limbs, eyes, as well as soft robotics and human–machine interfaces [93–95]. Recent progress on the strategies to realize large deformability or stretchability of devices has been leveraged on significant breakthroughs in the use of novel materials and structures [96–100]. The advent of novel deformable and transparent concepts has propelled a significant demand for the fabrication of soft, transparent and stretchable materials that can sustain their functionality under extreme deformations for potential application in the next-generation electronic devices and energy devices [101, 102].

One of the major challenges in the use of an electronic conductor in stretchable electronics is the significant increase in the resistance upon deformation [34, 61, 68, 69]. For example, at 100% strain, the resistance of silver nanowires (AgNWs) embedded in elastomers increases by a factor of 20–35 [34, 61]. Similarly, at 120% strain, the resistance of carbon nanotube (CNT) film embedded in elastomer increases by a factor of 8 [68, 69].

Earlier works on metal grids on flexible substrates have shown to form flexible conductors on plastics. However, the metal grid approach does not meet the needs and requirements for stretchable conductors. Therefore, Ag NWs have been extensively studied as stretchable conductors due to the high electrical conducting properties of Ag NW network that could withstand stretchability [103]. The adhesion between metal NWs and the elastomer substrate is important during the fabrication procedure (as shown in Fig. 6.12) to achieve electrical stability under deformation. Thus, modifications of the elastomer surface have been studied to create hydrophilic surfaces and improve the adhesion with metal NWs.

Despite the outstanding properties of Ag NW-based stretchable conductors, their relatively poor stability has remained a challenge for practical applications.

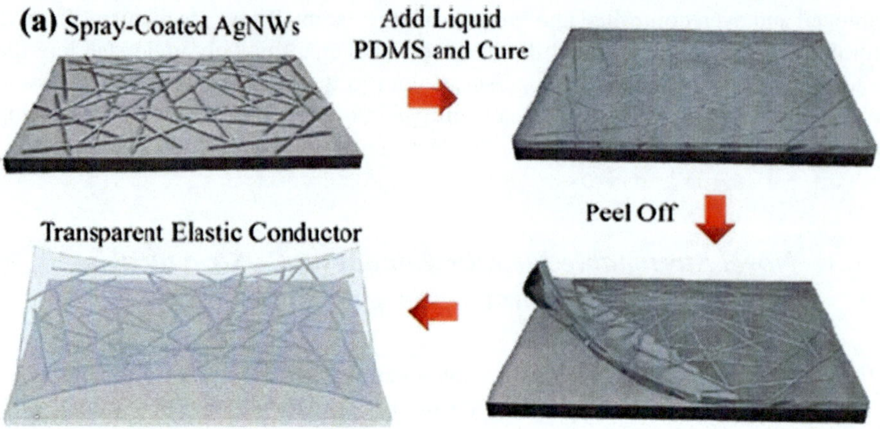

Fig. 6.12 Fabrication procedure of Ag nanowire-based stretchable conductor [61]

In addition, their application in soft energy devices, especially electrochemical storage/conversion devices, has been restricted, because Ag NW is easily oxidized in water and numerous electrolytes. Recently, the electrochemical and air stability of Ag NWs can be improved by Ag@Au core–shell structures and have been attempted in supercapacitors [104]. Furthermore, our group (PSL) synthesized Ag@Ni core–shell structures and found that percolated Ag@Ni NW networks showed reversible stretchability and good electrochemical stability in stretchable transparent supercapacitors [62]. There are some recent works focusing on Cu-based stretchable conductors in view of the lower cost and high conductivity. Nevertheless, oxidation issues and stability of Cu nanowires or Cu nanoparticles in air and electrochemical environment remain unresolved. Therefore, further ongoing effort is necessary to develop the highly stable stretchable conductors for various soft energy and electronics device applications, as summarized in a recent review paper [105].

Alternative materials configuration in quasi-planar form could overcome the strain limits or space requirements posed by the pre-strained approach or buckled conductors that are stretchable. Therefore, in this proposal, we identify the nanolaminates of bimetallic structures and their atomic reconfigurability to be highly promising for realizing a pragmatic and novel stretchable conductor.

6.2.4 Probing Atomic Reconfigurations Using In-Situ *Synchrotron X-ray Nanodiffraction*

X-ray diffraction remains one of the principal techniques to solve crystal structures. It was also recognized from the early days of X-ray crystallography that it is a very sensitive tool to analyze deviations from perfection in crystals of known structure. Defects (point defects, dislocations planar faults, etc.) yield very specific diffuse patterns in the diffraction space and elastic strains give rise to Bragg peak shifts that can be very sensitively measured [106]. Measuring a large enough number of independent Bragg peaks yield the full strain tensor. Multilayers are some kind of synthetic crystals in one dimension and some Bragg peaks exhibit coherent features related to the structure factor of one single bilayer [107–110]. This latter one is extremely sensitive to the phase shift between the two layers and hence to atomic reconfigurations at the interface. In real multilayers, deviations from an average structure may be important, but it is the strength of X-ray diffraction to yield with great accuracy the average structure but also the standard deviation from it [107]. Furthermore, a critical parameter about the interface structure is the relative orientation of the two layers, which can be easily characterized via texture measurements. All these measurements can be easily performed *in-situ* either during mechanical testing [111] or even annealing. It is worth noting that it is possible nowadays to perform *in-situ* texture measurements on a synchrotron beamline using a 2D detector [112, 113].

While X-ray diffraction is well known for its sensitivity to crystal defects and strains, it generally lacks spatial resolution. The last 15 years have witnessed a real revolution in the development of small X-ray beams. Progress in X-ray focusing optics allows nowadays scanning X-ray diffraction mapping to be performed with a resolution in the 50–100 nm range [114, 115]. Full field X-ray microscopy is improving a lot too with resolutions in the 100 nm range [116]. By far the best spatial resolution, on the other hand, is obtained with coherent diffraction imaging, which is a lens-less imaging technique, with a typical resolution of 8–10 nm [117, 118]. In addition, the comparatively large space available at synchrotron beamlines as well as the penetrating power of X-rays, in comparison to transmission electron microscopes, has paved the way to the development of various *in-situ* micro- and nano-mechanical testing devices [116] in the recent past. The use of micron-sized white beams to perform local Laue diffraction was pioneered at the Advanced Photon Source (APS) in the late 90 s and early 2000s [119–122]. It was soon applied to the evaluation of defects in crystals [121] and extended from a 2D projection to a full 3D investigation [122]. Spurred by the need for stress mapping at the micron and submicron scale in microelectronics components, Laue microdiffraction was implemented at Advanced Light Source (ALS) in Berkeley during the same period [123]. Very quickly, innovative results on stress distributions in metal lines and thin films were obtained [124–126] showing that very large stress inhomogeneities can be expected in these polycrystalline structures [127–129]. The development of Laue microdiffraction came to Europe 5 years later with a first publication from the Swiss Light Source in 2006 [130]. During the same year, a station dedicated to Laue microdiffraction started working at the BM32 beamline at ESRF in Grenoble [131].

Laue diffraction is based on polychromatic X-ray beams covering a large energy range (typically 5–25 keV). With monochromatic X-ray, the Ewald sphere passes only through a limited number of reciprocal lattice points making it necessary to rotate the sample to measure different Bragg reflections. However, when employing a broad spectrum of X-rays, i.e. a polychromatic beam, a large number of reciprocal lattice points are accessible simultaneously. Thanks to the large number of diffraction spots recorded in one pattern, the crystalline structure as well as the orientation of the crystal is readily available. From the refinement of the respective positions of the Laue spots, the local deviatoric strain tensor in the crystal may be obtained with a typical strain resolution of the order of 10^{-4} (Fig. 6.13).

Because no sample rotation under the beam is required, Laue microdiffraction is very adaptable to various *in-situ* environments and in particular to mechanical testing. For example, our group (IM2NP) demonstrated three-point bending tests on Au nanowires in combination with Laue microdiffraction at the BM32 beamline at ESRF [133]. The great sensitivity of Laue spot position to crystal rotation allows a detailed monitoring of the wire bending (and additional torsion). This kind of experiment, however, typically concentrates on one single position in an extended object due to the fact that any movement of diffractometer motors can induce vibrations, which eventually lead to the damaging of the nano-object under investigation. In order to circumvent this restriction, Leclere et al. recently demonstrated a novel beam scanning technique, which allows for mapping a sample in two dimensions during mechanical deformation [111, 134]. This technique gives access to the profile of the complete nanowire under load as well as to its boundary conditions, i.e. whether the wire is thoroughly clamped or simply supported. This beam scanning approach bases on the lateral and/or angular displacement of the Kirkpatrick-Baez mirrors used for focusing the incident white X-ray beam and, thus displacing the X-ray beam focus on the sample surface. It allows a complete mapping of strain and defects in the nano-object under load thus enabling the monitoring of the very first stages of plasticity.

With the advent of third-generation synchrotron sources and the increase of both the brilliance and the coherence of the provided synchrotron X-ray beams as well

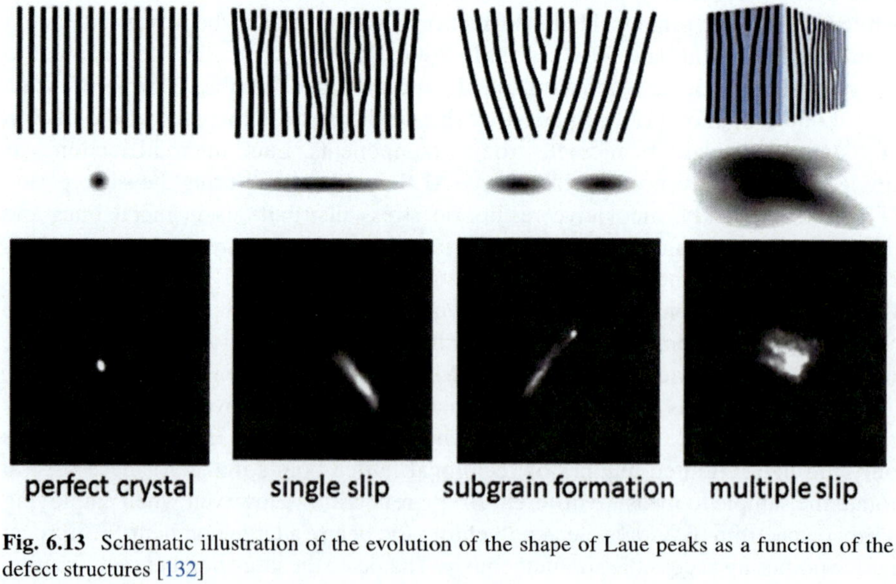

Fig. 6.13 Schematic illustration of the evolution of the shape of Laue peaks as a function of the defect structures [132]

as the progress on focusing optics, a new lens-less imaging technique, the Bragg coherent X-ray diffraction imaging (BCDI) method, started to being developed. In 2003, the first successful reconstruction of an Au nanocrystal from experimentally obtained coherent X-ray diffraction data was reported [135]. The first successful phase retrieval was demonstrated on Pb nanocrystals in 2006 [117]. Since these pioneering works, BCDI has been steadily improving over the last 10 years demonstrating its capability to image strain fields and single defects in 3D with very high spatial resolution (<10 nm), high strain sensitivity (10^{-4}) and unchallenged displacement field resolution (1 pm). For example, Labat et al. demonstrated the reconstruction of GaN nanowires containing inversion domain boundaries measuring five independent Bragg peaks of the same nanorod [136]. The authors reported a displacement of 8 pm along the c-direction with an accuracy of 1 pm demonstrating the very high accuracy of measuring displacement fields by BCDI (see Fig. 6.14). BCDI is perfectly adapted to *in-situ* or *operando* studies like, for example, *in-situ* mechanical testing. Single grain behavior in supported polycrystalline Au films has been investigated as a function of temperature allowing a detailed analysis of inter- and intra-grain interactions [137, 138]. *In-situ* nanoindentation combined with BCDI on Au nanocrystals was recently reported by Dupraz et al.[115] The Au nanocrystal was indented using a dedicated AFM (developed at IM2NP) and 2D coherent X-ray diffraction patterns were recorded simultaneously. After persistent changes were observed in the diffraction patterns, the Au crystal was completely unloaded and a Bragg coherent X-ray diffraction pattern was recorded. The reconstructed electron density and phase revealed a prismatic dislocation loop with a diameter of ~ 30 nm in the center of the particle.

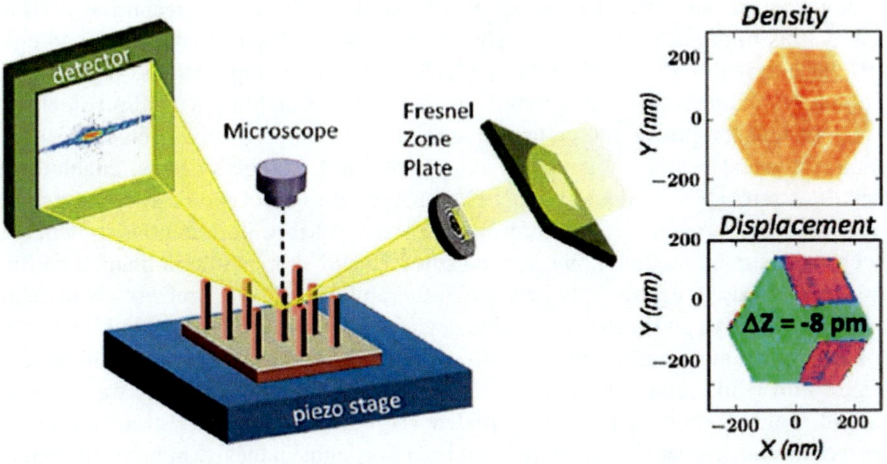

Fig. 6.14 Schematic illustration of the experimental configuration for BCDI on a single GaN nanopillar containing inversion domain boundaries. The green and violet parts are Ga-polar and N-polar zones within the nanopillar which are shifted along z by (c/2 – 8 pm) with respect to each other [136]

Monochromatic nanodiffraction has greatly benefited from the development of focusing optics. Nowadays, 50 nm hard X-ray beams are routinely available on the long nanofocusing beamlines existing at various facilities worldwide. New experimental strategies using continuous motor scans and on-the-fly image acquisitions allow for fast scanning X-ray microscopy where reciprocal space is measured as a function of position in the sample allowing for a detailed evaluation of local strain and defects. The spatial resolution is related to the beam size and the future use of multilayer Laue lenses should bring it down to few 10 nm in the coming years. The *in-situ* combination with nanoindentation or tensile testing has been successfully demonstrated and the scan of the energy of the incident X-ray beam enabled the measurement of the three-dimensional intensity distribution in reciprocal space during mechanical deformation without the need of moving the sample [116].

6.2.5 Enabling Novel Stretchable Metallic Conductor Materials for Emerging Technologies

In this section, we propose to harness the unique capacity for interfacial sliding and thus the atomic reconfigurability (instead of the "structural" reconfigurations in practically all existing metallic stretchable conductor materials/technologies) in FCC/BCC metallic nanolayers, such as Cu/Nb to enable novel, highly stretchable metallic conductors (high conductivity and other electrical performances) with enhanced reliability and robustness (against fatigue and other mechanical/cyclic loading). The approach is highly original, innovative and thus potentially groundbreaking to enable truly high performance, metallic stretchable technology. Ultimately, this high performance, stretchable metallic conductor material technology will have other widespread technologically important applications, such as in biomimicry, human–machine interactions to wearables and biomedical/prosthetics.

Given that the power source unit is an essential component of stretchable electronics devices/systems, soft energy conversion and storage systems, enabled by metallic stretchable conductors must be considered to meet the design and power needs of fully stretchable and deformable systems. More importantly, stretchable electronics can be mounted onto curved surfaces and skin and be implanted within the human body to enable light or thermal therapy or controlled release of medication. Therefore, to power the stretchable devices, power source units should not only possess multifunctional properties with a small form factor but also accommodate large strain while retaining their performance and remaining optically transparent. Recent progress on the strategies to realize large deformability or stretchability of devices has been leveraged on significant breakthroughs in the use of novel materials and structures. The advent of such novel deformable concepts have propelled a significant demand for the fabrication of metallic stretchable materials that can sustain their functionality under extreme deformations for potential application in the next-generation electronic devices and energy devices. To realize industrial applications, such new series of stretchable electrodes need to maintain a low sheet resistance, as well as possess mechanical deformability and low material/processing costs. In addition, they must meet some additional requirements for energy device applications, e.g. long-term electrical/electrochemical stability and interface compatibility with active materials. Thus, the material selection of metallic stretchable conductors is crucial for soft energy devices, because it can affect the overall cell efficiency and mechanical/electrical performances.

References

1. Wegst, U.G., Bai, H., Saiz, E., Tomsia, A.P., Ritchie, R.O.: Bioinspired structural materials. Nat. Mater. **14**(1), 23–36 (2015)
2. Yang, W., Chen, I.H., Gludovatz, B., Zimmermann, E.A., Ritchie, R.O., Meyers, M.A.: Natural flexible dermal armor. Adv. Mater. **25**(1), 31–48 (2013)
3. Curry, J.D.: The Structure of Bone Tissue. Bones: Structure and Mechanics (2006)
4. Patek, S.N., Caldwell, R.L.: Extreme impact and cavitation forces of a biological hammer: strike forces of the peacock mantis shrimp Odontodactylus scyllarus. J. Exp. Biol. **208**(19), 3655–3664 (2005)
5. Weaver, J.C., Milliron, G.W., Miserez, A., Evans-Lutterodt, K., Herrera, S., Gallana, I., Mershon, W.J., Swanson, B., Zavattieri, P., DiMasi, E., Kisailus, D.: The stomatopod dactyl club: a formidable damage-tolerant biological hammer. Science **336**(6086), 1275–1280 (2012)
6. Ashby, M.F., Gibson, L.J., Wegst, U., Olive, R.: The mechanical properties of natural materials. I. Material property charts. Proc. R. Soc. Lond. Series A Math. Phys. Sci. **450**(1938), 123–140 (1995)
7. Taylor, J.R., Scott, N.I., Rouse, G.W.: Evolution of mantis shrimp telson armour and its role in ritualized fighting. J. R. Soc. Interface **16**(157), 20190203 (2019)
8. Huang, W., Restrepo, D., Jung, J.Y., Su, F.Y., Liu, Z., Ritchie, R.O., McKittrick, J., Zavattieri, P., Kisailus, D.: Multiscale toughening mechanisms in biological materials and bioinspired designs. Adv. Mater. **31**(43), 1901561 (2019)
9. Cheng, L., Thomas, A., Glancey, J.L., Karlsson, A.M.: Mechanical behavior of bio-inspired laminated composites. Compos. A Appl. Sci. Manuf. **42**(2), 211–220 (2011)
10. Mencattelli, L. and Pinho, S.T.: Realising bio-inspired impact damage-tolerant thin-ply CFRP Bouligand structures via promoting diffused sub-critical helicoidal damage. Compos. Sci. Technol. **182**, 107684 (2019)
11. Tan, T., Ribbans, B.: A bioinspired study on the compressive resistance of helicoidal fibre structures. Proc. R. Soc. A Math. Phys. Eng. Sci. **473**(2206), 20170538 (2017)
12. Ribbans, B., Li, Y., Tan, T.: A bioinspired study on the interlaminar shear resistance of helicoidal fiber structures. J. Mech. Behav. Biomed. Mater. **56**, 57–67 (2016)
13. Ginzburg, D., Pinto, F., Iervolino, O., Meo, M.: Damage tolerance of bio-inspired helicoidal composites under low velocity impact. Compos. Struct. **161**, 187–203 (2017)
14. Apichattrabrut, T., Ravi-Chandar, K.: Helicoidal composites. Mech. Adv. Mater. Struct. **13**(1), 61–76 (2006)
15. Grunenfelder, L.K., Suksangpanya, N., Salinas, C., Milliron, G., Yaraghi, N., Herrera, S., Evans-Lutterodt, K., Nutt, S.R., Zavattieri, P., Kisailus, D.: Bio-inspired impact-resistant composites. Acta Biomater. **10**(9), 3997–4008 (2014)
16. Agarwal, K., Sahay, R., Baji, A., Budiman, A.S.: Biomimetic tough helicoidally structured material through novel electrospinning based additive manufacturing. MRS Adv. **4**(43), 2345–2354 (2019)
17. Agarwal, K., Zhou, Y., Anwar Ali, H.P., Radchenko, I., Baji, A., Budiman, A.S.: Additive manufacturing enabled by electrospinning for tougher bio-inspired materials. Adv. Mater. Sci. Eng. 2018 (2018)

18. Chen, R., Liu, J., Yang, C., Weitz, D.A., He, H., Li, D., Chen, D., Liu, K., Bai, H.: Transparent impact-resistant composite films with bioinspired hierarchical structure. ACS Appl. Mater. Interfaces **11**(26), 23616–23622 (2019)
19. Thomas, D., Cebe, P.: Self-nucleation and crystallization of polyvinyl alcohol. J. Therm. Anal. Calorim. **127**(1), 885–894 (2017)
20. Wang, X., Xiao, C., Liu, H., Chen, M., Hao, J., Wu, Y.: A study on fabrication of PVDF-HFP/PTFE blend membranes with controllable and bicontinuous structure for highly effective water-in-oil emulsion separation. RSC Adv. **8**(49), 27754–27762 (2018)
21. Li, N., Xiao, C., An, S., Hu, X.: Preparation and properties of PVDF/PVA hollow fiber membranes. Desalination **250**(2), 530–537 (2010)
22. Xu, W., Liu, C., Xue, X., Su, Y., Lv, Y., Xing, W., Lu, T.: New proton exchange membranes based on poly (vinyl alcohol) for DMFCs. Solid State Ionics **171**(1–2), 121–127 (2004)
23. Yang, C.C., Lee, Y.J., Yang, J.M.: Direct methanol fuel cell (DMFC) based on PVA/MMT composite polymer membranes. J. Power Sources **188**(1), 30–37 (2009)
24. Fratzl, P., Weinkamer, R.: Nature's hierarchical materials. Prog. Mater Sci. **52**(8), 1263–1334 (2007)
25. Sahay, R., et al.: Helicoidally arranged polyacrylonitrile fiber-reinforced strong and impact-resistant thin polyvinyl alcohol film enabled by electrospinning-based additive manufacturing. In: Polymers 12.10, p. 2376 (2020)
26. Agarwal, K., et al.: Impact-resistant and tough helicoidally aligned ribbon reinforced composites with tunable mechanical properties via integrated additive manufacturing methodologies. ACS Appl. Polym. Mater. **2**(8), 3491–3504 (2020)
27. Budiman, A.S., Anbazhagan, S., Illya, G., Song, W.J.R., Sahay, R., Tippabhotla, S.K., Tay, A.A.O.: Enabling curvable silicon photovoltaics technology using polycarbonate-sandwiched laminate design. Sol. Energy **220**, 462–472 (2021)
28. Bruce, P.G., Scrosati, B., Tarascon, J.M.: Angew. Chemie Int. Ed **47**(16), 2930–3029 (2008)
29. Song, W.J.R., Tippabhotla, S.K., Tay, A.A.O., Budiman, A.S.: Numerical simulation of the evolution of stress in solar cells during the entire manufacturing cycle of a conventional silicon wafer-based photovoltaic laminate. IEEE J. Photovolt. **8**(1), 210–217 (2018)
30. Chan, C.K., et al.: Nat. Nanotechnol **3**(1), 31–35 (2008)
31. Ali, I., et al.: Front. Energy Res. **6** (2018)
32. Tippabhotla, S.K., Radchenko, I., Stan, C.V., Tamura, N., Budiman, A.S.: J. Mater. Res **34**(9), 1622–1631 (2019)
33. Agarwal, K., Sahay, R., Baji, A.: Polymers (Basel) **12**(5), 1–11 2020 (2020)
34. Hu, L., Yuan, W., Brochu, P., Gruner, G. and Pei, Q.: Highly stretchable, conductive, and transparent nanotube thin films. Appl. Phys. Lett. **94**(16), 161108 (2009)
35. Liu, K., Sun, Y., Liu, P., Lin, X., Fan, S., Jiang, K.: Cross-stacked superaligned carbon nanotube films for transparent and stretchable conductors. Adv. Func. Mater. **21**(14), 2721–2728 (2011)
36. Zhang, Y., Sheehan, C.J., Zhai, J., Zou, G., Luo, H., Xiong, J., Zhu, Y.T., Jia, Q.X.: Polymer-embedded carbon nanotube ribbons for stretchable conductors. Adv. Mater. **22**(28), 3027–3031 (2010)
37. Sacco, J.E., Sokoloff, J.B.: Free sliding in lattices with two incommensurate periodicities. Phys. Rev. B **18**(12), 6549–6559 (1978)
38. Hirano, M., Shinjo, K.: Atomistic locking and friction. Phys. Rev. B **41**(17), 11837–11851 (1990)
39. Sorensen, M.R., Stoltze, P., Jacobsen, K.: Phys. Rev. B B1990,41, 2101–2113.
40. He, G., Müser, M.H., Robbins, M.O.: Adsorbed layers and the origin of static friction. Science **284**(5420), 1650–1652 (1999)
41. Lançon, F., Pénisson, J.M., Dahmen, U.: Quasicrystalline gold interface with a hypo-friction property. EPL (Europhysics Letters) **49**(5), 603–609 (2000)
42. Müser, M.H., Robbins, M.O.: Conditions for static friction between flat crystalline surfaces. Phys. Rev. B **61**(3), 2335–2342 (2000)
43. SER, M.H.M., Urbakh, M., Robbins, M.O.: Statistical mechanics of static and low-velocity kinetic friction. Editor. Board, 187–272 (2003)

44. Sutton, A.P.: Interfaces in crystalline materials. Monogr. Phys. Chem. Mater., 414–423 (1995)
45. Lançon, F., Ye, J., Caliste, D., Radetic, T., Minor, A.M., Dahmen, U.: Superglide at an internal incommensurate boundary. Nano Lett. **10**(2), 695–700 (2010)
46. Marks, L.D., Warren, O.L., Minor, A.M., Merkle, A.P.: Tribology in full view. MRS Bull. **33**(12), 1168–1173 (2008)
47. Siemens, M.E., Li, Q., Yang, R., Nelson, K.A., Anderson, E.H., Murnane, M.M., Kapteyn, H.C.: Quasi-ballistic thermal transport from nanoscale interfaces observed using ultrafast coherent soft X-ray beams. Nat. Mater. **9**(1), 26–30 (2010)
48. Wang, J., Hoagland, R.G., Hirth, J.P., Misra, A.: Atomistic modeling of the interaction of glide dislocations with “weak” interfaces. Acta Mater. **56**, 3109–3119 (2008)
49. Beyerlein, I.J., Mara, N.A., Carpenter, J.S., Nizolek, T., Mook, W.M., Wynn, T.A., McCabe, R.J., Mayeur, J.R., Kang, K., Zheng, S., Wang, J.: Interface-driven microstructure development and ultra high strength of bulk nanostructured Cu-Nb multilayers fabricated by severe plastic deformation. J. Mater. Res. **28**(13), 1799–1812 (2013)
50. Misra, A., Hirth, J.P., Hoagland, R.G.: Length-scale-dependent deformation mechanisms in incoherent metallic multilayered composites. Acta Mater. **53**(18), 4817–4824 (2005)
51. Misra, A., Hoagland, R.G., Kung, H.: Thermal stability of self-supported nanolayered Cu/Nb films. Phil. Mag. **84**(10), 1021–1028 (2004)
52. Misra, A., Demkowicz, M.J., Zhang, X., Hoagland, R.G.: The radiation damage tolerance of ultra-high strength nanolayered composites. Jom **59**(9), 62–65 (2007)
53. Demkowicz, M.J., Wang, Y.Q., Hoagland, R.G., Anderoglu, O.: Mechanisms of He escape during implantation in CuNb multilayer composites. Nucl. Instrum. Methods Phys. Res. Sect. B **261**(1–2), 524–528 (2007)
54. Budiman, A.S., Narayanan, K.R., Li, N., Wang, J., Tamura, N., Kunz, M., Misra, A.: Plasticity evolution in nanoscale Cu/Nb single-crystal multilayers as revealed by synchrotron X-ray microdiffraction. Mater. Sci. Eng. A **635**, 6–12 (2015)
55. Budiman, A.S., Han, S.M., Li, N., Wei, Q.M., Dickerson, P., Tamura, N., Kunz, M., Misra, A.: Plasticity in the nanoscale Cu/Nb single-crystal multilayers as revealed by synchrotron Laue x-ray microdiffraction. J. Mater. Res. **27**(3), 599–611 (2012)
56. Radchenko, I., Anwarali, H.P., Tippabhotla, S.K., Budiman, A.S.: Effects of interface shear strength during failure of semicoherent metal–metal nanolaminates: an example of accumulative roll-bonded Cu/Nb. Acta Mater. **156**, 125–135 (2018)
57. Ali, H.P.A., Radchenko, I., Li, N., Budiman, A.: The roles of interfaces and other microstructural features in Cu/Nb nanolayers as revealed by in situ beam bending experiments inside an scanning electron microscope (SEM). Mater. Sci. Eng. A **738**, 253–263 (2018)
58. Han, W.Z., Cerreta, E.K., Mara, N.A., Beyerlein, I.J., Carpenter, J.S., Zheng, S.J., Trujillo, C.P., Dickerson, P.O., Misra, A.: Deformation and failure of shocked bulk Cu–Nb nanolaminates. Acta Mater. **63**, 150–161 (2014)
59. Carpenter, J.S., Zheng, S.J., Zhang, R.F., Vogel, S.C., Beyerlein, I.J., Mara, N.A.: Thermal stability of Cu–Nb nanolamellar composites fabricated via accumulative roll bonding. Phil. Mag. **93**(7), 718–735 (2013)
60. Liang, J., Li, L., Niu, X., Yu, Z., Pei, Q.: Elastomeric polymer light-emitting devices and displays. Nat. Photon. **7**(10), 817–824 (2013)
61. Wang, J., Yan, C., Kang, W., Lee, P.S.: High-efficiency transfer of percolating nanowire films for stretchable and transparent photodetectors. Nanoscale **6**(18), 10734–10739 (2014)
62. Park, S., Tan, A.W.M., Wang, J., Lee, P.S.: Coaxial Ag–base metal nanowire networks with high electrochemical stability for transparent and stretchable asymmetric supercapacitors. Nanoscale Horiz. **2**(4), 199–204 (2017)
63. Vohra, A., Imin, P., Imit, M., Carmichael, R.S., Meena, J.S., Adronov, A., Carmichael, T.B.: Transparent, stretchable, and conductive SWNT films using supramolecular functionalization and layer-by-layer self-assembly. RSC Adv. **6**(35), 29254–29263 (2016)
64. Feng, C., Liu, K., Wu, J.S., Liu, L., Cheng, J.S., Zhang, Y., Sun, Y., Li, Q., Fan, S., Jiang, K.: Flexible, stretchable, transparent conducting films made from superaligned carbon nanotubes. Adv. Func. Mater. **20**(6), 885–891 (2010)

65. Kim, K.S., Zhao, Y., Jang, H., Lee, S.Y., Kim, J.M., Kim, K.S., Ahn, J.H., Kim, P., Choi, J.Y., Hong, B.H.: Large-scale pattern growth of graphene films for stretchable transparent electrodes. Nature **457**(7230), 706–710 (2009)
66. Kim, R.H., Bae, M.H., Kim, D.G., Cheng, H., Kim, B.H., Kim, D.H., Li, M., Wu, J., Du, F., Kim, H.S., Kim, S.: Stretchable, transparent graphene interconnects for arrays of microscale inorganic light emitting diodes on rubber substrates. Nano Lett. **11**(9), 3881–3886 (2011)
67. Chen, T., Xue, Y., Roy, A.K., Dai, L.: Transparent and stretchable high-performance supercapacitors based on wrinkled graphene electrodes. ACS Nano **8**(1), 1039–1046 (2014)
68. Chen, T., Peng, H., Durstock, M., Dai, L.: High-performance transparent and stretchable all-solid supercapacitors based on highly aligned carbon nanotube sheets. Sci. Rep. **4**(1), 3612 (2014)
69. Liang, J., Li, L., Tong, K., Ren, Z., Hu, W., Niu, X., Chen, Y., Pei, Q.: Silver nanowire percolation network soldered with graphene oxide at room temperature and its application for fully stretchable polymer light-emitting diodes. ACS Nano **8**(2), 1590–1600 (2014)
70. Wang, J., Hoagland, R.G., Hirth, J.P., Misra, A.: Atomistic simulations of the shear strength and sliding mechanisms of copper–niobium interfaces. Acta Mater. **56**(13), 3109–3119 (2008)
71. Beyerlein, I.J., Mara, N.A., Wang, J., Carpenter, J.S., Zheng, S.J., Han, W.Z., Zhang, R.F., Kang, K., Nizolek, T., Pollock, T.M.: Structure–property–functionality of bimetal interfaces. Jom **64**(10), 1192–1207 (2012)
72. Saito, Y., Utsunomiya, H., Tsuji, N., Sakai, T.: Novel ultra-high straining process for bulk materials—development of the accumulative roll-bonding (ARB) process. Acta Mater. **47**(2), 579–583 (1999)
73. Hirano, M., Shinjo, K., Kaneko, R., Murata, Y.: Observation of superlubricity by scanning tunneling microscopy. Phys. Rev. Lett. **78**(8), 1448–1451 (1997)
74. Ko, J.S., Gellman, A.J.: Friction anisotropy at Ni (100)/Ni (100) interfaces. Langmuir **16**(22), 8343–8351 (2000)
75. Dienwiebel, M., Verhoeven, G.S., Pradeep, N., Frenken, J.W., Heimberg, J.A., Zandbergen, H.W.: Superlubricity of graphite. Phys. Rev. Lett. **92**(12), 126101 (2004)
76. Socoliuc, A., Bennewitz, R., Gnecco, E. and Meyer, E.: Transition from stick-slip to continuous sliding in atomic friction: entering a new regime of ultralow friction. Phys. Rev. Lett. **92**(13), 134301 (2004)
77. Park, J.Y., Ogletree, D.F., Salmeron, M., Ribeiro, R.A., Canfield, P.C., Jenks, C.J., Thiel, P.A.: High frictional anisotropy of periodic and aperiodic directions on a quasicrystal surface. Science **309**(5739), 1354–1356 (2005)
78. Friedel, J., De Gennes, P.G.: Friction between incommensurate crystals. Phil. Mag. **87**(1), 39–49 (2007)
79. Wang, J., Zhou, Q., Shao, S., Misra, A.: Strength and plasticity of nanolaminated materials. Mater. Res. Lett. **5**(1), 1–19 (2017)
80. Jaya, B.N., Jayaram, V.: Fracture testing at small-length scales: From plasticity in Si to brittleness in Pt. JOM **68**(1), 94–108 (2016)
81. Wang, J., Misra, A., Hoagland, R.G., Hirth, J.P.: Slip transmission across fcc/bcc interfaces with varying interface shear strengths. Acta Mater. **60**(4), 1503–1513 (2012)
82. Hoagland, R.G., Kurtz, R.J., Henager, C.H., Jr.: Slip resistance of interfaces and the strength of metallic multilayer composites. Scripta Mater. **50**(6), 775–779 (2004)
83. Misra, A., Hoagland, R.G.: Dekker encyclopedia of nanoscience and nanotechnology, pp. 1–10 (2005)
84. Misra, A., Verdier, M., Lu, Y.C., Kung, H., Mitchell, T.E., Nastasi, M., Embury, J.D.: Structure and mechanical properties of Cu-X (X= Nb, Cr, Ni) nanolayered composites. Scripta Mater. **39**(4–5), 555–560 (1998)
85. Lee, S.B., LeDonne, J.E., Lim, S.C.V., Beyerlein, I.J., Rollett, A.D.: The heterophase interface character distribution of physical vapor-deposited and accumulative roll-bonded Cu–Nb multilayer composites. Acta Mater. **60**(4), 1747–1761 (2012)
86. Zheng, S.J., Beyerlein, I.J., Wang, J., Carpenter, J.S., Han, W.Z., Mara, N.A.: Deformation twinning mechanisms from bimetal interfaces as revealed by in situ straining in the TEM. Acta Mater. **60**(16), 5858–5866 (2012)

87. Li, T., Huang, Z.Y., Xi, Z.C., Lacour, S.P., Wagner, S., Suo, Z.: Delocalizing strain in a thin metal film on a polymer substrate. Mech. Mater. **37**(2–3), 261–273 (2005)
88. Demkowicz, M.J., Thilly, L.: Structure, shear resistance and interaction with point defects of interfaces in Cu–Nb nanocomposites synthesized by severe plastic deformation. Acta Mater. **59**(20), 7744–7756 (2011)
89. Zhang, R., Wang, J., Liu, X., Beyerlein, I.J. and Germann, T.C.: Nonequilibrium molecular dynamics simulations of shock wave propagation in nanolayered Cu/Nb nanocomposites. In: AIP Conference Proceedings, vol. 1426, no. 1, pp. 1251–1254. American Institute of Physics (2012)
90. Snel, J., Monclús, M.A., Castillo-Rodriguez, M., Mara, N., Beyerlein, I.J., Llorca, J., Molina-Aldareguia, J.M.: Deformation mechanism map of Cu/Nb nanoscale metallic multilayers as a function of temperature and layer thickness. Jom **69**(11), 2214–2226 (2017)
91. Rogers, J.A., Someya, T., Huang, Y.: Materials and mechanics for stretchable electronics. Science **327**(5973), 1603–1607 (2010)
92. Kim, D.H., Rogers, J.A.: Stretchable electronics: materials strategies and devices. Adv. Mater. **20**(24), 4887–4892 (2008)
93. Kim, D.H., Ghaffari, R., Lu, N., Rogers, J.A.: Flexible and stretchable electronics for biointegrated devices. Annu. Rev. Biomed. Eng. **14**, 113–128 (2012)
94. Kim, D.H., Lu, N., Ghaffari, R., Rogers, J.A.: Inorganic semiconductor nanomaterials for flexible and stretchable bio-integrated electronics. NPG Asia Mater. **4**(4), e15–e15 (2012)
95. Someya, T.: Stretchable Electronics. Wiley (2012)
96. Savagatrup, S., Printz, A.D., O'Connor, T.F., Zaretski, A.V., Lipomi, D.J.: Molecularly stretchable electronics. Chem. Mater. **26**(10), 3028–3041 (2014)
97. Sun, B., Long, Y.Z., Chen, Z.J., Liu, S.L., Zhang, H.D., Zhang, J.C., Han, W.P.: Recent advances in flexible and stretchable electronic devices via electrospinning. J. Mater. Chem. C **2**(7), 1209–1219 (2014)
98. Wang, Q., Robinson, D., Zhao, X.: On-demand hierarchical patterning with electric fields. Appl. Phys. Lett. **104**(23), 231605 (2014)
99. Yao, S., Zhu, Y.: Nanomaterial-enabled stretchable conductors: strategies, materials and devices. Adv. Mater. **27**(9), 1480–1511 (2015)
100. Nassar, J.M., Rojas, J.P., Hussain, A.M., Hussain, M.M.: From stretchable to reconfigurable inorganic electronics. Extrem. Mech. Lett. **9**, 245–268 (2016)
101. Kim, K., Kim, J., Hyun, B.G., Ji, S., Kim, S.Y., Kim, S., An, B.W., Park, J.U.: Stretchable and transparent electrodes based on in-plane structures. Nanoscale **7**(35), 14577–14594 (2015)
102. Lee, H., Kim, M., Kim, I., Lee, H.: Flexible and stretchable optoelectronic devices using silver nanowires and graphene. Adv. Mater. **28**(22), 4541–4548 (2016)
103. Yang, C., Gu, H., Lin, W., Yuen, M.M., Wong, C.P., Xiong, M., Gao, B.: Silver nanowires: from scalable synthesis to recyclable foldable electronics. Adv. Mater. **23**(27), 3052–3056 (2011)
104. Lee, H., Hong, S., Lee, J., Suh, Y.D., Kwon, J., Moon, H., Kim, H., Yeo, J., Ko, S.H.: Highly stretchable and transparent supercapacitor by Ag–Au core–shell nanowire network with high electrochemical stability. ACS Appl. Mater. Interfaces **8**(24), 15449–15458 (2016)
105. Park, S., Parida, K., Lee, P.S.: Deformable and transparent ionic and electronic conductors for soft energy devices. Adv. Energy Mater. **7**(22), 1701369 (2017)
106. Noyan, I.C., Cohen, J.B.: Measurement by Diffraction and Interpretation, Residual Stress. Springer, New York, NY (1987)
107. Fullerton, E.E., Schuller, I.K., Vanderstraeten, H., Bruynseraede, Y.: Structural refinement of superlattices from x-ray diffraction. Phys. Rev. B **45**(16), 9292 (1992)
108. Labat, S., Gergaud, P., Thomas, O., Gilles, B., Marty, A.: Interdependence of elastic strain and segregation in metallic multilayers: an X-Ray diffraction study of (111) Au/Ni multilayers. J. Appl. Phys. **87**(3), 1172–1181 (2000)
109. Bigault, T., Bocquet, F., Labat, S., Thomas, O., Renevier, H.: Interfacial structure in (111) Au: Ni multilayers investigated by anomalous x-ray diffraction. Phys. Rev. B **64**(12), 125414 (2001)

110. Benoudia, M.C., Gao, F., Roussel, J.M., Labat, S., Gailhanou, M., Thomas, O., Beke, D.L., Erdélyi, Z., Langer, G., Csik, A., Kis-Varga, M.: Thermoelasticity and interdiffusion in CuNi multilayers. Phys. Rev. B **85**(23), 235404 (2012)
111. Cornelius, T.W., Thomas, O.: Progress of in situ synchrotron X-ray diffraction studies on the mechanical behavior of materials at small scales. Prog. Mater Sci. **94**, 384–434 (2018)
112. Mocuta, C., Richard, M.I., Fouet, J., Stanescu, S., Barbier, A., Guichet, C., Thomas, O., Hustache, S., Zozulya, A.V., Thiaudiere, D.: Fast pole figure acquisition using area detectors at the DiffAbs beamline–Synchrotron SOLEIL. J. Appl. Crystallogr. **46**(6), 1842–1853 (2013)
113. Richard, M.I., Fouet, J., Texier, M., Mocuta, C., Guichet, C., Thomas, O.: Continuous and collective grain rotation in nanoscale thin films during silicidation. Phys. Rev. Lett. **115**(26), 266101 (2015)
114. Spolenak, R., Brown, W.L., Tamura, N., MacDowell, A.A., Celestre, R.S., Padmore, H.A., Valek, B., Bravman, J.C., Marieb, T., Fujimoto, H., Batterman, B.W.: Local plasticity of Al thin films as revealed by X-ray microdiffraction. Phys. Rev. Lett. **90**(9), 096102 (2003)
115. Chahine, G.A., Richard, M.I., Homs-Regojo, R.A., Tran-Caliste, T.N., Carbone, D., Jacques, V.L.R., Grifone, R., Boesecke, P., Katzer, J., Costina, I., Djazouli, H.: Imaging of strain and lattice orientation by quick scanning X-ray microscopy combined with three-dimensional reciprocal space mapping. J. Appl. Crystallogr. **47**(2), 762–769 (2014)
116. Simons, H., King, A., Ludwig, W., Detlefs, C., Pantleon, W., Schmidt, S., Stöhr, F., Snigireva, I., Snigirev, A., Poulsen, H.F.: Dark-field X-ray microscopy for multiscale structural characterization. Nat. Commun. **6**(1), 6098 (2015)
117. Pfeifer, M.A., Williams, G.J., Vartanyants, I.A., Harder, R., Robinson, I.K.: Three-dimensional mapping of a deformation field inside a nanocrystal. Nature **442**(7098), 63–66 (2006)
118. Minkevich, A.A., Gailhanou, M., Micha, J.S., Charlet, B., Chamard, V., Thomas, O.: Inversion of the diffraction pattern from an inhomogeneously strained crystal using an iterative algorithm. Phys. Rev. B **76**(10), 104106 (2007)
119. Chung, J.S., Ice, G.E.: Automated indexing for texture and strain measurement with broad-bandpass x-ray microbeams. J. Appl. Phys. **86**(9), 5249–5255 (1999)
120. Larson, B.C., Ice, G.E., Tamura, N., Yang, W., Tischler, J.Z., Chung, J.S., Budai, J.D., Chung, K.S., Weiland, H., Lowe, W.P.: pp. 247 (2001)
121. Barabash, R., Ice, G.E., Larson, B.C., Pharr, G.M., Chung, K.S., Yang, W.: White microbeam diffraction from distorted crystals. Appl. Phys. Lett. **79**(6), 749–751 (2001)
122. Larson, B.C., Yang, W., Ice, G.E., Budai, J.D., Tischler, J.Z.: Three-dimensional X-ray structural microscopy with submicrometre resolution. Nature **415**(6874), 887–890 (2002)
123. Dowell, A., Celestre, R., Tamura, N., Spolenak, R., Bravman, J., .Padmore, H., Batterman, B., Patel, J.: Nucl. Inst. Meth. A, 467–468 (2000)
124. Tamura, N., MacDowell, A.A., Celestre, R.S., Padmore, H.A., Valek, B., Bravman, J.C., Spolenak, R., Brown, W.L., Marieb, T., Fujimoto, H., Batterman, B.W.: High spatial resolution grain orientation and strain mapping in thin films using polychromatic submicron x-ray diffraction. Appl. Phys. Lett. **80**(20), 3724–3726 (2002)
125. Valek, B.C., Bravman, J.C., Tamura, N., MacDowell, A.A., Celestre, R.S., Padmore, H.A., Spolenak, R., Brown, W.L., Batterman, B.W., Patel, J.R.: Electromigration-induced plastic deformation in passivated metal lines. Appl. Phys. Lett. **81**(22), 4168–4170 (2002)
126. Budiman, A.S., Nix, W.D., Tamura, N., Valek, B.C., Gadre, K., Maiz, J., Spolenak, R., Patel, J.R.: Crystal plasticity in Cu damascene interconnect lines undergoing electromigration as revealed by synchrotron x-ray microdiffraction. Appl. Phys. Lett. **88**(23), 233515 (2006)
127. Budiman, A.S., Besser, P.R., Hau-Riege, C.S., Marathe, A., Joo, Y.C., Tamura, N., Patel, J.R., Nix, W.D.: Electromigration-induced plasticity: texture correlation and implications for reliability assessment. J. Electron. Mater. **38**(3), 379–391 (2009)
128. Budiman, A.S., Hau-Riege, C.S., Baek, W.C., Lor, C., Huang, A., Kim, H.S., Neubauer, G., Pak, J., Besser, P.R., Nix, W.D.: Electromigration-induced plastic deformation in Cu interconnects: effects on current density exponent, n, and implications for EM reliability assessment. J. Electron. Mater. **39**(11), 2483–2488 (2010)

129. Budiman, A.S., Shin, H.A.S., Kim, B.J., Hwang, S.H., Son, H.Y., Suh, M.S., Chung, Q.H., Byun, K.Y., Tamura, N., Kunz, M., Joo, Y.C.: Measurement of stresses in Cu and Si around through-silicon via by synchrotron X-ray microdiffraction for 3-dimensional integrated circuits. Microelectron. Reliab. **52**(3), 530–533 (2012)
130. Maaβ, R., Grolimund, D., Van Petegem, S., Willimann, M., Jensen, M., Van Swygenhoven, H., Lehnert, T., Gijs, M.A.M., Volkert, C.A., Lilleodden, E.T,. Schwaiger, R.: Defect structure in micropillars using x-ray microdiffraction. Appl. Phys. Lett. **89**(15), 151905 (2006)
131. Ulrich, O., Biquard, X., Bleuet, P., Geaymond, O., Gergaud, P., Micha, J.S., Robach, O., Rieutord, F.: A new white beam x-ray microdiffraction setup on the BM32 beamline at the European Synchrotron Radiation Facility. Rev. Sci. Instrum. **82**(3), 033908 (2011)
132. Kirchlechner, C., Keckes, J., Motz, C., Grosinger, W., Kapp, M.W., Micha, J.S., Ulrich, O., Dehm, G.: Mechanische in kleinen Dimensionen. Erlangen-Nürnberg Univ. (2010)
133. Leclere, C., Cornelius, T.W., Ren, Z., Davydok, A., Micha, J.S., Robach, O., Richter, G., Belliard, L., Thomas, O.: In situ bending of an Au nanowire monitored by micro Laue diffraction. J. Appl. Crystallogr. **48**(1), 291–296 (2015)
134. Leclere, C., Cornelius, T.W., Ren, Z., Robach, O., Micha, J.S., Davydok, A., Ulrich, O., Richter, G., Thomas, O.: KB scanning of X-ray beam for Laue microdiffraction on accelerophobic samples: application to in situ mechanically loaded nanowires. J. Synchrotron Radiat. **23**(6), 1395–1400 (2016)
135. Robinson, I.K., Vartanyants, I.A., Williams, G.J., Pfeifer, M.A., Pitney, J.A.: Reconstruction of the shapes of gold nanocrystals using coherent x-ray diffraction. Phys. Rev. Lett. **87**(19), 195505 (2001)
136. Labat, S., Richard, M., Dupraz, M., Gailhanou, M., Cornelius, F.T., Schülli, T., Eymery, J., Thomas, O.: ACS Nano **9**, 9210–9216 (2015)
137. Vaxelaire, N., Labat, S., Cornelius, T.W., Kirchlechner, C., Keckes, J., Schulli, T., Thomas, O.: New insights into single-grain mechanical behavior from temperature-dependent 3-D coherent X-ray diffraction. Acta Mater. **78**, 46–55 (2014)
138. Yau, A., Cha, W., Kanan, M.W., Stephenson, G.B., Ulvestad, A.: Bragg coherent diffractive imaging of single-grain defect dynamics in polycrystalline films. Science **356**(6339), 739–742 (2017)

Chapter 7
Conclusion

Ever since the Iron Ages, materials design has been mostly done through "trial and error" long before much was understood about them. This was the same for nanoscale metal–metal multilayered composites, which started with Koehler's theoretical attempt to design a strong solid just by stacking multiple layers of two dissimilar crystalline metallic thin films in 1970 [1]. Random attempts of synthesis, characterization and simulation of these nanocomposites were done over many decades, the interactions at the nanolayer interface are only understood in detail recently for their critical roles at determining the novel elastic and plastic responses of the materials. The roles of new experimental probing techniques as described in Chap. 3 of this book and the advent of nanotechnologies, including the ability for atomic engineering of interfaces in the nanoscales, along with recent advances in the technology of additive manufacturing approaches cannot be understated. The case with nanoscale metallic multilayered materials [2] is indeed an excellent illustration of not only how typical materials science discovery had happened up until very recently, and why an update of the classical laws of mechanics of materials (elasticity, plasticity and fracture mechanics) may be due today. Chapter 2 of this book represents an early attempt to provide updated accounts of the further evolution of the classical elasticity, plasticity and fracture mechanisms and laws as we had known them from textbook in the past decades.

Societies and cultures have been defined by materials for ages. From the Stone Age to the more recent Copper Age, and now we are in Silicon Age, our history has recorded how materials and their mechanical properties have always been the enabling factor in the technological revolutions of the days. Swordsmiths in the Middle Ages did not fully understand the metallurgical theory behind their sword-making until the nineteenth century except that quenching the red-hot iron (with a little carbon in it) in water made it hard and thus more lethal a weapon. In the today's Silicon Age, the inability to plasma-etch copper called for a drastic rethinking of the metal patterning process, and the result of this rethinking was a process referred to as an *additive patterning*, also known as a "Damascene" or "dual-Damascene" process

A. S. Budiman, *Principles of Extreme Mechanics (XM) in Design for Reliability (DfR)*, Engineering Materials,
https://doi.org/10.1007/978-981-15-6720-9_7

by analogy to the traditional technique of metal inlaying originating in the Persian city of Damascus. Chapter IV of this book highlighted the roles of the newly found insights of extended elastic regime and extreme deformability in such advanced copper interconnects enabling the wonders of silicon-based nanoelectronics technologies epitomized by the uses of smart phones, computers and interconnectivity in our modern world today. A stretch of several cities in southern part of the San Francisco Bay Area in Northern California is now called "Silicon Valley" signifying the unique roles silicon material has in our present world and culture. The principles of eXtreme Mechanics (XM) were part of that story as illustrated by the Through-Silicon Via (TSV) and the advanced 32 nm Cu interconnect schemes in Sects. 4.1 and 4.4, respectively.

The silicon story does not end there in this book. Chapter V in fact further elucidates the key, enabling roles XM play in extending the silicon story to next-generation energy technologies—in Lithium Ion Battery (LIB) for energy storage in Sect. 5.1, and in enabling novel and innovative silicon-based photovoltaics (PV) design for energy production in Sects. 5.2 and 5.3. By the time a reader of this book has reached these sections in Chap. 5, it would have been abundantly clear that XM has indeed extended the designer's palettes for reliability of the emerging technologies in the silicon story. This is the emerging transformation of "Design for Reliability" or DfR allowing the further extension of the Silicon Age. Not only for silicon-based semiconductor technologies that are the brick and mortar of our present lifestyles with hyperconnectivity and intense social media blurring the real world into the digital realms. But also for silicon-based next-generation energy technologies enabling the next industrial revolution based on clean technology!

While the preceding chapters have illustrated how XM enables mostly DfR (Design for Reliability) of important and strategic technologies of the future, Chap. 6 explores even further the unique roles XM may play in enabling design for novel functionalities. First, novel 3D architectures enabled by additive manufacturing approaches were shown here to have potential for next-generation LIB anode design with next levels of kinetics (charging rate) and performance (energy density) allowing for the vision of the electrification of transportation or electric vehicle (EV). Next, the unravelling of the interfacial sliding mechanisms through atomic reconfigurations in the FCC/BCC nanolayered materials may have opened up uncharted territories in the design and development of novel structural materials and their unprecedented limits and new functionalities as novel metallic stretchable materials [3]. The roles of the eXtreme Mechanics (XM) and the advanced experimentation techniques to enable the probing of the new mechanics aspects at the nano or mesoscales cannot be understated here. The real excitement here is the further potential that XM is not only enabling extreme reliability or DfXR (Design for Extreme Reliability) now, but also finally stretching the functional capabilities of the materials beyond our imagination. The interfacial sliding is after all an interface-mediated shear deformation at the extremes!

Finally, as a conclusion, in this book, we have captured a portion of the broad spectrum of research, representing some of the most important and recent advances in the probing of the principles of XM and in extending the DfR capabilities along with enabling extreme design and innovations. This book provides readers up-to-date information on the impact of these recent advances on XM principles and DfR innovations. We hope the readers will find this text to be a significant collection of important research highlights representing the rapid advancements in these exciting research fields.

References

1. Koehler, J.: Attempt to design a strong solid. Phys. Rev. B **2**, 547 (1970)
2. Misra, A., Thilly, L.: Structural metals at extremes. MRS Bull. **35**, 965–981 (2010)
3. Radchenko, I., Anwar Ali, H.P., Tippabhotla, S.K., Budiman, A.S.: Effects of interface shear strength during failure of semicoherent metal-metal nanolaminates: an example of accumulative roll-bonded Cu/Nb. Acta Materialia. **156**, 125–135 (2018). https://doi.org/10.1016/j.actamat.2018.06.023.